T0293847

Machine Learning in Bioinformatics of Protein Sequences

Algorithms, Databases
and Resources for
Modern Protein
Bioinformatics

Machine Learning in Bioinformatics of Protein Sequences

Algorithms, Databases and Resources for Modern Protein Bioinformatics

Editor

Lukasz Kurgan

NEW JERSEY · LONDON · SINGAPORE · BEIJING · SHANGHAI · HONG KONG · TAIPEI · CHENNAI · TOKYO

Published by

World Scientific Publishing Co. Pte. Ltd.

5 Toh Tuck Link, Singapore 596224

USA office: 27 Warren Street, Suite 401-402, Hackensack, NJ 07601

UK office: 57 Shelton Street, Covent Garden, London WC2H 9HE

Library of Congress Cataloging-in-Publication Data
Names: Kurgan, Lukasz, editor.
Title: Machine learning in bioinformatics of protein sequences : algorithms, databases and
    resources for modern protein bioinformatics / editor, Lukasz Kurgan.
Description: First edition. | New Jersey : World Scientific, [2023] |
    Includes bibliographical references and index.
Identifiers: LCCN 2022029972 | ISBN 9789811258572 (hardcover) |
    ISBN 9789811258589 (ebook for institutions) | ISBN 9789811258596 (ebook for individuals)
Subjects: LCSH: Bioinformatics. | Amino acid sequence--Data processing. | Machine learning. |
    Artificial intelligence--Biological applications. | Sequence alignment (Bioinformatics)
Classification: LCC QH324.25 .M33 2023 | DDC 570.285--dc23/eng/20220718
LC record available at https://lccn.loc.gov/2022029972

British Library Cataloguing-in-Publication Data
A catalogue record for this book is available from the British Library.

For any available supplementary material, please visit
https://www.worldscientific.com/worldscibooks/10.1142/12899#t=suppl

Desk Editor: Vanessa Quek

Typeset by Stallion Press
Email: enquiries@stallionpress.com

Dedicated to my beloved Family, talented
Magdalena and incredible Aleksander

Preface

As of mid-2022, we have access to over 230 million unique protein sequences, significant majority of which have unknown structure and functions. This number continues to grow rapidly and has more than tripled compared to just five years ago [1,2]. Structural biologists and bioinformaticians face an enormous challenge to structurally and functionally characterize these hundreds of millions of sequences. Experimental approaches to decipher protein structure and function do not keep up with this rapid expansion of the protein sequence space, motivating the development of fast computational tools that can help in filling the gap. Machine learning (ML) plays a vital role in the development of these tools since it provides a framework where the limited amount of experimentally solved data is used to design, develop and benchmark predictive algorithms that are later used to make predictions for the millions of uncharacterized sequences.

Hundreds of ML-based methods for the prediction of protein structure and function from the sequence were designed and released over the last few decades. They target a variety of structural aspects of proteins including prediction of tertiary structure, secondary structure, residue contacts, torsion angles, solvent accessibility, intrinsic disorder and flexibility. These methods also address numerous functional characteristics, such as prediction of binding to nucleic acids, proteins and lipids and identification of putative catalytic, cleavage and post-translational modification sites. Recent years have witnessed two significant advances in this area, the development of modern deep neural networks and the innovative representations of the protein sequences that that draw from the natural language processing (NLP) [3,4]. These advances have fueled the modern era in the sequence-based prediction of protein structure and function, resulting in the release of accurate and impactful solutions, such as AlphaFold for the tertiary structure prediction [5], SignalP for the signal peptide prediction [6] and flDPnn for the disorder and disorder function prediction [7]. Besides having free and convenient access to these predictive resources, nowadays, users also benefit from useful resources including large databases of protein structure and function predictions, such as AlphaFold DB [8], MobiDB [9], and DescribePROT [10], and user-friendly platforms that ease the development of new predictive tools, such as iLearnPlus [11] and BioSeq-BLM [12].

This book aims to guide students and scientists working with proteins through the exciting and rapidly advancing world of modern ML tools and resources for the efficient and accurate prediction and characterization of functional and structural aspects of proteins. This edited volume includes a mixture of introductory and advanced chapters written by well-published and accomplished experts. It covers a broad spectrum of predictions including tertiary structure, secondary structure, residue contacts, intrinsic disorder, protein, peptide and nucleic acids-binding sites, hotspots, post-translational modification sites, and protein function. It spotlights cutting-edge topics, such as deep neural networks, NLP-based sequence embedding, and prediction of residue contacts, tertiary structure, and intrinsic disorder. Moreover, it introduces and discusses several practical resources that include databases of predictions and software platforms for the development of novel predictive tools, providing a holistic coverage of this vibrant area.

The book includes 13 chapters that are divided into four parts. The **first part** focuses on ML algorithms. It features Chapter 1 (lead author: Dr. Hongbin Shen) that introduces key types of deep neural networks, such as convolutional, recurrent, and transformer topologies. The authors describe different ways to encode protein sequences and overview applications of deep learners across several types of protein structure predictions including secondary structure, contact maps and tertiary structure.

The **second part** addresses the inputs for the ML models and comprises of four chapters. It focuses on recently developed NLP-based approaches with three chapters that provide background and detail different applications. Chapter 2 (lead author: Dr. Daisuke Kihara) provides a comprehensive introduction to the popular NLP-inspired sequence embedding approaches including Word2Vec, UDSMProt, UniRep, SeqVec (ELMo), ESM-1b and BERT. The authors cover relevant databases and applications of these embeddings to the residue contact, secondary structure, and function predictions. Chapter 3 (lead author: Dr. Bin Liu) expands on this topic by discussing the applications of the NLP-based embedding to the prediction of protein folds, intrinsic disorder and protein-protein and protein-nucleic acids binding. It also introduces several popular ML-based predictors for these applications. Chapter 4 (lead author: Dr. Dukka KC) provides further details on several popular embedding methods and highlights their applications to the prediction of the post-translational modification sites and protein function. The key strength of this chapter is its comprehensive coverage of many ML algorithms in these two application areas. Finally, Chapter 5 (lead author: Dr. Shandar Ahmad) describes more traditional approaches to encoding protein sequences that include evolutionary profiles, biophysical properties and predicted structural features. It also sets these encodings in a practical context of methods for the prediction of protein-protein and protein-nucleic acids interactions.

The **third part** consists of six chapters that explain and describe sequence-based predictors for specific protein structure and function characteristics. Starting with Chapter 6 (lead author: Dr. Liam J. McGuffin), it introduces the contact prediction area, which is arguably one of the main innovations behind the success of the AlphaFold method. This chapter defines residue contacts, reviews the history of this topic, outlines its importance and growing influence, and describes key methods for predicting protein contacts and distance maps. The authors cover different ML algorithms, including hidden Markov models, support vector machines, random forests, Bayesian methods, and modern end-to-end deep neural networks. Chapter 7 (lead author: Dr. Dong-Jun Yu) provides an in-depth treatment of the residue contacts prediction area, focusing on algorithmic details of selected recent ML-based predictors, while assuming basic knowledge of this area and ML algorithms. It also surveys popular predictors, particularly focusing on the deep learning-based models. Subsequently, Chapter 8 (lead author: Dr. Lukasz Kurgan) focuses on the intrinsic disorder prediction area. It provides historical perspective and introduces numerous useful resources including commonly used and accurate disorder and disorder function predictors. In particular, the authors focus on the deep network-based solutions, databases of disorder predictions, webservers, and methods that provide quality assessment of disorder predictions. Chapter 9 (lead author: Dr. Yuedong Yang) defines protein-protein and protein-peptide interactions and introduces the corresponding prediction area. The authors survey related databases, sequence encodings, and predictive models. They also discuss popular assessment schemes and metrics for the evaluation of predictive performance, which they utilize to describe, compare, and recommend several leading methods in this area. Chapter 10 (lead authors: Drs Min Li and Lukasz Kurgan) describes the area concerned with the prediction of the protein-DNA and protein-RNA interactions. This chapter provides a comprehensive review of the current predictors of the protein-nucleic acids binding. Additionally, the authors recommend useful methods and discuss the importance of the underlying structural states of the nucleic acids binding regions. Lastly, Chapter 11 (lead author: Dr. Michael Gromiha) explores databases and ML methods for identifying cancer causing mutations. These resources have direct implication in the context of the development of precision medicine solutions. The authors emphasize on recently developed methods that rely on large scale data in order to produce practical and accurate results.

Part four centers on practical resources that support the development of new ML-based tools and that provide convenient access to the predictions generated by the methods described in the earlier chapters. Chapter 12 (lead authors: Drs Jiangning Song and Lukasz Kurgan) describes an innovative software platform, iLearnPlus, that can be used to analyze structural and functional characteristics of the DNA, RNA and protein sequences and to efficiently conceptualize, design, implement and comparatively evaluate ML-based predictors of these characteristics. The authors use an example application to demonstrate how easy it is to utilize iLearnPlus to design and evaluate a new and accurate predictor of the lysine malonylation sites. Chapter 13 (lead author: Dr. Lukasz Kurgan) highlights convenient ways to obtain pre-computed predictions of protein structure and function from large-scale databases, such as MobiDB, D^2P^2 and DescribePROT. The authors expand on their broad coverage of key characteristics, such as domains, secondary structure, solvent accessibility, intrinsic disorder, posttranslational modification sites, protein/DNA/RNA-binding, disordered linkers, and signal peptides. They also concisely discuss modern predictive webservers that should be used when users want to collect predictions for proteins that are not included in these databases.

Altogether, this book provides a comprehensive perspective on the concepts, methods and resources in the area of the ML-based prediction of protein structure and function from protein sequences. It introduces modern predictive models, including a variety of deep neural networks; provides in-depth treatment of cutting-edge methods for encoding of the input protein sequences; describes state-of-the art predictors for major structural and functional characteristics of proteins; and highlights useful resources that facilitate building new ML models and provide easy access to the predictions. This edited volume serves as a definitive reference for both budding and experienced developers and users of the ML models in this area.

Lukasz Kurgan

References

[1] UniProt C. UniProt: the universal protein knowledgebase in 2021. *Nucleic Acids Research*, 2021. **49**(D1): D480–D489.

[2] Li W, *et al.* RefSeq: expanding the Prokaryotic Genome Annotation Pipeline reach with protein family model curation. *Nucleic Acids Research*, 2021. **49**(D1): D1020–D1028.

[3] Ofer D, Brandes N and Linial M. The language of proteins: NLP, machine learning & protein sequences. *Computational and Structural Biotechnology Journal*, 2021. **19**: 1750–1758.

[4] LeCun Y, Bengio Y, and Hinton G. Deep learning. *Nature*, 2015. **521**(7553): 436–44.

[5] Jumper J, *et al.* Highly accurate protein structure prediction with AlphaFold. *Nature*, 2021. **596**(7873): 583–589.

[6] Teufel F, *et al.* SignalP 6.0 predicts all five types of signal peptides using protein language models. *Nature Biotechnology*, 2022. **40**: 1023–1025.

[7] Hu G, *et al.* flDPnn: Accurate intrinsic disorder prediction with putative propensities of disorder functions. *Nature Communications*, 2021. **12**(1): 4438.

[8] Varadi M, *et al.* AlphaFold Protein Structure Database: massively expanding the structural coverage of protein-sequence space with high-accuracy models. *Nucleic Acids Research*, 2021. **50**(D1): D439–D444.

[9] Piovesan D, *et al.* MobiDB: intrinsically disordered proteins in 2021. *Nucleic Acids Research*, 2021. **49**(D1): D361–D367.

[10] Zhao B, *et al.* DescribePROT: database of amino acid-level protein structure and function predictions. *Nucleic Acids Research*, 2021. **49**(D1): D298–D308.

[11] Chen Z, *et al.* iLearnPlus: a comprehensive and automated machine-learning platform for nucleic acid and protein sequence analysis, prediction and visualization. *Nucleic Acids Research*, 2021. **49**(10): e60.

[12] Li HL, Pang YH and Liu B. BioSeq-BLM: a platform for analyzing DNA, RNA and protein sequences based on biological language models. *Nucleic Acids Research*, 2021. **49**(22): e129.

Acknowledgments

First and foremost, I thank my wife, Dr. Magdalena Adamek, and my son, Aleksander Kurgan, for their unwavering support and permitting me to devote many long evening hours to this project. I had you in my thoughts from the start to the finish of this rewarding journey.

I am in great debt to the remarkable groups of students and researchers who contributed, worked and graduated from my lab. These are the people who designed, built, tested and deployed many of the tools and resources that are described in this book. I would not be able to do any research at all without your hard work, ingenuity and commitment. The key people include (in alphabetical order) Dr. Amita Barik, Mr. Balint Biro, Dr. Ke Chen, Dr. Xiao Fan, Mr. Sina Ghadermarzi, Mrs. Leila Homaeian, Dr. Akila Katuwawala, Dr. Fanchi Meng, Mrs. Fatemeh Miri, Dr. Marcin Mizianty, Dr. Christopher Oldfield, Dr. Zhenling Peng, Dr. Chen Wang, Dr. Jing Yan, Mr. Fuhao Zhang, Dr. Hua Zhang, Dr. Tuo Zhang, and Dr. Bi Zhao. I also acknowledge with big thanks my long-term collaborators, Dr. A. Keith Dunker, Dr. Jianzhao Gao, Dr. Gang Hu, Dr. Jishou Ruan, Dr. Vladimir Uversky, Dr. Kui Wang, Dr. Zhonghua Wu and Dr. Jian Zhang. I hope that I did not miss anyone, and I deeply apologize if I did.

I am very grateful to Dr. Shandar Ahmad, Dr. KC Dukka, Dr. Michael Gromiha, Dr. Daisuke Kihara, Dr. Min Li, Dr. Bin Liu, Dr. Liam McGuffin, Dr. Hong-Bin Shen, Dr. Jiangning Song, Dr. Yuedong Yang, and Dr. Dong-Jun Yu for contributing chapters to this book. It was very enjoyable and gratifying to work together.

Finally, I thank Xiao Ling, Vanessa Quek ZhiQin and Joy Quek for managing the process of the book production and to the World Scientific for publishing this volume.

About the Editor

Dr. Kurgan received his Ph.D. degree in Computer Science from the University of Colorado at Boulder, U.S.A. in 2003. He joined Computer Science Department at the Virginia Commonwealth University in 2015 as the Robert J. Mattauch Endowed Professor. Before that, he was a Professor at the Electrical and Computer Engineering Department at the University of Alberta. His research focuses on applications of modern machine learning in structural bioinformatics and spans structural genomics, protein function prediction, prediction of protein-drug interactions and characterization of intrinsic disorder. His work was featured in high-impact venues including *Chemical Reviews, Nature Communications, Nucleic Acids Research, Cellular and Molecular Life Sciences* and *Science Signaling*. He is known for the development of accurate bioinformatics methods for the intrinsic disorder and disorder function predictions including MFDp that was ranked third in the CASP10 assessment of disorder predictions [*Proteins* 2014 82(Suppl 2):127–3] and flDPnn that won the CAID experiment [*Nat Methods* 2021 18:454–5; *Nat Methods* 2021 18:472–81]. One of his early machine learning inventions, the CAIM discretization algorithm, was incorporated into several popular platforms, such as MATLAB, R and KEEL, and inspired many derived algorithms. Dr. Kurgan also spearheaded the development of large bioinformatics databases, such as PDID (1.1 million protein-drug interactions) and DescribePROT (13.5 billion protein structure and function predictions). Dr. Kurgan was recognized as the member of the European Academy of Sciences and Arts in 2022, Fellow of the Asia-Pacific Artificial Intelligence Association (AAIA) in 2021, Fellow of the American Institute for Medical and Biomedical Engineering (AIMBE) in 2018, and Fellow of the Kosciuszko Foundation Collegium of Eminent

Scientists in 2018. He was also inducted as a member of Faculty Opinions in the "Bioinformatics, Biomedical Informatics and Computational Biology" area in 2021. His research was and is supported by numerous competitive grants from the National Science Foundation, Canadian Institutes of Health Research, Natural Sciences and Engineering Research Council of Canada, Alberta Cancer Foundation and Canadian Arthritis Network. Dr. Kurgan is the Associate Editor-in-Chief for *Biomolecules*, Structural Bioinformatics Section Editor for *BMC Bioinformatics* and serves on the Editorial Boards for *Bioinformatics* and several other journals. He has reviewed applications and articles for over a dozen funding agencies and close to 100 different journals and serves on the Steering Committee for the International Conference on Machine Learning and Applications (ICMLA) and as Program Co-Chair for BIBM'21, ICMLA'09, ICMLA'08, ICMLA'06, and ICMLA'05 conferences. However, he is particularly proud of the over 20 postdocs, doctoral and master's students who he mentored and who graduated from his lab. More details about him and his laboratory are available at http://biomine.cs.vcu.edu/.

Contents

Part I

Machine Learning Algorithms

Chapter 1

Deep Learning Techniques for *De novo* Protein Structure Prediction

Chunqiu Xia and Hong-Bin Shen*

*Institute of Image Processing and Pattern Recognition,
Shanghai Jiao Tong University, and
Key Laboratory of System Control and Information Processing,
Ministry of Education of China,
Shanghai, China 200240
corresponding author: hbshen@sjtu.edu.cn

De novo tertiary protein structure prediction is a highly complex and long-standing problem in structural bioinformatics. Traditional methods divide the structure prediction procedure into several more manageable sub-tasks, such as local structure prediction, contact map prediction, fragment assembling, refinement and quality assessment. In recent years, deep learning techniques have been introduced to the protein structure prediction and have achieved great success in some of the subtasks, such as the contact map and secondary structure predictions. High-quality predicted contact maps and their variants, such as distance maps and orientation maps, help to improve the performance of the tertiary structure prediction. In the latest CASP competition, end-to-end deep neural network models have significantly enhanced quality of the predicted protein

structures, suggesting that deep learning techniques provide a promising
direction for solving this complex problem. In this chapter, we introduce
progress of the development and applications of deep learning techniques to the protein structure prediction and discuss potential reasons
for their effectiveness.

1. Introduction

Proteins are important biological macromolecules, composed of one or
multiple chains of amino acids (AAs). These polypeptides fold into a variety of structures in the three-dimensional (3D) space. Considering that
protein structures determine their functions and interactions with other
molecules to a large extent, many experimental methods are designed and
used to solve the protein structures, such as X-ray crystallography, NMR
spectroscopy and cryo-electron microscopy (cryo-EM). However, these
methods are usually expensive, time-consuming and labor-intensive.
Determining protein structure computationally, which is also known as the
de novo protein structure prediction, provides low-cost information that
complements and supports these experimental efforts. The major goal of
the protein structure prediction is to derive the 3D coordinate of each
heavy atom in a given protein from its sequence. Its theoretical basis is the
thermodynamic hypothesis that the native structure of a protein can be
determined by its AA sequence under the standard physiological environment [1]. Although some proteins show switched folds [2,3] and some are
intrinsically disordered [4,5], arguably most of proteins follow the above
paradigm.

Many efforts over decades have been dedicated to protein structure
prediction. Most of the approaches can be roughly divided into two types:
template-based modeling (TBM) and free modeling (FM) [6]. TBM
requires the query protein, which has homologous proteins with known
structures. FM, also known as the *ab initio* modeling, does not require very
similar structure templates. As more hybrid algorithms that combine the
two types of modelling approaches are proposed, the boundary between
TBM and FM is blurred [7–9]. The procedure adopted by many approaches
to the protein structure prediction usually consists of several modules, such
as the local structure prediction, contact map prediction, fragment assembling, refinement, quality assessment, and etc. (Fig. 1). Here, local

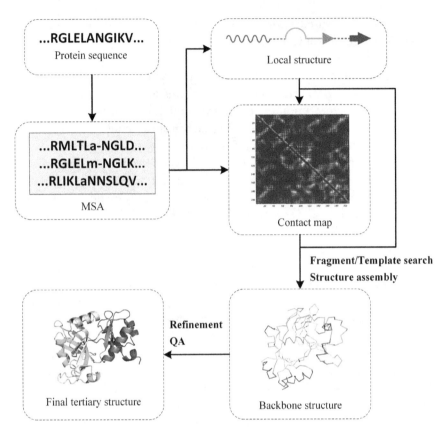

Figure 1. An illustration of relatively independent procedures of protein structure prediction before the application of end-to-end deep neural networks. For a given protein sequence, sequence alignment tools are used to obtain its MSA by searching against sequence databases. Then, local structures and contact maps are predicted based on the MSA and can be further used to assemble the backbone structure. Refinement would be used to add side chains and remove the structural violation and clashes. Quality assessment (QA) will be applied to estimate the accuracy of decoys and determine the final tertiary structure.

structures refer to the geometric features of a continuous fragment of a polypeptide, such as the secondary structures. Contact maps describe the Euclidean distance between each residue pairs in a protein. If local structures and contact maps are accurately predicted, the conformation search space of the corresponding structure is largely reduced. Then, heuristic algorithms can be used to assemble the fragments according to physical principles or empirical rules. This procedure had been used as a common pipeline until end-to-end deep learning techniques were introduced.

Deep learning is a subfield of machine learning and representation learning [10,11]. It extracts high-level features and performs prediction simultaneously while traditional shallow models rely heavily on feature engineering. Therefore, deep learning algorithms are also considered as the end-to-end models. The core of deep learning techniques is the artificial neural network (ANN) which is biologically inspired by the human brain [12]. From the view of machine learning, it can be seen as a multilayer perceptron (MLP) with nonlinear activation function and error backpropagation. Features learned by multiple nonlinear transformation are more discriminative and compact compared with those handcraft features [10]. Since AlexNet won the ImageNet Challenge in 2012 [16], deep learning has shown very promising performance in speech recognition, image recognition, and natural language processing (NLP) [13–15].

With the development of high-throughput sequencing techniques, genetic and protein sequences are accumulated rapidly. There are about 2.5 billion sequences which are derived from metagenomes in the largest protein sequence dataset, BFD [17,18]. Researchers have applied deep learning techniques to molecular biology, especially structural biology, in the past few years [19,20]. The success of RaptorX-Contact in the Critical Assessment of protein Structure Prediction (CASP) competition showed that deep learning performed better on the protein contact map prediction when large-scale protein sequence data are available, which spurred growing interest in this technology [21,22]. Afterwards, more deep learning-based methods were proposed for a variety of sequence-based protein structure prediction subtasks [23–25]. Compared with traditional machine learning methods, higher prediction accuracy achieved by deep learning-based methods may be due to their powerful representation ability. In CASP13, AlphaFold which applied deep neural networks to contact map and torsion angle prediction, showed promising results [26]. The end-to-end recurrent geometric network (RGN) was also proposed in CASP13 [27].

Deep neural networks require large-scale experimentally annotated data for training and evaluation. However, there are only about 181,000 proteins in Protein Data Bank (PDB) by 2021 [28]. Lack of structurally annotated sequence data raises challenges for the supervised end-to-end models. To partially overcome this problem, self-supervised language models are applied to the protein sequence representation task since 2019 [14].

Protein language models are used to learn evolutionary information from unsupervised AA type prediction tasks [29]. The learned protein sequence embeddings are found more effective for many downstream structure prediction tasks than features directly derived from sequence profiling or direct coupling analysis (DCA) [30,31]. In CASP 14, an end-to-end model AlphaFold2 makes use of unlabeled and labeled data by self-supervised learning and data augmentation [32]. The improvement brought by AlphaFold2 is tremendous and demonstrates that deep learning is a promising tool for the protein structure prediction in the future.

In this chapter, we review recent development of deep learning models for sequence-based protein structure prediction. We first introduce several architectures of deep neural networks which have been widely used in the structural bioinformatics. Then, we discuss the latest protein language modeling and its effects on protein structure prediction tasks. Next, we compare and analyze existing supervised secondary structure prediction models, contact map prediction models, and end-to-end tertiary structure prediction models in detail. Finally, we examine potential future direction of deep learning in the structural bioinformatics field.

2. Architectures of deep neural networks

Neural networks usually stack multiple modules/layers. Each of these modules could be modeled as a simple nonlinear function. Theoretically, a multi-layer dense network (i.e., MLP) can approximate any continuous function [33]. However, this is challenging to achieve in real-world applications because the data is limited and the target functions are complicated. Therefore, several more effective neural network architectures with stronger inductive bias are designed for processing data that has specific format. For protein sequence data, convolutional neural networks (CNNs), recurrent neural networks (RNNs) and attention-based neural networks are among the most popular architectures. We briefly introduce them in this section.

2.1 *Convolutional neural networks*

The architecture of CNNs is related to the *Neocognitron* and the time delay neural network (TDNN) [34,35]. Modern CNNs use backpropagation for network optimization [36]. CNNs are designed for data with structure, such as sequences (1D data), images (2D data) and point clouds (3D data). For example, any pixel in an image is more correlated to its spatially neighboring pixels and pixels in the same neighborhood can form a local pattern. These patterns are unrelated to their locations in the image and can further form larger or higher-level patterns. Convolutional layers and pooling layers in the CNNs capture both low-level and high-level patterns more easily than MLPs.

The implementation of convolution and pooling is similar to a sliding window or a filter. The difference between them is the function to process the data in the window. Taking the image processing as an example, when a batch of images which are represented as a tensor $\mathbf{I} \in \mathbb{R}^{(N \times W \times H \times D_{in})}$ are input into a convolutional layer, the output tensor $\mathbf{O} \in \mathbb{R}^{(N \times W \times H \times D_{out})}$ can be derived as follows:

$$\mathbf{O}_{i,j} = \sum_{d=1}^{D_{in}} \mathbf{W}_d * \mathbf{I}_{i,d} + b_j \tag{1}$$

where N denotes the batch size, W and H denote the size of images, D_{in} and D_{out} denote the number of input and output channels respectively, $\mathbf{I}_{i,d}$ denotes the dth channel of the ith input image, and $\mathbf{O}_{i,j}$ denotes the jth channel of the ith output image. $\mathbf{W}_d$ is the dth learnable weight matrix, b_j is the jth learnable bias, and * is the valid 2D cross-correlation operator. The pooling operation substitutes the weighted sum function with a maximum or average function and usually has a larger stride.

CNNs stack multiple convolutional layers and pooling layers to learn features. When the network is deeper, its capacity is growing but its performance may degrade due to problems, such as overfitting and gradient vanishing. A variety of methods are proposed to mitigate these problems, such as dropout, batch normalization, and identical map [13,37,38]. These techniques are also used in other types of deep neural networks.

2.2 *Recurrent neural networks*

RNNs are effective frameworks for processing sequence data. Compared with MLPs, RNNs have a memory cell to record the state of the previous time step. A simple RNN can be formulated as follows [39]:

$$\mathbf{h}_t = \sigma_h(\mathbf{W}_h \mathbf{x}_t + \mathbf{U} \mathbf{h}_{t-1} + b_h) \tag{2}$$

$$\mathbf{y}_t = \sigma_y(\mathbf{W} y \mathbf{h}_t + b_y) \tag{3}$$

where $\mathbf{x}_t$, $\mathbf{y}_t$, and $\mathbf{h}_t$ is the input embedding, the output embedding, and hidden state vector at the tth time step, respectively. $\mathbf{W}$, $\mathbf{U}$, and b are learnable parameters and σ is the activation function. When RNNs are unrolled along the long sequences, they can be seen as deep feedforward neural networks. Therefore, RNNs also suffer from gradient vanishing. To solve the problem, long short-term memory (LSTM) neural networks are proposed, which add a memory cell and implement gate control [40]. For any input vector $\mathbf{x}_t$ at the tth time step, the output vector $\mathbf{h}_t$ (i.e., the hidden state vector) and the memory cell $\mathbf{c}_t$ are derived as follows:

$$\mathbf{f}_t = \sigma_g(\mathbf{W}_f\mathbf{x}_t + \mathbf{U}_f\mathbf{h}_{t-1} + b_f) \tag{4}$$

$$\mathbf{i}_t = \sigma_g(\mathbf{W}_i\mathbf{x}_t + \mathbf{U}_i\mathbf{h}_{t-1} + b_i) \tag{5}$$

$$\mathbf{o}_t = \sigma_g(\mathbf{W}_o\mathbf{x}_t + \mathbf{U}_o\mathbf{h}_{t-1} + b_o) \tag{6}$$

$$\tilde{\mathbf{c}}_t = \sigma_c(\mathbf{W}_c\mathbf{x}_t + \mathbf{U}_c\mathbf{h}_{t-1} + b_c) \tag{7}$$

$$\mathbf{c}_t = \mathbf{f}_t \circ \mathbf{c}_{t-1} + \mathbf{i}_t \circ \tilde{\mathbf{c}}_t \tag{8}$$

$$\mathbf{h}_t = \mathbf{o}_t \circ \sigma_t(\mathbf{c}_t) \tag{9}$$

where $\mathbf{f}_t$, $\mathbf{i}_t$, $\mathbf{o}_t$ and $\tilde{\mathbf{c}}_t$ are forget gate, input gate, output gate and cell input respectively. $\mathbf{W}$, $\mathbf{U}$ and b are learnable parameters. σ_g, σ_c and σ_t are sigmoid, tanh and tanh function respectively. The operator $\circ$ is the element-wise product. LSTM learns long-term dependency because memory cells can store information from many time steps before. There have been various successful applications of LSTMs, such as speech recognition, machine translation, robot control and protein secondary structure and function prediction [41–45].

2.3 *Attention-based neural networks*

Although LSTM provides a promising way to alleviate the gradient vanishing problem, it has problems when processing very long sequences. For example, RNN-based encoder-decoder approaches were once popular for machine translation [43,46]. However, when the length of an input sentence increases, the performance of these approaches degrades. The reason is that the decoder only uses the fixed-length context vector generated at the last time step of the decoder, which leads to an information loss.

The attention mechanism is proposed to solve this problem by aggregating all hidden state vectors from the encoder [47].

If a simple RNN, like in equations (2) and (3), is used as an encoder, an attention-based decoder can be constructed as follows:

$$\mathbf{s}_i = f(\mathbf{s}_{i-1}, \mathbf{y}_{i-1}, \mathbf{c}_i) \tag{10}$$

where $\mathbf{s}_i$, and $\mathbf{c}_i$ denote the hidden state vector and the context vector at the ith time step of the decoder. f is a nonlinear function which is similar to those used in the encoder but adds a context vector. $\mathbf{y}_{i-1}$ denotes the output word probability vector at the $(i-1)^{\text{th}}$ time step. If $\mathbf{c}_i = h_T$, the model becomes a simple RNN encoder-decoder, where T is the length of the input sentence [43]. In an attention-based model, $\mathbf{c}_i$ can be derived as follows:

$$\mathbf{c}_i = \sum_{t=1}^{T} a_{it} \mathbf{h}_t \tag{11}$$

$$a_{it} = \left. exp\left(e_{it}\right) \middle/ \sum_{j=1}^{T} exp\left(e_{ij}\right) \right. \tag{12}$$

$$e_{it} = \mathbf{v}_a^T \sigma_a (\mathbf{W}_a \mathbf{s}_{i-1} + \mathbf{U}_a \mathbf{h}_t) \tag{13}$$

where $\mathbf{W}_a$, $\mathbf{U}_a$ and $\mathbf{v}_a$ are learnable parameters. σ_a is the activation function tanh. a_{it} is normalized to $(0,1)$ and $\mathbf{c}_i$ is a weighted sum of $\mathbf{h}_t$. Each output word learns from all hidden states from the encoder by applying attention a_{it}.

Combining RNNs and attention mechanism significantly improves the performance of the encoder-decoder approaches [47]. However, any hidden state in an RNN has to receive information from the previous time step [48,49]. Moreover, it is difficult to parallelize RNNs for the same reason. Therefore, a purely attention-based neural network, named Transformer, is proposed and has been proven to be the effective models for NLP [50]. The core of the transformer is the self-attention module, the scaled dot-product attention, which is formulated as follows:

$$A(\mathbf{Q}, \mathbf{K}, \mathbf{V}) = softmax\left(\frac{\mathbf{Q}\mathbf{K}^T}{\sqrt{d_k}}\right)\mathbf{V} \tag{14}$$

where $\mathbf{Q}$, $\mathbf{K}$, $\mathbf{V}$ and d_k denote the query matrix, key matrix, value matrix and the length of the row vectors in the key matrix, respectively. The transformer is also an encoder-decoder approach. In the encoder,

$\mathbf{Q} = \mathbf{K} = \mathbf{V} \in \mathbb{R}^{T \times d_k}$ and each row of them represents the hidden state at the corresponding time step. The scaled dot-product of $\mathbf{Q}$ and $\mathbf{K}$ is the attention matrix and will be sent to the decoder. Besides self-attention, several tricks including multi-head attention, positional encoding and masking are also used to enhance the representation ability of the network.

3. Self-supervised protein sequence representation

Searching query sequence by alignment against large sequence databases is often used to generate multiple sequence alignment (MSA) [51]. Evolutionary statistics can be extracted from MSA and are widely used for the *in silico* protein structure, function and interaction prediction [8,52,53]. However, it is difficult to directly apply machine learning methods to MSA if it is binarized. Thus, it is necessary to transform high-dimensional MSA to a more compact and dense representation. Position-specific scoring matrices (PSSMs) derived from PSI-BLAST during MSA generation reflect conservation information and are suitable for machine learning models [54]. The reason is that the shape of a PSSM is $L \times 20$, which makes it independent on the number of sequences in the MSA, where L is the length of the query sequence. Direct coupling analysis (DCA), also called Potts model, extracts coevolutionary information from MSA [31].

In the recent years, many self-supervised models are proposed for learning word embedding from natural language and achieve great success [14,48,49,55]. These models are also applied to protein sequence for AA residue representation. The input of protein language models is a single sequence or its corresponding MSA. Structure information is not involved in the training procedure. Thus, millions of sequences are used to train a model with high capacity. The resulting sequence representation is also effective for downstream tasks, such as contact map prediction and secondary structure prediction [56].

3.1 *Single-sequence-based protein sequence representation*

Single-sequence-based models learn protein representation from only a one-hot encoding of protein sequence. The language models they use are categorized into autoregressive (AR) models and autoencoding (AE) models. As shown in Figure 2, the pretext tasks for AR and AE protein language models are the next AA prediction and masked AA prediction, respectively.

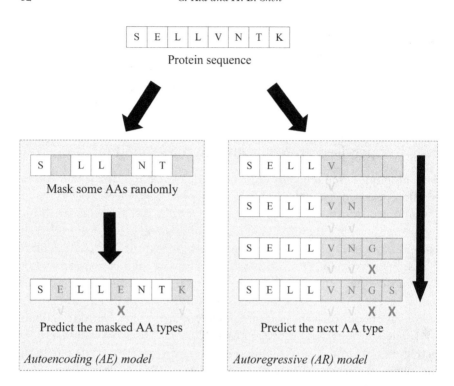

Figure 2. Pretext tasks for AR and AE protein language modeling. For a given protein sequence, AE models will mask some amino acids (AAs) randomly and then predict the masked AA types according to the remaining known AAs. AR models will predict the next AA type according to all the AAs before it iteratively.

Both RNNs and attention-based neural networks can serve as the encoder of AR models while the latter can be further used for AE models due to their different pretext tasks. Initially, AR models access contexts from one direction during training. Later, modified models, such as XLNet, have been proposed to let AR learn from bidirectional contexts in NLP, but there is still much room to further improve these methods' modeling of the protein sequence data [49]. On the contrary, AE models learn from bidirectional contexts but may suffer from a pretrain-finetune discrepancy due to introducing [MASK] token.

In Table 1, we summarize part of the representative self-supervised protein language models for protein sequence representation. TAPE method introduces the Transformer to protein sequence representation

Table 1. Some self-supervised protein language models for protein sequence representation.

Method	Input	LM[a]	Encoder	Training dataset[b]
TAPE [57]	Single sequence	AE	Transformer [50]	Pfam [61]
SeqVec [62]	Single sequence	AR	ELMo [48]	UniRef50 [63]
UniRep [58]	Single sequence	AR	mLSTM [64]	UniRef50
ESM-1b [29]	Single sequence	AE	Transformer	Uniparc [65]
ESM-MSA-1b [56]	MSA	AE	Transformer	UniRef50

[a]Abbreviation for Language model.
[b]Pfam, UniRef50, and Uniparc contain about 31 million, 24 million, and 250 million protein sequences, respectively.

and provides several benchmarks for evaluating protein language models [57]. The results show that the performance of TAPE and UniRep is close to each other on several supervised downstream tasks [58]. A possible reason could be that the training sets of TAPE and UniRep are not large enough. ESM-1b also applies Transformer but has 650 million parameters and is trained on 250 million protein sequences [29]. Features derived from it also show promising results on the secondary structure prediction and contact map prediction tasks [29,59,60].

3.2 MSA-based protein sequence representation

The input of an MSA-based model is generally the MSA of the target sequence rather than the sequence itself. The pretext task of an MSA-based model is to predict randomly masked AAs in the MSA. It should be noted that the masked AAs can be predicted not only from the residues in the context of the same sequence, but also from the residue at the same position from other sequences in the MSA [56]. If we simply concatenate all sequences in an MSA, the size of attention matrix will be M^2L^2 where M is the number of sequences in the MSA and L is the length of each sequence. The space and time complexity of the operation could be prohibitive. In ESM-MSA-1b, a variant of axial attention is applied to reduce the computational cost [56,66,67]. As shown in Figure 3, untied row attention helps to reduce the attention cost to $O(ML^2)$ and $O(M^2L)$. Moreover, tied row attention makes different sequences share the same attention matrix and further reduce the attention cost to $O(L^2)$, which makes full use of the evolutionary information in the MSA.

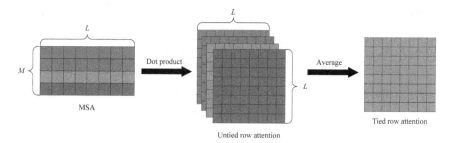

Figure 3.   Untied and tied row attention used in the ESM-MSA-1b. M is the number of sequences in the MSA and L is the length of each sequence.

4. Secondary structure prediction

Secondary structure prediction (SSP) is important for the tertiary structure prediction. Secondary structures can be used for template search in threading algorithms [68,69]. Early methods for SSP are based on statistical analysis and were developed in 1970s [70–73]. These methods mainly consider the AA types of the target single residue or a triplet centered on it. Thus, the Q3 accuracy (the fraction of correctly predicted residues for 3-class prediction) of them is generally not very high, i.e., ~60% [74]. The sliding window strategy is then used to extract longer continuous fragments as the input, based on the assumption that the secondary structure of the target residue is highly related to the residues around it. Moreover, as more protein structures are determined, powerful machine learning models applied to SSP of the introduction of evolutionary information improves the Q3 accuracy to over 70% [74]. Artificial neural networks are currently one of the most used and the state-of-the-art machine learning models for SSP and further improve the prediction accuracy.

4.1 *Neural networks used for local structure prediction*

MLPs (i.e., dense neural networks or fully connected neural networks) used for SSP often contain one or two hidden layers. Early MLP-based approaches accept a sliding window of one-hot encoding AA fragments as their input and achieve ~65% Q3 accuracy [75]. Then, evolutionary features were one of the popular encoding approaches. For example, in PSIPRED, the normalized PSSMs were used as the input of the first MLP and its Q3 accuracy is ~76% [76]. Other similar methods also demonstrated the effectiveness of evolutionary features [77,78]. Besides SSP, MLPs are also used for other local properties of protein backbones. The initial version of SPIDER predicts the θ angle (i.e., the angle formed by three consecutive C_α atoms) and τ angle (i.e., the dihedral angle formed by four consecutive C_α atoms) based on an MLP with three hidden layers [79], where its input consists of predicted secondary structures and predicted solvent accessible surface area from SPINE-X [80]. The improved version SPIDER2 additionally predicts secondary structures, solvent accessible surface area and torsion angles [81].

In general, MLPs use the sliding window strategy to derive fixed-length input. The size of sliding window is an important hyper-parameter and needs careful adjustment [82]. Too large sliding windows may lead to overfitting or introduce noise while too small sliding windows may lose useful information. RNNs are designed for sequential data and can accept variable-length sequences. RNN-based models often use bidirectional RNNs to extract context from both sides of the target residue. SSpro and its improved version Porter which achieve ~79% Q3 accuracy are composed of an ensemble of two-stage bidirectional RNNs (BRNNs) [83,84]. SPIDER3 changes its architecture to bidirectional LSTMs (BiLSTMs) from MLPs and achieves ~84% Q3 accuracy [44].

CNNs are usually used for 2D data, such as images, but have also been applied to SSP and perform as well as other neural network architectures [85,86]. DeepCNF substitutes the MLPs in conditional random fields (CRF) with CNNs which can capture complex relationships between input features and output labels [85]. The experiments show that the improvement of DeepCNF is mainly brought by deep CNNs [85], which has also been applied in other models [86].

4.2 *State-of-the-art SSP approaches benefit from larger data, deeper networks, and better evolutionary features*

When investigating the difference between some non-consensus SSP approaches and their updated versions, the results summarized in Table 2 suggest that the improvement of SSP may be related to several factors. We note that this table aims to compare different versions of the same method, not different methods with each other. After introducing sequence profile from PSI-BLAST, HHblits and MMseqs2 [54,59,87], most of the deep learning-based SSP approaches can achieve over 80% Q3 accuracy [88,89]. Then, these approaches use more annotated data for training and change

Table 2.   Comparison between PSI-PRED, Porter, SPIDER, NetSurfP and their corresponding updated versions.

Method	N_{train}[a]	Architecture	Features	Q3 accuracy[f]
PSIPRED [76]	~1100[b]	1-layer MLPs[c]	PSSM	76%
PSIPRED4 [91, 92]		2-layer MLPs	PSSM	84.2%
Porter [84]	2171	BRNNs	PSSM	79%
Porter4.0 [93]	7522	BRNNs	PSSM	82.2%
SPIDER2 [81]	4590	3-layer MLPs	PSSM,PP[d]	82%
SPIDER3 [44]	4590	BiLSTMs	PSSM, PP, HMM[e]	84%
NetSurfP1.0 [94]	2085	1-layer MLPs	PSSM	81%
NetSurfP2.0 [90]	10337	HNNs	HMM, MMseqs2, one-hot	85%

[a]The number of proteins used for network training.
[b]There are three overlapped training sets of PSIPRED, each of which contains ~1100 proteins
[c]N-layer MLP means the MLP contains N hidden layers.
[d]PP: Physicochemical properties.
[e]HMM: HMM profile derived from HHblits [59].
[f]All Q3 accuracy is derived from their original papers.

shallow MLPs to deeper neural networks, including MLPs, RNNs and hybrid neural networks (HNNs) that are composed of different types of layers. For example, Netsurf2.0 implements a very deep HNN which is composed of CNNs and RNNs and uses over 10k protein sequences for training [90]. It gains 4% improvement in the Q3 accuracy compared with its single-layer MLP version. Moreover, better MSA tools and fast accumulated unannotated protein sequence data also make sequence profiling more precise. When substituting the HMM profile used in the Netsurf2.0 with the representation from the ESM-MSA-1b, the Q8 accuracy (the fraction of correctly predicted residues for 8-class prediction) increases by 2% [56]. The results indicate that better evolutionary features from self-supervised neural networks can help to enhance SSP.

5. Contact map prediction

One of the most recent and successful applications of deep learning to protein structural bioinformatics is the contact/distance map prediction.

When contact/distance maps are determined accurately, protein structures can be reconstructed directly or indirectly with improvements (e.g., as an energy term for heuristic algorithms) [95,96].

5.1 *Neural networks used for contact map prediction*

The major goal of contact map prediction (CMP) is to figure out whether each residue pair in a protein is in contact or not. The most widely used definition of *contact*, which is also accepted by CASP, is that the distance between the C_{β} of two AA residues (C_{α} for Glycine) is smaller than 8Å. CMP methods can be generally categorized into two types: statistical models and machine learning models. Covariance-based methods were one of early successful statistical models. They are established based on a biological premise that when an AA residue mutate, its interacting residue would be mutated simultaneously to maintain the structure and function of the corresponding protein [97]. Thus, two interacting residues coevolve and their AA types are highly correlated. The correlation between a residue pair can be approximately derived from MSA by calculating mutual information (MI) and Pearson correlation coefficient (PCC) [98,99]. DCA is another type of statistical method used to avoid introducing the transitive noise involved in the traditional covariance methods [30], which can be considered as the unsupervised CMP methods.

Supervised machine learning methods usually rely on the output of DCA and sequence profiles. Most of early methods are shallow models, such as SVM and random forests [100,101]. Although MLPs also have been applied to CMP, their architectures are hardly suitable for processing 2D coevolutionary data and are relatively shallow. For example, MetaPSICOV is a two-stage MLP with one hidden layer [102], which extracts the context of a residue pair by a sliding window strategy due to the architecture of MLPs. In contrast, CNNs are designed for 2D data, such as images, and can be transferred to predict protein CMP by treating the coevolutionary features as images [103]. Convolutional kernels are similar to 2D sliding windows and can extract multiscale features by stacking multiple convolutional layers. CNN-based CMP methods' basic pipeline is shown in Figure 4. Using deeper neural networks or ensemble learning helps to increase the capacity of the models. For instance, DNCON2 used an ensemble of five CNNs at different distance thresholds and SPOT-Contact combines ResNet and BiLSTMs [13,104,105].

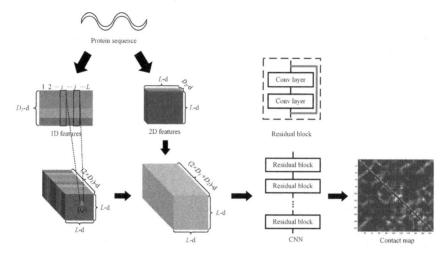

Figure 4. A basic pipeline of CNN-based protein contact map prediction approaches. The 1D features (e.g., PSSMs) and 2D features (e.g., coevolutionary features) are extracted from a protein sequence of length L. The number of types of 1D features and 1D features is D_1 and D_2, respectively. Then, 1D features are concatenated into 2D and further concatenated with 2D features. Finally, a residual neural network (i.e., a CNN with identical maps) is used to predict the contact map based on the input concatenated features.

5.2 Novel strategies used in state-of-the-art CMP approaches

Information loss is inevitable when binarizing a distance matrix into a contact map with a fixed distance threshold. To further improve the accuracy of CMP, top methods in CASP13 discretize inter-residues distance into more than two bins. In modified RaptorX-Contact, the number of distance bins is 25 (<4.5, 4.5–5, 5–5.5, ..., 15–15.5, 15.5–16, and >16) [106]. Similar strategies are also applied to TripletRes, trRosetta and AlphaFold [26,107,108]. Apart from distance prediction, searching against larger metagenomic databases to generate high-quality MSA is also meaningful. In MetaPSICOV, it was found that the number of hit sequences doubles when additionally using metagenomic databases, and the overall precision was improved [109].

Coevolutionary information is important for supervised CMP [107]. The latest self-supervised learning models also provide promising discriminative features for various downstream tasks. For example, the attention maps generated by ESM-MSA-1b are also useful for CMP task [56]. These results suggest that constructing an end-to-end tertiary structure prediction model is feasible.

6. End-to-end tertiary structure prediction

The traditional protein structure prediction procedure is divided into several parts which are shown in Figure 1. Deep learning methods applied to some of them have significantly improved the prediction accuracy [107]. Based on these experiments and observations, end-to-end paradigm of applying deep learning methods for this task has been gradually developed, such as the end-to-end tertiary structure prediction through RGN [27]. It contains three main stages: computation, geometry and assessment. In the first stage, a BiLSTM is used to predict three torsion angles from the input PSSM of the protein sequence. Then, torsion angles of each residue are transformed to the corresponding Cartesian coordinate by Natural Extension Reference Frame [110]. At last, the distance-based root mean square deviation (dRMSD) metric is used to measure the similarity between predicted and experimental structures. The whole procedure is differentiable and can be optimized by a gradient descent algorithm.

AlphaFold2 is another state-of-the-art end-to-end protein structure prediction model [32]. Compared with other approaches, AlphaFold2's networks are deeper and more sophisticated. This predictor consists of two main modules: Evoformer and Structure module [32]. The Evoformer is used to learn residue representation based on self-attention mechanism. Its input is the MSA and template of the query sequence, which are more informative than sequence profiles. Six large sequence databases (i.e., BFD, UniRef90, MGnify, Uniclust30, PDB70, and PDB) are used to generate high-quality MSA [18,28,63,111,112]. The Structure module is designed to predict rotation and translation for backbones and χ angles for side chains by invariant point attention (IPA). There are a total of 48 Evoformer blocks and 8 Structure modules in AlphaFold2. Considering that limited known protein structures may not support the large model, 350k high-confidence structures predicted by itself are also used as the training data. RoseTTAFold's architecture consists of two major stages [113], where in the first stage, the MSA and template information is used to generate the sequence representation and then, SE(3)-equivariant Transformer is used to generate atomic coordinates [114]. The prediction accuracy of

AlphaFold2 (e.g., the highest all-atom accuracy in CASP14) and RoseTTAFold demonstrate that end-to-end deep learning methods are promising directions for the tertiary structure prediction [32,113].

7. Conclusions

Protein structure prediction has been dramatically improved by deep learning techniques. Several factors are closely related to the success of deep learning-based methods, such as the rapidly accumulated unlabeled protein sequence data and effective network architectures. Different network architectures have been developed for protein structure prediction task, such as CNN, RNN, and more recent attention-based networks, such as BERT and Transformer. To train these large models, self-supervised learning techniques and data augmentation are generally applied. Several topics are worth further investigation. For instance, steric clashes still exist in the predicted structures, which need further refinement, and current major deep neural networks are black boxes which lack interpretability. Additionally, more efforts are still needed to find solutions to these problems. It can be expected that many downstream tasks would benefit from accurately predicted structures, such as protein docking, protein-ligand binding site prediction, and protein function prediction. These various applications will overall improve our understandings of protein structures and their complex functions.

References

[1]   Anfinsen CB. Principles that govern the folding of protein chains. *Science (New York, N.Y.)*, 1973. **181**(4096): 223–230.

[2]   Porter LL and Looger LL. Extant fold-switching proteins are widespread. *Proceedings of the National Academy of Sciences of the United States of America*, 2018. **115**(23): 5968–5973.

[3]   Pauwels K, *et al.* Chaperoning Anfinsen: The steric foldases. *Molecular Microbiology*, 2007. **64**(4): 917–922.

[4]   Oldfield CJ, *et al.* Introduction to intrinsically disordered proteins and regions. In *Intrinsically Disordered Proteins*, Salvi N (ed.), Academic Press, 2019. 1–34.

[5]   Lieutaud P, *et al.* How disordered is my protein and what is its disorder for? A guide through the "dark side" of the protein universe. *Intrinsically Disordered Proteins*, 2016. **4**(1): e1259708.

[6] Zhang Y. Progress and challenges in protein structure prediction. *Current Opinion in Structural Biology*, 2008. **18**(3): 342–348.

[7] Zhang Y. Template-based modeling and free modeling by I-TASSER in CASP7. *Proteins-Structure Function and Bioinformatics*, 2007. **69**: 108–117.

[8] Yang J, *et al.* The I-TASSER suite: Protein structure and function prediction. *Nature Methods*, 2015. **12**(1): 7–8.

[9] Pearce R and Zhang Y. Toward the solution of the protein structure prediction problem. *Journal of Biological Chemistry*, 2021. **297**(1): 100870.

[10] Bengio Y, *et al.* Representation learning: A review and new perspectives. *IEEE Transactions on Pattern Analysis and Machine Intelligence*, 2013. **35**(8): 1798–1828.

[11] LeCun Y, *et al.* Deep learning. *Nature*, 2015. **521**(7553): 436–444.

[12] Alom MZ, *et al.* The history began from AlexNet: A comprehensive survey on deep learning approaches. arXiv, 2018.

[13] He KM, *et al.* Deep residual learning for image recognition. *2016 IEEE Conference on Computer Vision and Pattern Recognition (CVPR)*, Las Vegas, IEEE, 2016. 770–778.

[14] Devlin J, *et al.* BERT: Pre-training of deep bidirectional transformers for language understanding. *Proceedings of NAACL-HLT*, Minneapolis, ACL, 2019. 4171–4186.

[15] Amodei D, *et al.* Deep speech 2: End-to-end speech recognition in English and Mandarin. *Proceedings of The 33rd International Conference on Machine Learning*, Balcan MF & Weinberger KQ (eds.), PMLR: Proceedings of Machine Learning Research, 2016. 173–182.

[16] Krizhevsky A, *et al.* ImageNet classification with deep convolutional neural networks. *International Conference on Neural Information Processing Systems*, 2012. Lake Tahoe, Nevada: Curran Associates Inc., 1097–1105.

[17] Steinegger M and Soding J. Clustering huge protein sequence sets in linear time. *Nature Communications*, 2018. **9**: 1–8.

[18] Steinegger M, *et al.* Protein-level assembly increases protein sequence recovery from metagenomic samples manyfold. *Nature Methods*, 2019. **16**(7): 603.

[19] Pan X, *et al.* IPMiner: Hidden ncRNA-protein interaction sequential pattern mining with stacked autoencoder for accurate computational prediction. *BMC Genomics*, 2016. **17**: 582.

[20] Alipanahi B, *et al.* Predicting the sequence specificities of DNA- and RNA-binding proteins by deep learning. *Nature Biotechnology*, 2015. **33**(8): 831–838.

[21] Wang S, *et al.* Accurate de novo prediction of protein contact map by ultra-deep learning model. *PLOS Computational Biology*, 2017. **13**(1): e1005324.

[22] Wang S, *et al.* Analysis of deep learning methods for blind protein contact prediction in CASP12. *Proteins-Structure Function and Bioinformatics*, 2018. **86**: 67–77.

[23] Xia Y, *et al.* GraphBind: Protein structural context embedded rules learned by hierarchical graph neural networks for recognizing nucleic-acid-binding residues. *Nucleic Acids Research*, 2021. **49**(9): e51.

[24] Feng SH, *et al.* Topology prediction improvement of alpha-helical transmembrane proteins through helix-tail modeling and multiscale deep learning fusion. *Journal of Molecular Biology*, 2020. **432**(4): 1279–1296.

[25] Katuwawala A, *et al.* DisoLipPred: Accurate prediction of disordered lipid binding residues in protein sequences with deep recurrent networks and transfer learning. *Bioinformatics*, 2021. **38**(1): 115–124.

[26] Senior AW, *et al.* Improved protein structure prediction using potentials from deep learning. *Nature*, 2020. **577**(7792): 706.

[27] AlQuraishi M. End-to-end differentiable learning of protein structure. *Cell Systems*, 2019. **8**(4): 292–301.e3.

[28] Berman HM, *et al.* The protein data bank. *Nucleic Acids Research*, 2000. **28**(1): 235–242.

[29] Rives A, *et al.* Biological structure and function emerge from scaling unsupervised learning to 250 million protein sequences. *Proceedings of the National Academy of Sciences of the United States of America*, 2021. **118**(15): e2016239118.

[30] Weigt M, *et al.* Identification of direct residue contacts in protein-protein interaction by message passing. *Proceedings of the National Academy of Sciences of the United States of America*, 2009. **106**(1): 67–72.

[31] Morcos F, *et al.* Direct-coupling analysis of residue coevolution captures native contacts across many protein families. *Proceedings of the National Academy of Sciences of the United States of America*, 2011. **108**(49): E1293–E1301.

[32] Jumper J, *et al.* Highly accurate protein structure prediction with AlphaFold. *Nature*, 2021. **96**(7873): 583–589.

[33] Hornik K, *et al.* Multilayer feedforward networks are universal approximators. *Neural Networks*, 1989. **2**(5): 359–366.

[34] Fukushima K. Neocognitron: A self-organizing neural network model for a mechanism of pattern recognition unaffected by shift in position. *Biological Cybernetics*, 1980. **36**: 193–202.

[35] Waibel A, *et al.* Phoneme recognition using time-delay neural networks. *IEEE Transactions on Acoustics, Speech, and Signal Processing*, 1989. **37**: 328–339.

[36] LeCun Y, *et al.* Backpropagation applied to handwritten zip code recognition. *Neural Computation*, 1989. **1**: 541–551.

[37] Srivastava N, *et al.* Dropout: A simple way to prevent neural networks from overfitting. *Journal of Machine Learning Research*, 2014. **15**: 1929–1958.

[38] Ioffe S and Szegedy C. Batch normalization: Accelerating deep network training by reducing internal covariate shift. *Proceedings of the 32nd*

International Conference on Machine Learning, Francis B & David B (eds.), PMLR: Proceedings of Machine Learning Research, 2015. 448–456.

[39]   Elman JL. Finding structure in time. *Cognitive Science*, 1990. **14**: 213–252.

[40]   Hochreiter S and Schmidhuber J. Long short-term memory. *Neural Computation*, 1997. **9**(8): 1735–1780.

[41]   Berner C, *et al*. Dota 2 with large scale deep reinforcement learning. arXiv, 2019.

[42]   Graves A, *et al*. Speech recognition with deep recurrent neural networks. *2013 IEEE International Conference on Acoustics, Speech and Signal Processing*, 2013. IEEE, 6645–6649.

[43]   Cho K, *et al*. Learning phrase representations using RNN encoder-decoder for statistical machine translation. *Proceedings of the 2014 Conference on Empirical Methods in Natural Language Processing (EMNLP)*, Moschitti A, *et al*. (eds.), ACL, 2014. 1724–1734.

[44]   Heffernan R, *et al*. Capturing non-local interactions by long short-term memory bidirectional recurrent neural networks for improving prediction of protein secondary structure, backbone angles, contact numbers and solvent accessibility. *Bioinformatics*, 2017. **33**(18): 2842–2849.

[45]   Lv Z, *et al*. Protein function prediction: From traditional classifier to deep learning. *Proteomics*, 2019. **19**(14): e1900119.

[46]   Sutskever I, *et al*. Sequence to sequence learning with neural networks. *International Conference on Neural Information Processing Systems*, Ghahramani Z, *et al*. (eds.), Montreal, MIT Press, 2014. 3104–3112.

[47]   Bahdanau D, *et al*. Neural machine translation by jointly learning to align and translate. *International Conference on Learning Representations*, Bengio Y & LeCun Y (eds.), San Diego, OpenReview, 2015.

[48]   Peters ME, *et al*. Deep contextualized word representations. *Proceedings of the 2018 Conference of the North American Chapter of the Association for Computational Linguistics: Human Language Technologies, NAACL-HLT*, Walker MA, *et al*. (eds.), Association for Computational Linguistics, 2018. 2227–2237.

[49]   Yang Z, *et al*. XLNet: Generalized autoregressive pretraining for language understanding. *Proceedings of the 33rd International Conference on Neural Information Processing Systems, NeurIPS*, Wallach HM, *et al*. (eds.), Vancouver, Curran Associates, 2019. 5754–5764.

[50]   Vaswani A, *et al*. Attention is all you need. *International Conference on Neural Information Processing Systems*, Guyon I, *et al*. (eds.), Long Beach, Curran Associates, 2017. 5998–6008.

[51]   Hu G and Kurgan L. Sequence similarity searching. *Current Protocols in Protein Science*, 2019. **95**(1): e71.

[52] Yang J, *et al.* Protein-ligand binding site recognition using complementary binding-specific substructure comparison and sequence profile alignment. *Bioinformatics*, 2013. **29**(20): 2588–2595.

[53] Gong Q, *et al.* GoFDR: A sequence alignment based method for predicting protein functions. *Methods*, 2016. **93**: 3–14.

[54] Altschul SF, *et al.* Gapped BLAST and PSI-BLAST: A new generation of protein database search programs. *Nucleic Acids Research*, 1997. **25**(17): 3389–3402.

[55] Lan Z, *et al.* ALBERT: A lite BERT for self-supervised learning of language representations. *ICLR*, 2020. OpenReview.net.

[56] Rao R, *et al.* MSA transformer. arXiv, 2021.

[57] Rao R, *et al.* Evaluating protein transfer learning with TAPE. *Proceedings of the 33rd International Conference on Neural Information Processing Systems, NeurIPS*, Wallach HM, *et al.* (eds.), Vancouver, Curran Associates, 2019. 9686–9698.

[58] Alley EC, *et al.* Unified rational protein engineering with sequence-only deep representation learning. *Nature Methods*, 2019. **16**: 1315–1322.

[59] Remmert M, *et al.* HHblits: Lightning-fast iterative protein sequence searching by HMM-HMM alignment. *Nature Methods*, 2011. **9**(2): 173–175.

[60] Seemayer S, *et al.* CCMpred — fast and precise prediction of protein residue-residue contacts from correlated mutations. *Bioinformatics*, 2014. **30**(21): 3128–3130.

[61] El-Gebali S, *et al.* The Pfam protein families database in 2019. *Nucleic Acids Research*, 2019. **47**(D1): D427–D432.

[62] Heinzinger M, *et al.* Modeling aspects of the language of life through transfer-learning protein sequences. *BMC Bioinformatics*, 2019. **20**(1): 723.

[63] Suzek BE, *et al.* UniRef clusters: A comprehensive and scalable alternative for improving sequence similarity searches. *Bioinformatics*, 2015. **31**(6): 926–932.

[64] Krause B, *et al.* Multiplicative LSTM for sequence modelling. *ICLR (Workshop)*, 2017. OpenReview.net.

[65] UniProt Consortium. The universal protein resource (UniProt). *Nucleic Acids Research*, 2007. **35**(Database issue): D193–D197.

[66] Ho J, *et al.* Axial attention in multidimensional transformers. arXiv, 2019.

[67] Child R, *et al.* Generating long sequences with sparse transformers. arXiv, 2019.

[68] Gront D, *et al.* BioShell threader: Protein homology detection based on sequence profiles and secondary structure profiles. *Nucleic Acids Research*, 2012. **40**(Web Server issue): W257–W262.

[69] Zheng W, *et al.* LOMETS2: Improved meta-threading server for fold-recognition and structure-based function annotation for distant-homology proteins. *Nucleic Acids Research*, 2019. **47**(W1): W429–W436.

[70] Chou PY and Fasman GD. Conformational parameters for amino acids in helical, beta-sheet, and random coil regions calculated from proteins. *Biochemistry*, 1974. **13**(2): 211–222.

[71] Nagano K. Triplet information in helix prediction applied to the analysis of super-secondary structures. *Journal of Molecular Biology*, 1977. **109**(2): 251–274.

[72] Wu TT and Kabat EA. An attempt to evaluate the influence of neighboring amino acids (n–1) and (n+1) on the backbone conformation of amino acid (n) in proteins. Use in predicting the three-dimensional structure of the polypeptide backbone of other proteins. *Journal of Molecular Biology*, 1973. **75**(1): 13–31.

[73] Chou PY and Fasman GD. Prediction of protein conformation. *Biochemistry*, 1974. **13**(2): 222–245.

[74] Torrisi M, *et al.* Deep learning methods in protein structure prediction. *Computational and Structural Biotechnology Journal*, 2020. **18**: 1301–1310.

[75] Qian N and Sejnowski T. Predicting the secondary structure of globular proteins using neural network models. *Journal of Molecular Biology*, 1988. **202**(4): 865–884.

[76] Jones DT. Protein secondary structure prediction based on position-specific scoring matrices. *Journal of Molecular Biology*, 1999. **292**(2): 195–202.

[77] Rost B and Sander C. Prediction of protein secondary structure at better than 70% accuracy. *Journal of Molecular Biology*, 1993. **232**(2): 584–599.

[78] Cuff JA and Barton GJ. Application of multiple sequence alignment profiles to improve protein secondary structure prediction. *Proteins*, 2000. **40**(3): 502–511.

[79] Lyons J, *et al.* Predicting backbone Calpha angles and dihedrals from protein sequences by stacked sparse auto-encoder deep neural network. *Journal of Computational Chemistry*, 2014. **35**(28): 2040–2046.

[80] Faraggi E, *et al.* SPINE X: Improving protein secondary structure prediction by multistep learning coupled with prediction of solvent accessible surface area and backbone torsion angles. *Journal of Computational Chemistry*, 2012. **33**(3): 259–267.

[81] Heffernan R, *et al.* Improving prediction of secondary structure, local backbone angles, and solvent accessible surface area of proteins by iterative deep learning. *Scientific Reports*, 2015. **5**: 11476.

[82] Chen K, *et al.* Optimization of the sliding window size for protein structure prediction. *Proceedings of the 2006 IEEE Symposium on Computational Intelligence in Bioinformatics and Computational Biology*, Toronto, IEEE, 2006.

[83] Baldi P, *et al.* Exploiting the past and the future in protein secondary structure prediction. *Bioinformatics*, 1999. **15**(11): 937–946.

[84] Pollastri G and McLysaght A. Porter: A new, accurate server for protein secondary structure prediction. *Bioinformatics*, 2005. **21**(8): 1719–1720.

[85] Wang S, *et al.* Protein secondary structure prediction using deep convolutional neural fields. *Scientific Reports*, 2016. **6**: 18962.

[86]   Wang S, *et al.* RaptorX-Property: A web server for protein structure property prediction. *Nucleic Acids Research*, 2016. **44**(W1): W430–W435.

[87]   Steinegger M and Soding J. MMseqs2 enables sensitive protein sequence searching for the analysis of massive data sets. *Nature Biotechnology*, 2017. **35**(11): 1026–1028.

[88]   Zhang H, *et al.* Critical assessment of high-throughput standalone methods for secondary structure prediction. *Briefings in Bioinformatics*, 2011. **12**(6): 672–688.

[89]   Yang Y, *et al.* Sixty-five years of the long march in protein secondary structure prediction: The final stretch? *Briefings in Bioinformatics*, 2018. **19**(3): 482–494.

[90]   Klausen MS, *et al.* NetSurfP-2.0: Improved prediction of protein structural features by integrated deep learning. *Proteins-Structure Function and Bioinformatics*, 2019. **87**(6): 520–527.

[91]   Buchan DWA and Jones DT. The PSIPRED protein analysis workbench: 20 years on. *Nucleic Acids Research*, 2019. **47**(W1): W402–W407.

[92]   Bryson K, *et al.* Protein structure prediction servers at University College London. *Nucleic Acids Research*, 2005. **33**: W36–W38.

[93]   Mirabello C and Pollastri G. Porter, PaleAle 4.0: High-accuracy prediction of protein secondary structure and relative solvent accessibility. *Bioinformatics*, 2013. **29**(16): 2056–2058.

[94]   Petersen B, *et al.* A generic method for assignment of reliability scores applied to solvent accessibility predictions. *BMC Structural Biology*, 2009. **9**: 51.

[95]   Vassura M, *et al.* Reconstruction of 3D structures from protein contact maps. *IEEE/ACM Transactions on Computational Biology and Bioinformatics*, 2008. **5**(3): 357–367.

[96]   Nugent T and Jones DT. Accurate de novo structure prediction of large transmembrane protein domains using fragment-assembly and correlated mutation analysis. *Proceedings of the National Academy of Sciences of the United States of America*, 2012. **109**(24): E1540–E1547.

[97]   Kortemme T, *et al.* Computational redesign of protein-protein interaction specificity. *Nature Structural & Molecular Biology*, 2004. **11**(4): 371–379.

[98]   Gobel U, *et al.* Correlated mutations and residue contacts in proteins. *Proteins*, 1994. **18**(4): 309–317.

[99]   Gloor GB, *et al.* Mutual information in protein multiple sequence alignments reveals two classes of coevolving positions. *Biochemistry*, 2005. **44**(19): 7156–7165.

[100] Cheng JL and Baldi P. Improved residue contact prediction using support vector machines and a large feature set. *BMC Bioinformatics*, 2007. **8**: 113.

[101] Li YQ, *et al.* Predicting residue-residue contacts using random forest models. *Bioinformatics*, 2011. **27**(24): 3379–3384.

[102] Jones DT, *et al.* MetaPSICOV: Combining coevolution methods for accurate prediction of contacts and long range hydrogen bonding in proteins. *Bioinformatics*, 2015. **31**(7): 999–1006.

[103] Yang J and Shen HB. MemBrain-contact 2.0: A new two-stage machine learning model for the prediction enhancement of transmembrane protein residue contacts in the full chain. *Bioinformatics*, 2018. **34**(2): 230–238.

[104] Hanson J, *et al.* Accurate prediction of protein contact maps by coupling residual two-dimensional bidirectional long short-term memory with convolutional neural networks. *Bioinformatics*, 2018. **34**(23): 4039–4045.

[105] Adhikari B, *et al.* DNCON2: Improved protein contact prediction using two-level deep convolutional neural networks. *Bioinformatics*, 2018. **34**(9): 1466–1472.

[106] Xu JB and Wang S. Analysis of distance-based protein structure prediction by deep learning in CASP13. *Proteins-Structure Function and Bioinformatics*, 2019. **87**(12): 1069–1081.

[107] Li Y, *et al.* Deducing high-accuracy protein contact-maps from a triplet of coevolutionary matrices through deep residual convolutional networks. *PLOS Computational Biology*, 2021. **17**(3): e1008865.

[108] Yang JY, *et al.* Improved protein structure prediction using predicted inter-residue orientations. *Proceedings of the National Academy of Sciences of the United States of America*, 2020. **117**(3): 1496–1503.

[109] Kandathil SM, *et al.* Prediction of interresidue contacts with DeepMetaPSI-COV in CASP13. *Proteins-Structure Function and Bioinformatics*, 2019. **87**(12): 1092–1099.

[110] Parsons J, *et al.* Practical conversion from torsion space to Cartesian space for in silico protein synthesis. *Journal of Computational Chemistry*, 2005. **26**(10): 1063–1068.

[111] Mitchell AL, *et al.* MGnify: The microbiome analysis resource in 2020. *Nucleic Acids Research*, 2020. **48**(D1): D570–D578.

[112] Mirdita M, *et al.* Uniclust databases of clustered and deeply annotated protein sequences and alignments. *Nucleic Acids Research*, 2017. **45**(D1): D170–D176.

[113] Baek M, *et al.* Accurate prediction of protein structures and interactions using a three-track neural network. *Science*, 2021. **373**(6557): 871–876.

[114] Fuchs F, *et al.* SE(3)-transformers: 3D roto-translation equivariant attention networks. *Advances in Neural Information Processing Systems*, Curran Associates, Inc., 2020. 1970–1981.

Part II
Inputs for Machine Learning Models

https://doi.org/10.1142/9789811258589_0002

Chapter 2

Application of Sequence Embedding in Protein Sequence-Based Predictions

Nabil Ibtehaz* and Daisuke Kihara*,†,‡

*Department of Computer Science, Purdue University,
West Lafayette, IN, United States
†Department of Biological Sciences, Purdue University,
West Lafayette, IN, United States
‡corresponding author: dkihara@purdue.edu

Conventionally, in sequence-based predictions, an input sequence is represented by a multiple sequence alignment (MSA) or a representation derived from MSA, such as a position-specific scoring matrix. Recently, inspired by the development in natural language processing, several applications of sequence embedding have been developed and published. Here, we review different approaches of protein sequence embeddings and their applications including protein contact prediction, secondary structure prediction, and function prediction.

1. Introduction

Proteins are the fundamental building blocks of life, driving the entire network of interconnected and intercorrelated functional mechanisms in an organism [1]. Proteins are involved in almost every cellular function, including signaling pathway, DNA repair, transmembrane transport, catalytic activity and transporter activity. The tertiary structures of proteins and probably also protein interactions are encoded in the sequence of amino acids. Thus, protein sequences are often termed as the *language of life* [2].

Analyzing protein sequences and inferring various functional and structural information have been one of the major goals and long-standing themes of the bioinformatics field. Next-generation sequencing technologies have led to an exponential increase in the size of protein databases, nearly doubling almost in every two years [3]. However, labeling them with valid and meaningful annotations requires an extensive amount of effort, expertise, experiments and expense. As a result, we collect orders of magnitudes more new proteins than it is viable to annotate manually. This discrepancy becomes more apparent when we observe that the sparsely annotated TrEMBL database contains 219 million sequences, whereas the manually curated SwissProt database contains only 565,000 proteins [3]. Thus, the so-called sequence-structure gap [4] is increasingly growing.

Bioinformatics researchers have devoted decades in developing various computational prediction methods of protein structures, features, and annotations from amino acid sequence information. Typically, in conventional prediction methods, sequence information of a target protein is provided in various forms including a single protein sequence, position-specific scoring matrix (PSSM) [5], Hidden Markov Model (HMM) [6], and k-grams. Often, physicochemical properties of amino acids, such as hydrophobicity, charge, and size information are also used, instead of, or in addition to, the amino acid sequence itself [7].

In recent years, the field of Natural Language Processing (NLP) has observed a radical paradigm shift by embracing pre-trained language models [9,10]. The trend has been to train a language model on a large corpus of unlabeled text data in an unsupervised or semi-supervised fashion [11], which enables the models to learn patterns and structures of the language. This pre-training provides us with a general knowledge of the language in form of embeddings, which are found to be effective in solving various downstream tasks, occasionally with the aid of some task-specific finetuning. These pre-trained embedding approaches significantly improved upon the earlier supervised methods trained on task-specific smaller datasets [9].

As a result of deriving motivations from the success of word embeddings and pretrained language models in NLP, they are gradually gaining popularity in protein sequence analysis. Several language models have been adopted and applied for proteins, for example, ProtVec [12], SeqVec [2], and ProtBERT [13].

In this chapter, we explore the application of sequence embedding in protein sequence-based predictions. We briefly explain some notable language models and how they are used in bioinformatics, facilitated by access to large-scale protein databases. We present the effectiveness of learned sequence embeddings in solving various problems on diverse topics.

2. A brief overview of language models and embeddings in Natural Language Processing

First, we briefly present a few prominent language models and corresponding embedding generation schemes used in Natural Language Processing (NLP). For a more elaborate explanation, readers are encouraged to consider broader surveys [9,10].

Distributed representations [14] and Neural Network-based language models [15] have a long history of gradual progressive development. The first significant breakthrough came from the works of Mikolov *et al.* [16,17]. They proposed a novel word embedding named word2vec, which represents the words as dense vectors in a relatively low dimensional space. Figure 1(A) presents a simplified representation of word2vec, where we show how the input-output can be represented and computed in the skip-gram model. The embeddings are learned from a shallow neural network, by analyzing the neighboring words using Log-Linear models such as Continuous Bag-of-Words, which tries to predict the word based on the context, or Skip-gram, which attempts to predict the context based on the word. Both these two approaches are based on the fact that not only the semantic but also the syntactic meaning of a word can be estimated by its neighboring words [17]. In this word embedding, similar words are projected nearby in the vector space, thus providing a means of comparing the words both syntactically and semantically, which turns out valuable in various downstream NLP tasks.

Despite the effectiveness of mapping similar and dissimilar words, one major problem the word2vec model faces is that the generated embeddings are context independent. In natural language, the same word can have multiple meanings based on the context it is used. Word2vec discards this

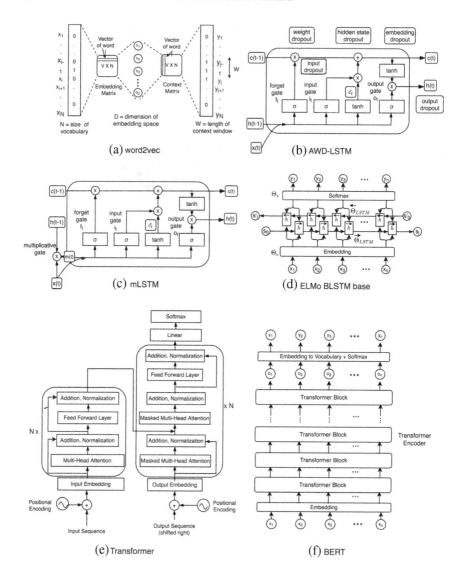

(a) word2vec

(b) AWD-LSTM

(c) mLSTM

(d) ELMo BLSTM base

(e) Transformer

(f) BERT

contextual and positional dependency of words. Various Long-Short Term Memory (LSTM)-based models, for example, ASGD Weight-Dropped LSTM (AWD-LSTM) [18] (Fig. 1B), multiplicative LSTM (mLSTM) [19] (Fig. 1C), and more specifically, ELMo (Embeddings from Language Models) [20] (Fig. 1D) target to solve this issue by treating the word embeddings not merely as a function of the words, but rather as a function of the entire sentence. LSTM, being a recurrent neural network, keeps track of the order of the words unlike the feed-forward networks used in word2vec. As a result, it manages to learn sentences, not just as a collection of words. AWD-LSTM investigates several strategies for regularizing and optimizing LSTM models, incorporating various levels of dropout at inputs, hidden layers, weights, embeddings, and outputs. Furthermore, the model is trained with a modified averaged stochastic gradient method, and it outperforms other models on word-level perplexities, a standard language model task, where a model's ability to compute the probability of unseen test sentences is evaluated on two datasets, Penn Treebank and WikiText-2 [18]. On the other hand, mLSTM combines the standard LSTM with multiplicative RNN architecture. This provides mLSTM with the ability to have different recurrent transition functions for different inputs and these increases expressivity. This enables mLSTM to consistently outperform

---

Figure 1.   Popular language models used in natural language processing. (A) word2vec being the simplest one, uses a shallow feed forward neural network to compute the word embeddings using long-linear models like continuous bag of words or skipgram. LSTM models consider the context and are capable of producing better embeddings. The vanilla LSTM model has been augmented with improved regularization and expressivity in the different variants. (B) AWD-LSTM aims at better regularization by employing different kinds of dropouts in inputs, outputs, hidden layers, weights, and embeddings. Furthermore, a modified averaged stochastic gradient descent is employed during training. (C) mLSTM on the other hand attempts at increasing the expressivity of LSTM, by learning different recurrent transformations for different inputs. (D) ELMo improves such models further by leveraging bidirectional LSTM layers which enables ELMo to produce context-sensitive embeddings for words, unlike word2vec. (E) Transformers dispense recurrent layers and use attention instead. This not only makes the computations faster but also makes the process parallelizable, which resulted in a huge leap in NLP over the time-consuming RNN model training. (F) BERT improves the state-of-the-art even further by employing stacks of bidirectional transformers. This deeper architecture accompanied by the availability of large-scale text corpora makes BERT the de facto standard in language modeling.

vanilla LSTM on character-level language modeling task. Among the various LSTM-based models, ELMo produces superior contextualized word representation, managing to capture the syntax and semantics of the words across various linguistic contexts. ELMo achieves this through a semi-supervised setup, by pretraining a Bidirectional Language Model (biLM) on large-scale datasets, later incorporating that with diverse NLP architectures. The biLM aims to model the probability of a token t_k, in a sequence $(t_1, t_2, t_3, ..., t_N)$ by considering both the history and the future contexts.

$$p(t, t, t, ..., t_N) = \begin{cases} \prod_{k=1}^{N} p(t_k \mid t, t, ..., t_{k-}), & \text{forward LM} \\ \prod_{k=1}^{N} p(t_k \mid t_{k+1}, t_{k+2}, ..., t_N), & \text{backward LM} \end{cases} \qquad (1)$$

In the process of learning these probabilities, each of the L layers of the BLSTMs outputs a context-dependent representation for a token t_k in both directions $\vec{h}_{k,j}^{LM}, \overleftarrow{h}_{k,j}^{LM}$, where $j = 1, 2, ..., L$. Combining these with a context-independent representation x_k^{LM}, ELMo embeddings are computed. By incorporating the context of the words, the issue of homonyms is thus solved by ELMo, as well as capturing high-level concepts from context. AWD-LSTM is significantly better compared to the traditional LSTM models. With the efficient utilization of the regularizations, it manages to reduce perplexity by 20%, despite requiring 1/3 of the parameters [18]. Although mLSTM decreases perplexity by 10% over a vanilla LSTM, it is outperformed by the AWD-LSTM model, due to it being a character-level architecture. On the other hand, as a result of using bidirectional LSTM layers, ELMo appears to be the most capable LSTM-based model, going beyond the state-of-the-art performance in six tasks, including question answering, named entity recognition, and sentiment analysis [20].

The LSTM-based models AWD-LSTM, mLSTM, and ELMo have been illustrated in Figures 1(B), 1(C), and 1(D) respectively. AWD-LSTM is defined by the repertoire of dropouts it employs, as demonstrated in Figure 1(B). mLSTM (Fig. 1C) on the other hand, disregards the typical hidden state in LSTM, and maps the hidden state $h(t-1)$ to an input dependent one $m(t)$ by suitable weighted multiplication with the input $x(t)$. The foundational component of ELMo (Fig. 1D) is the BLSTM base, which is accompanied by embeddings generated from character convolutions and suitable softmax top layers that address a particular downstream task.

ELMo surpasses word2vec-based embedding, but the next breakthrough came from Transformer-based architectures, most notably BERT (Bidirectional Encoder Representations from Transformers) [11]. Transformer is a novel network architecture based on attention [21], instead of traditional convolutional or recurrent operations. Discarding recurrent operations makes the transformer model parallelizable, which reduces the training time and improves the performance simultaneously. Transformer is composed of an encoder and a decoder, which themselves contain stacks of layers involving multi-head attention and position-wise feedforward networks, accompanied by residual connections and layer normalizations. A simplified schematic diagram of Transformer has been presented in Figure 1(E). Transformers have an encoder and a decoder component, and they comprise of generalized stackable blocks, making it feasible to experiment with deeper architectures. Both the inputs and outputs are translated to an embedding space, but for the lack of convolutional or recurrent operations, the sequential or ordering information gets lost, which is compensated by using positional encoding. BERT, the next frontier in language modeling, outperforms the state-of-the-art models on 11 different tasks, including language understanding, question answering, multi-genre natural language inference [11]. It employs fine-tuning, thereby making task-specific architectural modifications redundant. Internally, BERT is a multi-layer bidirectional Transformer encoder [21], which comprises stacks of transformer blocks as shown in Figure 1(F). Although transformers are general sequence transduction models, BERT is more oriented towards being a language model. The BERT model is pretrained on two different tasks, namely, masked language modeling and next sentence prediction. After developing the pretrained BERT model, it can be conveniently adapted for most downstream NLP tasks just by fine-tuning the model using the new task-specific input-output pair, without any major modification. Ever since its introduction, BERT has become the de facto standard of transfer learning in NLP problems. The state-of-the-art performance of BERT is the result of a few contributory factors such as using stacks of transformer blocks with almost double attention heads, leveraging bidirectional information, and the use of segment embeddings.

3. Protein databases facilitating language modeling

The primary goal of language model pretraining is to train it with a huge volume of data so that the model can learn diversified and distinct patterns in the language. Although managing such a high volume of data apparently seems challenging, the pretraining pipeline is simplified as it does not require any labeling of the data. This makes the task easier for NLP as we have huge volumes of unlabeled textual information sources like Wikipedia [22], which can be used for language model pretraining. Similarly, the combination of the diligent efforts of biologists over many decades with next-generation sequencing technologies has resulted in billions of protein sequence data. Thus, this opens up a huge scope for employing language models in protein sequence analysis.

Protein sequence databases, SwissProt [23], Pfam [24], and UniRef [25] are among the most notable ones. SwissProt is a manually curated protein sequence database, which provides high-quality annotations of 565,254 proteins. UniRef (UniProt Reference Clusters) provides clustering of sequences from the UniProtKB (UniProt Knowledgebase) and selected UniParc (UniProt Archive) records, i.e., a total of 216 million proteins, to obtain complete coverage as well as removing redundancy. The Pfam database is one of the most widely used resources to analyze protein families and domains, having a huge collection of 47 million proteins in 19,179 families.

Metagenome databases turned out to be very useful for training models due to their large size. BFD [26,27] is a metagenome database containing around 2,122 million protein sequences. BFD is the largest protein database at the time of writing, even eight times larger than the previous largest merged database [28]. MGnify is another metagenome resource, which contains over 1.1 billion sequences [81].

When training deep language models, the volume of the dataset plays an important role. For example, Heinzinger *et al.* [2] found that training an ELMo model with the SwissProt dataset resulted in less useful models, compared to training the same model with a larger UniRef50 dataset. For instance, when training the model on UniRef50 dataset a significant improvement was observed in downstream tasks such as the protein secondary structure prediction and localization prediction [2]. Elnaggar *et al.* [13] similarly investigated the impact of database size on performance using three datasets, namely UniRef100, UniRef50, and BFD. They observed that UniRef databases, particularly UniRef50 was sufficient and using BFD, which is 10 times larger than UniRef50, resulted in a minor improvement in classification accuracy and membrane protein prediction [13]. But the outcome of using UniRef50 and BFD seemed to be task dependent. For the secondary structure prediction, UniRef50 performed on par with an existing method, NetSurfP-2.0 [82]. On the other hand, UniRef50 declined the model's performance by about 2% on tasks such as subcellular localization prediction, and membrane protein prediction when a larger ProtT5-XXL model was trained in comparison with BFD [13].

4. Adapting language models for protein sequences

In this section, we highlight some of the major attempts in adapting language models for protein sequences to perform prediction tasks.

4.1 *ProtVec (word2vec)*

Asgari *et al.* [12], for the first time, applied the concept of word embedding in analyzing protein sequences. They used the skip-gram-based word2vec model to generate embeddings from 3-mers of amino acids. The word2vec model is trained on 546,790 manually annotated sequences from SwissProt. In order to tokenize the sequences, they took three shifted versions of the sequence and broke them into non-overlapping 3-mers. These 3-mers are thus treated as words and the broken sequences are treated as sentences, with which the word2vec model is trained using negative sampling. In addition to selecting the suitable contexts for a word, negative sampling randomly picks some contexts which are not related to that word, thus it enables the embeddings of similar words proximate and dissimilar words distant. The trained model, named as ProtVec, can embed 3-mers in 100-dimensional space. The learned protein space is also consistent in terms of the distribution of various biophysical and biochemical properties such as volume, polarity, hydrophobicity, etc.

The model is applied in two tasks, protein family classification, and disordered protein region prediction. They used the computed 100 dimensional 3-mer vectors, averaged over the entire sequence as the input to an SVM classifier. The family classification, applied on 324,018 protein sequences from SwissProt, spanning over 7,027 protein families, show an average accuracy of 93%, surpassing existing methods. The accuracy of disordered protein prediction was 99.8% and 100% on the two datasets they used [12]. ProtVec was later extended to ProtVecX [32], which can operate on variable-length segments of protein sequences. The extension of word2vec, doc2vec are applied for protein sequence analysis tasks, e.g., protein localization prediction [33].

Since this foundational work of Asgari *et al.* [12], word2vec has been used in diverse prediction tasks including protein-protein interaction binding sites [34], compound-protein interaction [35], protein Glycation sites [36], and generalized protein classification [37]. Word2vec embeddings are used as input for conventional machine learning models such as SVM, KNN, Random Forest [38–40] and also for deep learning models, e.g., CNN [36], RNN [41], and Transformers [35].

In some works, FastText skip-gram model [42], which represents each word as a bag of character n-gram, was also employed to generate embeddings from protein sequences, which was later used for different types of analysis [38,40,43]. Islam *et al.* [37] proposed m-NGSG, which modifies the k-skip-bi-gram model by employing a combination of n-grams and skip grams and demonstrated consistent improvement on tasks including locallization prediction, fluorescent protein prediction, and antimicrobial peptide prediction. Ideally, the embeddings are expected to work well in many prediction tasks, but there is a report that a network specifically trained on a particular task (kinase-substrate phosphorylation prediction) showed a better performance than ProtVec [44]. Ibtehaz *et al.* [45] analyzed this further and found that the ProtVec embeddings hardly correspond with similarity scores of the k-mers, i.e., the vector similarity (cosine similarity) of the embeddings of the k-mers correlates little with the similarity score (alignment score) of the k-mers (Pearson correlation coefficient of 0.226). Ibtehaz *et al.* proposed the Align-gram model [45], which modified the Skip-gram model to make it more aligned with protein analysis, making vector similarity and k-mer similarity equivalent (Pearson correlation coefficient of 0.902).

4.2 *UDSMProt (AWD-LSTM)*

Strodthoff *et al.* attempted to devise a single, universal model architecture to solve diverse problems related to proteins. The proposed architecture UDSMProt [29] is based on the AWD-LSTM language model [18], which is internally a 3-layer LSTM network, with different dropout regularizations (input, embedding, weight, hidden state, and output layer dropout). The model is pretrained on SwissProt.

UDSMProt operates on protein sequence data tokenized to the amino acid level. The pretraining aimed towards predicting the next token for a given sequence of tokens, implicitly learning the structure and semantics of the language i.e., protein sequences. During various downstream prediction tasks, the embeddings obtained from the model are compiled through a Concat-Pooling layer and some dense layers are added on top which are trained in the process of finetuning. The UDSMProt pipeline is evaluated on three different tasks, enzyme class prediction, gene ontology prediction, and remote homology and fold detection. With mere finetuning on the problem datasets, the proposed method performs on par with state-of-the-art algorithms that were tailored to those specific tasks, even surpassing them in two tasks [29].

UDSMProt was later used in the USMPEP [63] pipeline. State-of-the-art results are obtained on MHC class I binding prediction, using just a generic model without any domain-specific heuristics [63].

4.3 *UniRep (mLSTM)*

Alley *et al.* [31] trained an mLSTM model with 1,900 hidden units, UniRep, on around 24 million protein sequences in UniRef50. Despite being trained in an unsupervised manner, i.e., predicting the next amino acid from a sequence of amino acids, UniRep embeddings managed to create physicochemically meaningful clusters of amino acids and partition structurally similar protein sequences. The learned embeddings predict protein secondary structure, the stability of natural and de novo designed proteins, and the quantitative function of molecularly diverse mutants [31]. It was also shown that UniRep has the potential to enhance efficiency in protein engineering tasks, as demonstrated in predicting fluorescence in engineered proteins. The primary contribution of UniRep is extracting the fundamental protein features using unsupervised deep learning as fixed-length vectors which are both semantically rich and structurally, evolutionarily, and biophysically grounded.

In addition, UniRep embedding-based feature representation demonstrates improved performance in other tasks, such as anticancer peptides prediction [64], assessing disease risk of protein mutation [65], localizing sub-Golgi proteins [66], and Peroxisomal proteins [52]. UniRep embeddings are correlated with biological features important for protein expression in *B. subtilis* [67] and can be used to analyze interaction patterns between virus and human proteins [68].

4.4 *SeqVec (ELMo)*

As mentioned in earlier sections, word2vec embeddings ignore context, which can be resolved by using a complex language model like ELMo. Heinzinger *et al.* [2] proposed SeqVec, which is an ELMo model trained on protein sequences. The authors basically used the standard ELMo implementation. The two-layer ELMo model applies dropout and shared weights between forward and backward LSTMs to reduce overfitting. It is trained on the UniRef50 database. The SeqVec model takes a protein sequence and returns 3,076 features for each residue in the sequence.

The embeddings generated from SeqVec are evaluated in four different tasks, namely, secondary structure prediction, disorder prediction, localization prediction, and membrane prediction and show better performance than other sequence-based representations, such as one-hot encoding and ProtVec [2]. Moreover, SeqVec runs much faster than evolutionary methods e.g., HHBlits [46], and the speed is not affected by the size of the database, thus is massively scalable.

With the release of SeqVec [2], ELMo has been promptly received as a welcomed addition to the bioinformatics analysis toolbox. Zeng *et al.* [47] used ELMo to learn a context-dependent embedding of amino acids for MHC I class ligand prediction. Litmann *et al.* [48] demonstrated that by simply using ELMo and BERT-based embeddings, it is possible to almost reach the state-of-the-art in protein function prediction. Again, Villegas *et al.* [49] leveraged ELMo embedding as features for protein function prediction. SeqVec features is also applied to B-cell epitope prediction [50] and cross-species protein function prediction [51]. Moreover, Elmo embeddings are used as input for SVM [52] and Graph Neural Networks [53].

Apart from the standard ELMo architecture, general BLSTM networks are used in several other protein language modeling tasks. Bepler *et al.* [75] trained a multitask neural network to solve protein structural tasks, contact prediction and structural similarity prediction by training on protein structure information. In another work [76], the authors also experimented with introducing a two-stage feedback mechanism where they trained a BLSTM on protein sequences and contact map information and a proposed "soft symmetric alignment". Primarily, the encoder generates embeddings from the amino acid sequence. Subsequently, the embeddings are used to predict contact maps and compute L1 distance between pairs of proteins by the proposed soft symmetric alignment. These error terms are fed as feedback signals to the language model, thus making the embeddings more biologically driven. DeepBLAST [77] on the other hand, obtains alignments from embeddings learned from the protein language model in [75] and integrates them into an end-to-end differentiable alignment framework.

4.5 *ESM-1b (Transformer)*

Rives *et al.* [69] trained transformer models on 250 million protein sequences from UniParc [3]. Initially, transformer models with 100M parameters are trained and a systematic hyperparameter optimization is performed. After finalizing the suitable hyperparameter set, the model is scaled to 33 layers, having around 650M parameters. The trained ESM-1b transformer manages to learn the biochemical properties of the amino acids. The output embeddings allow the clustering of residues in several groups which are consistent with the hydrophobic, polar, and aromatic nature of amino acids. Furthermore, the molecular weight and charge information is also reflected across the amino acids. Moreover, the different biological variations are encoded in the representation space. Specifically, the embeddings without any explicit information cluster the orthologous genes together. Furthermore, the learned embeddings are suitable to be used as feature representations for various downstream tasks. The authors demonstrate applications of the trained model in remote homology prediction, secondary structure prediction, and contact prediction [69]. ESM-1b embeddings were used in protein function prediction [70], effects of mutations on protein function [71], contact map prediction [72], protein fitness prediction [73], and Lectin-Glycan Binding Prediction [74].

A recent work based on transformer language modeling, MSA transformer [83] uses multiple sequence alignment as inputs to a transformer and significantly improves the performance over ESM-1b in unsupervised contact prediction, increasing top-L long-range contact precision by 15

points. MSA transformer also outperforms NetSurfP-2.0 in secondary structure prediction by increasing in Q8 accuracy by 2% [84].

4.6 *ProtTrans (BERT)*

Since its introduction, BERT has become the de facto standard model for
solving various NLP problems. Elnaggar *et al.* [13] ported BERT for protein
sequence analysis. They trained two auto-regressive models (Transformer-XL,
XLNet) and four auto-encoder models (BERT, Albert, Electra, T5) on data
from UniRef and BFD, using roughly 2,122 million protein sequences. The
authors followed the standard implementations of the transformer models
and trained different instances on different datasets. Training such networks
on the astounding amount of data require the assistance of HPC (High-
Performance Computing), using 5,616 GPUs and TPU Pod up to 1,024 cores.

The embeddings capture various biophysical properties of the amino
acids, structure classes of proteins, domains of life and viruses, and protein
functions in conserved motifs. The embeddings are also evaluated on per
residue (protein secondary structure prediction) and per-protein (cellular
localization and membrane protein classification) levels. No task-specific
modifications were performed, rather, the models are used as static feature
extractors, by extracting embeddings derived from the hidden state of the
last attention layer. From the experiments, it is observed that for both
localization and secondary structure prediction, fine-tuning improved performance [13]. Impressively, embeddings from their trained ProtT5
model, for the first time, surpass the state-of-the-art methods in the secondary structure prediction task, without using any evolutionary information
[13]. The authors assessed the impact of database size. They observed that
models trained on UniRef50 are enough and adding the huge amount of
data from BFD hardly presents noticeable improvements.

Despite BERT being just recently adopted for proteins, it has rapidly
gained popularity. Hiranuma *et al.* [54] used ProtBERT embeddings along
with several structural features to guide and improve protein structure
refinement. Litmann *et al.* [48] investigated the effectiveness of BERT
embeddings in Gene Ontology prediction. Charoenkwan *et al.* [55] used
BERT embeddings to predict amino acid sequences of peptides that taste
bitter without using any structural information and greatly outperform the
existing works. Filipavicius *et al.* [56] pretrained a RoBERTa model on a
mixture of binding and random protein pairs and achieved enhanced

downstream protein classification performance for tasks such as homology prediction, localization prediction, protein-protein interaction prediction as compared to the ESM-1b transformer [69]. The application of BERT embeddings improves several peptide prediction tasks [57–59]. BERT embeddings are also effective as input representations for clustering algorithms [60] and graph neural networks [61,62], for tasks such as clustering protein functional families [60], predicting effects of mutation [61], and protein-protein interaction site prediction [62].

Vig *et al.* [78] analyzed underlying learned information of protein transformer models, utilizing attention mechanisms. They analyzed and experimented with transformer models from TAPE [30] and ProtTrans [13], with a specific focus on the attention mechanism. Their analysis revealed that attention can capture high-level structured properties of proteins, namely, amino acids that are nearby in the 3D structure, despite being further apart in the 1D sequence. Furthermore, they found that attention reflects binding sites, amino acid contact maps, and amino acid substitution matrices.

5. Conclusions

Various successful adaptations of the language models in bioinformatics have greatly benefitted the analysis of protein sequences and various kinds of predictions of protein structure and function. We show that in addition to using embedding learning on specific datasets and tasks [8,79,80], pre-training language models on millions of protein sequences can dramatically improve the performance of downstream tasks.

We review a total of 33 methods (as summarized in Table 1) that rely on sequence embeddings as input, moving away from the traditional bioinformatics pipeline of computing PSSM or HMM profiles. Among the various types of embeddings, ProtVec and ProtTrans are used most frequently in nine and eight methods respectively. For various prediction problems, such as secondary structure prediction and protein-protein interaction site prediction, previous state-of-the-art performance has already been surpassed by merely using embeddings learned in an unsupervised manner. We expect to observe that embedding techniques and pretrained embeddings will be applied in many other tasks and that they will make substantial improvements in the field.

Table 1. The list of methods reviewed in this article.

Embedding	Problem	Model	Additional Features	Dataset	Ref	Source Code/Server
ProtVec (word2vec)	Protein-protein interaction (PPI) binding sites prediction	CNN RNN fusion	HSP, PSSM, ECO, RSA, RAA, disorder, hydropathy, physicochemcial properties	Recent publications	[34]	https://github.com/lucian-ilie/DELPHI Server: https://delphi.csd.uwo.ca
	Compound–protein interaction prediction	Transformer	Atomic properties	Human dataset, Caenorhabditis elegans dataset, BindingDB dataset	[35]	https://github.com/lifanchen-simm/transformerCPI
	Protein glycation sites prediction	LSTM	—	3 Surveyed Datasets	[36]	Server: http://watson.ecs.baylor.edu/ngsg
	Protein classification	LR	—	Subchlo, osFP, iAMP-2L, Cypred and PredSTP, TumorHPD 1 and 2, HemoPI 1 and 2, IGPred and PVPred	[37]	https://bitbucket.org/sm_islam/mngsg/src/master/
	Transporter substrate specificities identification	SVM	—	Proteins involved in transporting ion/molecules, collected from UniProt (release 2018_10). Dataset available in : http://bio216.bioinfo.yzu.edu.tw/fasttrans/	[38]	Server: http://bio216.bioinfo.yzu.edu.tw/fasttrans
	Nuclear localization signal identification	Multivariate Analysis	Physicochemcial properties, disorder, PSSM	NLSdb 2003, NLSdb 2017, SeqNLS	[39]	Server: http://www.csbio.sjtu.edu.cn/bioinf/INSP/
	Tumor necrosis factors identification	SVM	—	106 protein from tumor necrosis factor family and 1023 sequences from other major cytokine families were collected from UniProt (release 2019_05)	[40]	https://github.com/khucnam/TNFPred

	Application	Embedding	ML method	Dataset	Reference
	MHC binding prediction	MHC Allele embedding	GRU	IEDB and recent publications	[41] https://github.com/cmb-chula/MHCSeqNet
	Nucleic acid-binding protein identification	RNA sequence embedding	NN	RNA compete dataset, PBM dataset and recent publications	[43] https://github.com/syang11/ProbeRating
UDSMProt (AWD-LSTM)	MHC binding prediction	—	LSTM	IEDB, HPV	[63] https://github.com/nstrodt/USMPep
UniRep. (mLSTM)	Anticancer peptide prediction	Pretrained SSA embedding	KNN, LDA, SVM, RF, LGBM, NB	AntiCP 2.0 datasets	[64] https://github.com/zhibinlv/iACP-DRLF
	Disease risk prediction	Hydrophilic properties	MLP	BRCA1, PTEN	[65] https://github.com/xzenglab/BertVS
	Sub-Golgi localization identification	—	SVM	Recent publications	[66] https://github.com/zhibinlv/isGP-DRLF Server: http://isgp-drlf.aibiochem.ne
	Peroxisomal proteins localization prediction	SeqVec embedding	SVM	Protein sequences for peroxisomal membrane and matrix proteins collected from UniprotKB/SwissProt database. Dataset available in: https://github.com/MarcoAnteghini/In-Pero/tree/master/Dataset	[52] https://github.com/MarcoAnteghini/In-Pero
SeqVec (ELMo)	MHC class I ligand prediction	One hot encoding, BLOSUM50	Residual Network	Recent publications	[47] https://github.com/gifford-lab/DeepLigand
	Protein function prediction	BERT embedding	Modified KNN	CAFA3	[48] https://github.com/Rostlab/goPredSim Server: https://embed.protein.properties

(Continued)

Table 1. (*Continued*)

Embedding	Problem	Model	Additional Features	Dataset	Ref	Source Code/Server
	Protein function prediction	KNN, LR, MLP, CNN, GCN	One hot encoding, k-mer, DeepFold features, Contact map	CAFA3	[49]	https://github.com/stamakro/GCN-for-Structure-and-Function
	Linear B-cell epitope prediction	NN	Amino acid embedding	IEDB Linear Epitope Dataset	[50]	https://github.com/mcollatz/EpiDope
	Protein function prediction	LR	—	SwissProt, cross-species datasets	[51]	
	Peroxisomal proteins localization prediction	SVM	UniRep embedding	Protein sequences for peroxisomal membrane and matrix proteins collected from UniprotKB/SwissProt database. Dataset available in : https://github.com/MarcoAnteghini/In-Pero/tree/master/Dataset	[52]	https://github.com/MarcoAnteghini/In-Pero
	Cofactor specificity of Rossmann-fold protein prediction	GCN	—	ECOD and literature datasets	[53]	https://github.com/labstructbioinf/rossmann-toolbox Server: https://lbs.cent.uw.edu.pl/rossmann-toolbox
ESM-1b (Transformer)	Protein function prediction	GAT	Inter-residue contact graphs	PDB-cdhit	[70]	
	Effect of mutation prediction	Transformer	—	41 deep mutational scans	[71]	

	Application	Method	Features	Dataset	Reference
	Contact map prediction	CNN	One hot encoding, SS3, SS8, ASA< HSE, protein backbone torsion angles	ProteinNet, CASP14-FM, SPOT-2018	[72] https://github.com/jaspreet/SPOT-Contact-Single Server: https://sparks-lab.org/server/spot-contact-single/
	Protein fitness prediction	Ridge Regression	One hot encoding, physicochemical representation	19 labelled mutagenesis datasets	[73]
	Lectin-Glycan binding prediction	MLP	SweetNet features	Dataset was curated from 3,228 glycan arrays from the Consortium for Functional Glycomics database and 100 glycan arrays from the Carbohydrate Microarray Facility of Imperial College London	[74] https://github.com/BojarLab/LectinOracle
ProtTrans (BERT)	Protein structure refinement	CNN	Distance maps, amino acid identities and properties, backbone angles, residue angular orientations, Rosetta energy terms, secondary structure information, MSA information	PISCES	[54] https://github.com/hiranumn/DeepAccNet
	Protein function prediction	Modified KNN	ELMo embedding	CAFA3	[48] https://github.com/Rostlab/goPredSim Server: https://embed.protein.properties

(*Continued*)

Table 1. (*Continued*)

Embedding	Problem	Model	Additional Features	Dataset	Ref	Source Code/Server
	Peptide binding site identification	Transformer	—	Peptide complex dataset	[57]	
	Signal peptide prediction	CRF	—	Extended previously published dataset with newly available sequences from UniProt, Prosite and TOPDB	[58]	Server: https://services.healthtech.dtu.dk/service.php?SignalP-6.0
	MHC-peptide class II interaction prediction	Transformer	—	IEDB and recent publications	[59]	https://github.com/s6juncheng/BERTMHC Server: https://bertmhc.privacy.nlehd.de
	Functional family clustering	DBSCAN	—	CATH	[60]	https://github.com/Rostlab/FunFamsClustering
	Effect of mutation prediction	LGBM	ProteiSolver features	ProTherm, SKEMPI	[61]	Server: http://elaspic.kimlab.org
	Protein-protein interaction (PPI) binding sites prediction	GCN	—	Recent publications	[62]	https://github.com/Sazan-Mahbub/EGRET

For each method, we list the problem solved, machine learning model, additional features, dataset, and software availability.

Abbreviations : CNN = Convolutional Neural Network, RNN = Recurrent Neural Network, LSTM = Long Short-Term Memory, LR = Logistic Regression, SVM = Support Vector Machine, GRU = Gated Recurrent Unit, NN = Neural Network, KNN = K-Nearest Neighbors, MLP = Multi-Layer Perceptrons, GCN = Graph Convolutional Network, DBSCAN = Density-Based Spatia˜ Clustering of Applications with Noise, CRF = Conditional Random Field, LGBM = Light Gradient Boosting Machine, LDA = Latent Dirichlet Allocation, RF = Random Forest, NB = Naive Bayes, GAT = Graph Attention

6. Acknowledgement

This work was partly supported by the National Institutes of Health (R01GM133840, R01GM123055, and 3R01GM133840-02S1) and the National Science Foundation (CMMI1825941, MCB1925643, and DBI2003635).

References

[1] Levitt M. Nature of the protein universe. *Proceedings of the National Academy of Sciences*, 2009. **106**(27): 11079–11084.

[2] Heinzinger M, *et al.* Modeling aspects of the language of life through transfer-learning protein sequences. *BMC Bioinformatics*, 2019. **20**: 723.

[3] UniProt Consortium. UniProt: A worldwide hub of protein knowledge. *Nucleic Acids Research*, 2019. **47**(D1): D506–D515.

[4] Rost B and Sander C. Bridging the protein sequence-structure gap by structure predictions. *Annual Review of Biophysics and Biomolecular Structure*, 1996. **25**(1): 113–136.

[5] Gribskov M, *et al.* Profile analysis: Detection of distantly related proteins. *Proceedings of the National Academy of Sciences*, 1987. **84**(13): 4355–4358.

[6] Koski T. Hidden Markov Models: an Overview. In *Hidden Markov Models for Bioinformatics*, Dress A (ed.), Kluwer Academic Publisher, 2001. 211–231.

[7] Jing X, *et al.* Amino acid encoding methods for protein sequences: a comprehensive review and assessment. *IEEE/ACM Transactions on Computational Biology and Bioinformatics*, 2019. **17**(6): 1918–1931.

[8] Jia J, *et al.* Identification of protein-protein binding sites by incorporating the physicochemical properties and stationary wavelet transforms into pseudo amino acid composition. *Journal of Biomolecular Structure and Dynamics*, 2016. **34**(9): 1946–1961.

[9] Qiu X, *et al.* Pre-trained models for natural language processing: A survey. *Science China Technological Sciences*, 2020. **63**: 1872–1897.

[10] Otter DW, *et al.* A survey of the usages of deep learning for natural language processing. *IEEE Transactions on Neural Networks and Learning Systems*, 2020. **32**(2): 604–624.

[11] Devlin J, *et al.* BERT: Pretraining of deep bidirectional transformers for language understanding. arXiv, 2018.

[12] Asgari E and Mofrad MRK. Continuous distributed representation of biological sequences for deep proteomics and genomics. *PloS One*, 2015. **10**(11): e0141287.

[13]   Elnaggar A, *et al.* ProtTrans: Towards cracking the language of life's code through self-supervised deep learning and high performance computing. arXiv, 2020.

[14]   Hinton G, *et al.* Distributed representations. In *Parallel Distributed Processing: Explorations in the Microstructure of Cognition*, Rumelhart DE & McClelland JL (eds.), MIT Press, Cambridge, MA, USA, 1986, 77–109.

[15]   Bengio Y, *et al.* A neural probabilistic language model. *The Journal of Machine Learning Research*, 2003. **3**: 1137–1155.

[16]   Mikolov T, *et al.* Distributed representations of words and phrases and their compositionality. *Advances in Neural Information Processing Systems*, 2013. **26**: 3111.

[17]   Mikolov T, *et al.* Efficient estimation of word representations in vector space. arXiv, 2013.

[18]   Merity S, *et al.* Regularizing and optimizing LSTM language models. arXiv, 2017.

[19]   Krause B, *et al.* Multiplicative LSTM for sequence modelling. arXiv, 2016.

[20]   Peters ME, *et al.* Deep contextualized word representations. arXiv, 2018.

[21]   Vaswani A, *et al.* Attention is all you need. *Advances in Neural Information Processing Systems*, 2017. **30**: 5998.

[22]   Wikimedia Foundation. Wikimedia downloads. https://dumps.wikimedia.org

[23]   Boutet E, *et al.* UniProtKB/Swiss-Prot, the manually annotated section of the UniProt KnowledgeBase: How to use the entry view. In *Plant Bioinformatics*, Edwards D (ed.), Springer, 2016. 23–54.

[24]   Mistry J, *et al.* Pfam: The protein families database in 2021. *Nucleic Acids Research*, 2021. **49**(D1): D412–D419.

[25]   Suzek BE, *et al.* UniRef clusters: A comprehensive and scalable alternative for improving sequence similarity searches. *Bioinformatics*, 2015. **31**(6): 926–932.

[26]   Steinegger M, *et al.* Protein-level assembly increases protein sequence recovery from metagenomic samples manyfold. *Nature Methods*, 2019. **16**(7): 603–606.

[27]   Steinegger M and Söding J. Clustering huge protein sequence sets in linear time. *Nature Communications*, 2018. **9**: 2542.

[28]   Madani A, *et al.* Progen: Language modeling for protein generation. arXiv, 2020.

[29]   Strodthoff N, *et al.* UDSMProt: Universal deep sequence models for protein classification. *Bioinformatics*, 2020. **36**(8): 2401–2409.

[30]   Rao R, *et al.* Evaluating protein transfer learning with TAPE. *Advances in Neural Information Processing Systems*, 2019. **32**: 9689.

[31]   Alley EC, *et al.* Unified rational protein engineering with sequence-based deep representation learning. *Nature Methods*, 2019. **16**(12): 1315–1322.

[32]   Asgari E, *et al.* Probabilistic variable-length segmentation of protein sequences for discriminative motif discovery (DiMotif) and sequence embedding (ProtVecX). *Scientific Reports*, 2019. **9**: 3577.

[33] Yang KK, *et al.* Learned protein embeddings for machine learning. *Bioinformatics*, 2018. **34**(15): 2642–2648.

[34] Li Y, *et al.* Delphi: Accurate deep ensemble model for protein interaction sites prediction. *Bioinformatics*, 2021. **37**(7): 896–904.

[35] Chen L, *et al.* TransformerCPI: Improving compound–protein interaction prediction by sequence-based deep learning with self-attention mechanism and label reversal experiments. *Bioinformatics*, 2020. **36**(16): 4406–4414.

[36] Chen J, *et al.* DeepGly: A deep learning framework with recurrent and convolutional neural networks to identify protein glycation sites from imbalanced data. *IEEE Access*, 2019. **7**: 142368–142378.

[37] Ashiqul Islam SM, *et al.* Protein classification using modified n-grams and skip-grams. *Bioinformatics*, 2018. **34**(9): 1481–1487.

[38] Nguyen T-T-D, *et al.* Using word embedding technique to efficiently represent protein sequences for identifying substrate specificities of transporters. *Analytical Biochemistry*, 2019. **577**: 73–81.

[39] Guo Y, *et al.* Discovering nuclear targeting signal sequence through protein language learning and multivariate analysis. *Analytical Biochemistry*, 2020. **591**: 113565.

[40] Nguyen T-T-D, *et al.* TNFPred: Identifying tumor necrosis factors using hybrid features based on word embeddings. *BMC Medical Genomics*, 2020. **13**: 155.

[41] Phloyphisut P, *et al.* MHCSeqNet: A deep neural network model for universal MHC binding prediction. *BMC Bioinformatics*, 2019. **20**: 270.

[42] Bojanowski P, *et al.* Enriching word vectors with subword information. *Transactions of the Association for Computational Linguistics*, 2017. **5**: 135–146.

[43] Yang S, *et al.* ProbeRating: A recommender system to infer binding profiles for nucleic acid-binding proteins. *Bioinformatics*, 2020. **36**(18): 4797–4804.

[44] Kirchoff KE and Gomez SM. Ember: Multi-label prediction of kinase-substrate phosphorylation events through deep learning. bioRxiv, 2020. 2020.02.04.934216.

[45] Ibtehaz N, *et al.* Align-gram: Rethinking the skip-gram model for protein sequence analysis. arXiv, 2020.

[46] Remmert M, *et al.* HHblits: Lightning-fast iterative protein sequence searching by HMM-HMM alignment. *Nature Methods*, 2012. **9**(2): 173–175.

[47] Zeng H and Gifford DK. DeepLigand: Accurate prediction of MHC class I ligands using peptide embedding. *Bioinformatics*, 2019. **35**(14): i278–i283.

[48] Littmann M, *et al.* Embeddings from deep learning transfer go annotations beyond homology. *Scientific Reports*, 2021. **11**: 1160.

[49] Villegas-Morcillo A, *et al.* Unsupervised protein embeddings outperform hand-crafted sequence and structure features at predicting molecular function. *Bioinformatics*, 2021. **37**(2): 162–170.

[50] Collatz M, *et al.* EpiDope: A deep neural network for linear B-cell epitope prediction. *Bioinformatics*, 2021. **37**(4): 448–455.

[51] van den Bent I, *et al.* The power of universal contextualised protein embeddings in cross-species protein function prediction. bioRxiv, 2021. 2021.04.19.440461.

[52] Anteghini M, *et al.* In-Pero: Exploiting deep learning embeddings of protein sequences to predict the localisation of peroxisomal proteins. *International Journal of Molecular Sciences*, 2021. **22**(12): 6409.

[53] Kaminski K, *et al.* Graph neural networks and sequence embeddings enable the prediction and design of the cofactor specificity of Rossmann fold proteins. bioRxiv, 2021. 2021.05.05.440912.

[54] Hiranuma N, *et al.* Improved protein structure refinement guided by deep learning based accuracy estimation. *Nature Communications*, 2021. **12**: 1340.

[55] Charoenkwan P, *et al.* Bert4Bitter: A bidirectional encoder representations from transformers (bert)-based model for improving the prediction of bitter peptides. *Bioinformatics*, 2021. **37**(17): 2556–2562.

[56] Filipavicius M, *et al.* Pre-training protein language models with label-agnostic binding pairs enhances performance in downstream tasks. arXiv, 2020.

[57] Abdin O, *et al.* PepNN: A deep attention model for the identification of peptide binding sites. bioRxiv, 2021. 2021.01.10.426132.

[58] Teufel F, *et al.* SignalP 6.0 achieves signal peptide prediction across all types using protein language models. bioRxiv, 2021. 2021.06.09.447770.

[59] Cheng J, *et al.* BERTMHC: Improves MHC-peptide class II interaction prediction with transformer and multiple instance learning. *Bioinformatics*, 2021. In press.

[60] Littmann M, *et al.* Clustering FunFams using sequence embeddings improves EC purity. *Bioinformatics*, 2021. In press.

[61] Strokach A, *et al.* ELASPIC2 (EL2): Combining contextualized language models and graph neural networks to predict effects of mutations. *Journal of Molecular Biology*, 2021. **433**(11): 166810.

[62] Mahbub S and Bayzid MS. EGRET: Edge aggregated graph attention networks and transfer learning improve protein-protein interaction site prediction. bioRxiv, 2020. 2020.11.07.372466.

[63] Vielhaben J, *et al.* USMPep: Universal sequence models for major histocompatibility complex binding affinity prediction. *BMC Bioinformatics*, 2020. **21**: 279.

[64] Lv Z, *et al.* Anticancer peptides prediction with deep representation learning features. *Briefings in Bioinformatics*, 2021. In press.

[65] Li K, *et al.* Predicting the disease risk of protein mutation sequences with pre-training model. *Frontiers in Genetics*, 2020. **11**: 1535.

[66] Lv Z, *et al.* Identification of sub-Golgi protein localization by use of deep representation learning features. *Bioinformatics*, 2020. **36**(24): 5600–5609.

[67] Martiny H-M, *et al.* Deep protein representations enable recombinant protein expression prediction. bioRxiv, 2021. 2021.05.13.443426.

[68] Dong TN and Khosla M. A multitask transfer learning framework for novel virus-human protein interactions. bioRxiv, 2021. 2021.03.25.437037.

[69] Rives A, *et al.* Biological structure and function emerge from scaling unsupervised learning to 250 million protein sequences. *Proceedings of the National Academy of Sciences*, 2021. **118**(15): e2016239118.

[70] Lai B and Xu J. Accurate protein function prediction via graph attention networks with predicted structure information. bioRxiv, 2021. 2021.06.16.448727.

[71] Meier J, *et al.* Language models enable zero-shot prediction of the effects of mutations on protein function. bioRxiv, 2021. 2021.07.09.450648.

[72] Singh J, *et al.* SPOT-contact-single: Improving single-sequence-based prediction of protein contact map using a transformer language model, large training set and ensembled deep learning. bioRxiv, 2021. 2021.06.19.449089.

[73] Hsu C, *et al.* Combining evolutionary and assay-labelled data for protein fitness prediction. bioRxiv, 2021. 2021.03.28.437402.

[74] Lundstrøm J, *et al.* LectinOracle — A generalizable deep learning model for lectin-glycan binding prediction. bioRxiv, 2021. 2021.08.30.458147.

[75] Bepler T and Berger B. Learning the protein language: Evolution, structure, and function. *Cell Systems*, 2021. **12**(6): 654–669.

[76] Bepler T and Berger B. Learning protein sequence embeddings using information from structure. arXiv, 2019.

[77] Morton J, *et al.* Protein structural alignments from sequence. bioRxiv, 2020.

[78] Vig J, *et al.* BERTology meets biology: Interpreting attention in protein language models. arXiv, 2020.

[79] ElAbd H, *et al.* Amino acid encoding for deep learning applications. *BMC Bioinformatics*, 2020. **21**: 235.

[80] Broyles BK, *et al.* Activation of gene expression by detergent-like protein domains. *Iscience*, 2021. **24**(9): 103017.

[81] Mitchell AL, *et al.* MGnify: The microbiome analysis resource in 2020. *Nucleic Acids Research*, 2020. **48**(D1): D570–D578.

[82] Klausen MS, *et al.* NetSurfP-2.0: Improved prediction of protein structural features by integrated deep learning. *Proteins*, 2019. **87**: 520–527.

[83] Rao R, et al. MSA Transformer. In *Proceedings of the 38th International Conference on Machine Learning*, Meila M & Zhang T (eds.), Proceedings of Machine Learning Research (PMLR), 2021. **139**: 8844–8856.

Chapter 3

Applications of Natural Language Processing Techniques in Protein Structure and Function Prediction

Bin Liu[*,‡], Ke Yan[*], Yi-He Pang[*], Jun Zhang[†],
Jiang-Yi Shao[*], Yi-Jun Tang[*] and Ning Wang[*]

*School of Computer Science and Technology,
Beijing Institute of Technology, Beijing 100081, China;
†School of Computer Science and Technology,
Harbin Institute of Technology, Shenzhen,
Guangdong, 518055, China
‡corresponding author: bliu@bliulab.net

Protein structure and function prediction are instrumental areas in the bioinformatics field. They are important for a number of applications in rational drug discovery, disease analysis, and many others. Protein sequences and natural languages share some similarities. Therefore, many techniques derived from natural language processing (NLP) have been applied to the protein structure and function prediction. In this chapter, we discuss sequence-based predictors of protein structure and function that utilize techniques derived from NLP field. We include methods that target protein sequence analysis, fold recognition, identification of intrinsically disordered regions/proteins, and prediction of protein-nucleic acids binding. The concepts and computational methods discussed in this chapter will be especially useful for the researchers who are working in the related field. We also aim to bring new computational NLP techniques into the protein structure and function prediction area.

1. Introduction

The genome is the "book of life," with protein sequences as languages. Natural languages and protein sequences share some similarities [1]. For example, peptide bonds connect amino acid residues together to form a protein with a certain structure and function, while words are combined by grammar and linguistic rules to form a natural sentence with a specific meaning. Based on these similarities, techniques derived from natural language processing (NLP) have been successfully applied to predict the structures and functions of proteins, indirectly uncovering the underlying meanings of the "book of life" (Fig. 1). Here we review selected applications of the NLP technologies to predict the structures and functions of proteins, such as fold recognition, identification of intrinsically disordered regions/proteins, protein-nucleic acids binding prediction. In particular, the recently constructed platform BioSeq-BLM [2] is introduced and discussed. This tool integrates a broad range of biological language models (BLMs) derived from the field of NLP for biological analysis. BLMs are able to represent the biological sequences based on language models. BioSeq-BLM is a powerful toolkit, which automatically constructs the BLMs, selects a predictor, evaluates the performance of this predictor, and analyzes the results. These functionalities are particularly useful for solving problems related to extracting linguistic features and designing techniques derived from NLP, providing a new approach to explore the meaning of "book of life".

2. Methods for protein sequence analysis

Most of the protein structure and function prediction methods are based on the Multiple Sequence Alignments (MSAs). The MSAs are generated by searching a series of homology sequences from a database like Uniprot [3]. Protein remote homology detection is one of the core tasks in protein sequence analysis, whose aim is to detect the remote homology relationships based on protein sequences sharing lower than 30% sequence similarity.

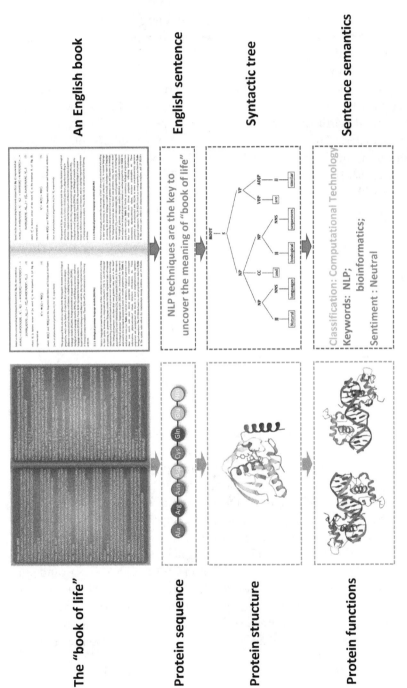

Figure 1.   Similarities between protein sequence and natural language sentence.

Given the importance of the remote homology detection, many computational methods have been proposed for this problem. For example, profile-profile alignment method, like HHblits [4], effectively search for homologous/similar sequences, but this search may lead to some false positives occurring for remote homologues. The remote homology relationship among proteins depends not only on the position information of amino acids contained in sequence, but also on some relevant physical and chemical properties. These properties play a fundamental role in some classification methods like SMI-BLAST [5] and ranking methods like PL-search [6]. Inspired by this, ProtRe-CN [7] combines the classification method with the protein network method, and feeds them into the Learning to Rank model [8], in which the protein network method contains an alignment method for building its protein network. This classification method is discriminative enough to identify the related proteins, while the protein network method determines the global relationship among proteins [9–14]. Refer to Figure 2 for details.

3. Computational prediction of protein structures

3.1 *Protein fold recognition*

Understanding the functions of proteins and protein-protein interactions rely on learning and analyzing the tertiary structures of proteins. Proteins within the same fold usually have similar structures and functions [15]. Protein fold classification is a typical taxonomy-based problem aiming to classify a query protein into one of the known fold types according to its sequence. Therefore, predicting the folds of proteins is critical to understanding their structures and functions [16].

Inspired by the self-attention mechanism in the neural machine translation (NMT), which captures potential associations and dependencies among words in a sentence and encodes the semantics of language, the SelfAT-Fold [17] applies the self-attention mechanism to protein fold recognition. The motif-based self-attention network (MSAN) and the residue-based self-attention network (RSAN) are constructed to capture the global associations among the structural motifs and residues along with the protein sequences, respectively. The fold-specific attention features captured by MSAN or RSAN are combined with Support Vector Machines (SVMs) to predict the protein fold type.

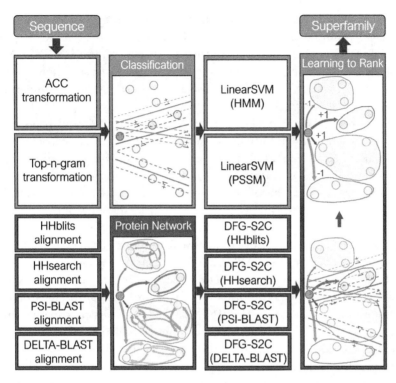

Figure 2. The workflow of ProtRe-CN. The ProtRe-CN first projects the sequence into the feature space through variety of representation methods (i.e., ACC, Top-n-gram transformation and HHblits, HHsearch, PSI-BLAST and DELTA-BLAST alignment method). Based on the feature representations of the sequences, we integrate the classification methods and the network methods into a ranking framework in a supervised manner. Finally, we select the highest scored superfamily as the predictive results from the output ranking list.

Similarity at the protein fold level among proteins could be seen as equivalent to the similarity at the topic level among book chapters. Similar to different sentences in the text concerning the same topic, different sequences may be categorized into the same fold type. Protein sequence is treated as the chapter text to solve the false positive issues in the predicted results by using the protein similarity network. FoldRec-C2C [18] was proposed based on the Cluster-to-Cluster model (C2C model) to prevent query protein sequences, belonging to the same fold type, from being predicted to be in different folds. The Cluster-to-Cluster model eliminates the false positives by constructing the cluster-to-cluster network, The ProtFold-DFG [19] is based on the Directed Fusion Graph (DFG), to fuse multiple ranking results from different sources. Based on such heterogeneous ranking results, the DFG network precisely measures similarities among protein sequences.

Moreover, protein sequences can be described by several properties, such as evolutionary information, secondary structure, physicochemical information, etc. MV-fold [20] relies on multi-view learning and regularized least square regression framework to utilize consistent information among different features in order to accurately predict the protein fold types. Multi-view learning model constructs the latent subspace shared with the multiple features using a subspace learning algorithm. MV-fold applies the $L_{2,1}$ norm regularization to extract the discriminative features from each view of data to construct the latent subspace. Please refer to Figure 3 for more details.

Because proteins in the same fold typically have similar structures and share similar evolutionary information, these proteins can be regarded as common semantic space in NLP. TSVM-fold [21] constructs semantic similarity features based on three tools including HHblits [4], SPARKS-X [22] and DeepFR [23]. HHblits utilizes evolutionary information based on the MSAs between different protein sequences. SPARKS-X considers estimated match probability between the predicted and actual structural properties. DeepFR utilizes fold-specific features using contact maps extracted from the protein sequences. TSVM-fold uses three tools to calculate sequence similarity between a target sequence and training dataset to develop scores used to construct semantic features. Then the semantic similarity features based on the sequence similarity are combined with the SVM to predict the protein fold types. In order to extract the principal features from the semantic features and decrease noise, MLDH-Fold [24] proposes a multi-view low-rank learning framework for the protein fold recognition. MLDH-Fold utilizes the low-rank decomposition framework to extract precise similarity scores from the semantic features and construct a latent subspace with the common information of different views to predict the target proteins.

3.2 Intrinsically disorder regions/proteins identification

Intrinsically disordered regions (IDRs) are sequence regions lacking stable three-dimensional structure under physiological conditions [25,26]. These

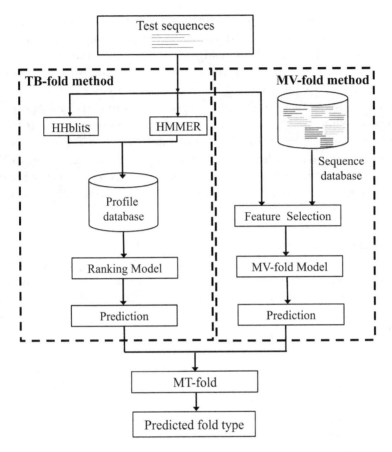

Figure 3.   The flowchart of the MT-fold method. The proposed method can be divided into two parts: dRHP-PseRA method (the left module in the flowchart) and MV-fold (the right module in the flowchart with darker background) method. The dRHP-PseRA method searches the query sequence against the dataset by HHblits and HMMER respectively. The MV-fold utilizes the multi-view learning model in different views features.

regions are very difficult to characterize experimentally, given that their backbones vary significantly over time and are inherently very flexible [27,28]. The emergence of IDRs challenges traditional point of view that sequences encode specific structures. At the same time, IDRs play essential role in biological functions [29,30], such as storage of small molecules, cellular signal transduction and protein phosphorylation, etc. IDRs are also related to many diseases, such as cancer [30], Alzheimer's disease [31,32] and cardiovascular diseases [33], etc. Therefore, accurate identification of IDRs is an important task for studying protein functions and disease discovery. Experimental methods for identifying IDRs include nuclear magnetic resonance [34], X-ray crystallography [35], and circular dichroism [35]. However, IDRs are hard to identify because of their intrinsic lack of a stable structure. While the 50-years-old grand challenge to solve protein structure prediction was arguably solved by Alphafold2, which is a computational method based on deep learning techniques proposed in 2020 [36], IDRs cannot be accurately detected by Alphafold2. In this regard, many predictors have been proposed to identify IDRs based on the protein sequences [25,37].

As mentioned earlier, protein sequences can be regarded as text in NLP. Disordered regions can be considered as keywords or key sentences in the text [38]. For example, IDRs are divided into long disordered regions (LDRs) and short disordered regions (SDRs). SDRs are discrete, short fragments not exceeding 30 in length distributed in the protein sequence, similar to the distribution characteristics of keywords in the text. LDRs are fragments with more than 30 residues distributed at N and C termini of the protein sequence, similar to the key sentences at the beginning and end of the text. In view of the high similarity between natural language and biological sequence, algorithms derived from NLP have been applied to the field of identifying IDRs. Previous studies have carried out related research. For example, conditional random fields (CRF) is a conditional probability distribution model of another set of output sequences given a set of input sequences, and has been widely used in named entity (named entities are words with specific name in text) recognition tasks [39]. IDP-CRF [40] applies CRF to identify IDRs, which effectively improves the predictive performance [40]. IDP-FSP [41] constructs two CRF models based on LDRs and SDRs by extracting discriminative features and achieves good performance by integrating the two models. Since Hinton and his two students used a deep convolution neural network (CNN) to achieve great success in the ImageNet in 2012, deep learning technology undergone rapid development. At the same time, the fast increase in the number of protein sequences also provides sufficient amount of data for the application of deep learning in protein sequence analysis. One-dimensional CNN extracts information between adjacent words in the text, which can also be used to capture information between adjacent residues when identifying IDRs, such as in AUCpreD [42]. Long short term network (LSTM) is usually used to capture global action information of protein sequences, such as in SPOT-Disorder [43]. RFPR-IDP [44] uses both CNN and LSTM models, which not only capture local information between adjacent residues, but also capture the non-local interaction features between protein residues, which improves the performance of prediction [44]. In addition to the use of general deep learning models, more-grained natural language models are employed to capture characteristics of different IDRs. Seq2Seq and attention mechanisms [45] are used to capture global characteristics of natural language, obtain new sequences based on the context information of text, and are widely used in machine translation. Similarly, IDP-Seq2Seq [38] uses this mechanism to capture characteristics of protein sequences to identify disordered regions. Since the characteristics of LDRs and SDRs are different, training the corresponding models separately is conducive to a more comprehensive acquisition of IDRs characteristics. The CAN [39] model uses convolutional attention network commonly used in the NLP keyword recognition to capture discretely distributed SDRs. The HAN [46] model uses a commonly used hierarchical attention mechanism for key sentence extraction to capture characteristics of LDRs. As the high-dimensional features obtained by these two models capture common feature and differences between LDRs and SDRs, these two models are fused to further improve predictive performance [47]. The above discussion shows that the natural language processing methods are useful in the field of the IDRs identification. Please refer to Figure 4 for more details.

4. Computational prediction of protein functions

4.1 *Prediction of functions of intrinsically disorder regions*

IDRs carry out many critical functions, such as transcriptions, signaling and aggregation and they host post-translational modification sites [29]. Experimental annotation of IDRs functions primarily relies on X-ray crystallography, NMR spectroscopy, and circular dichroism. However, experimental methods are expensive and time-consuming, which is not suitable for high-throughput analysis. Recent development of computational methods facilitated the understanding of intrinsic disorder and its functions. The information of disordered regions and functions are both encoded in their primary sequences, similar to the syntax and semantics of text. To this end, many technologies derived from NLP can be used to

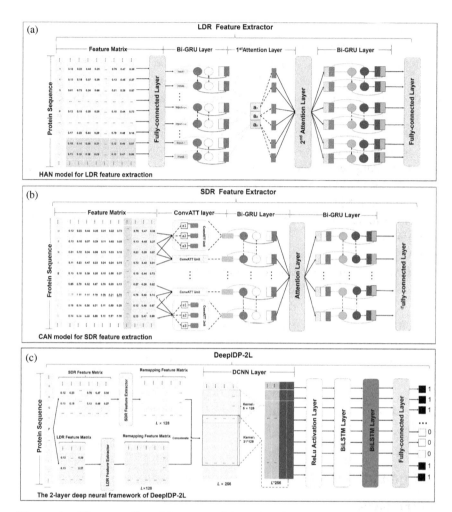

Figure 4. The overall architecture of DeepIDP-2L. DeepIDP-2L employs the 2-layer framework. In the first layer, the SDR features and LDR features were extracted by CAN and HAN, respectively, which were then concatenated and mapped into a new feature space. In the second layer, the Bi-LSTM was performed on this feature space to identify the IDRs by considering the global interactions among residues along the protein sequences.

uncover the semantics of disordered proteins. Here, we review some representative IDRs functional predictors especially related to NLP.

According to the primary IDRs functional annotated database (DisProt [48]), the functions of IDRs can be generally divided into two categories: entropic chains and binding-related functions. Entropic chain is the popular molecular function of IDRs where disordered flexible linkers (DFLs) contribute most. DFLs are the disordered regions, often located between domains, which are characterized by a high level of flexibility and long sequence. By treating the DFLs prediction as a classification task, DFLpred [49] is the first DFLs predictor that identifies the DFLs via combining the logistic regression (LR) and four sequence-based features including structure domain propensities, putative disordered regions, and two secondary structure propensities. Later, APOD[50] incorporates various sequence profile features, such as evolutionary conservation and relative solvent accessible area, into SVMs to further improve the predictive performance. IDRs interact with a wide range of partners, such as proteins, nucleic acids and lipids. For example, disordered residues act as recognition units to interact with proteins, called molecular recognition features (MoRFs) [51]. Most efforts have been made for this binding-based function of IDRs. However, the existing MoRF prediction methods have limited accuracy, largely due to the limited number of MoRF data for model training. Motivated by the similarities between disordered functional prediction and low resource machine translation, SPOT-MoRF [52] employs a pre-trained IDR predictor and transfers it into MoRF prediction, alleviating the problem of lack of MoRF data. Furthermore, MoRFs can be subdivided according to the specific types of functions they perform after binding to specific proteins [53,54]. However, so far, there is no predictor for identifying the specific functional MoRFs. Thus, further studies of this field are needed. Moreover, DisoRDPbind [55], DeepDISObind [56] and DisoLipPred [57] are the only methods that predict disordered RNA/DNA-binding residues and disordered lipids-bind residues. DisoRDPbind and DeepDISObind are integrated predictors based on logistic regression and deep convolutional neural network, respectively, which predict disordered nucleic acid binding and protein binding residues. DisoLipPred combines bi-directional recurrent network and transfer learning to capture the comprehensive sequence features for the prediction of disordered lipids-binding residues.

An important aspect of IDRs' functions is that residues in disordered regions have various functions but are in no way exclusive. In other words, one disordered residue can carry out multi-functions. The multi-functional IDRs are the so-called moonlighting regions. DMRpred [58] predictor employs a 2-layer framework model for disordered multi-functional region prediction, where the first layer is a feature extraction and the second layer is a random forest to predict the multifunctional residue tendency score. Although the DMRpred predictor [9] predicts whether a residue has multiple functions, it does not determine these specific functions. Therefore, some integrated methods have been proposed to realize the prediction of multiple functions of disordered proteins. DEPICTER [59] integrates many disordered function servers into one platform covering protein and nucleic acids binding, linkers, and moonlighting regions. Furthermore, flDPnn [60] incorporates the disordered region and function prediction into one computational architecture, which generates putative functions for the predicted IDRs covering the four most commonly annotated functions including protein-binding, DNA-binding, RNA-binding, and linkers. The integrated architecture of flDPnn improves performance of disordered region predictions and function predictions by sharing features between these sub-modules [60].

4.2 *Protein-nucleic acids binding prediction*

Interactions between proteins and nucleic acids are important for many cellular functions. Accurate identification of nucleic acid binding proteins and the corresponding binding residues is important for studying and characterizing interactions between proteins and nucleic acids [61,62]. This is of great significance in the applications related to gene expression, gene editing, treatment of genetic diseases, drug development, and others [63,64].

The nucleic acid binding residues in proteins are mainly detected by wet-lab experimental methods, such as X-ray crystallography and nuclear magnetic resonance. However, these methods are relatively expensive, slow, and not suitable for whole-genome scale analysis [65–67]. Therefore, only a small number of complexes of proteins and nucleic acids have been resolved and deposited in Protein Data Bank (PDB) [68]. It is important to study computational methods that predict the nucleic acid binding proteins and their binding residues based on the protein sequences or structures. The protein sequence is similar to the natural language. The methods and techniques derived from natural language processing provide an innovative way to analyze the interaction of proteins and nucleic acids.

4.2.1 *Nucleic acid binding protein prediction*

Recognition of nucleic acid binding proteins is similar to the classification of unlabeled text in the NLP field. The idea of text classification is to use N-Gram, TF-IDF, or other methods to encode the text, and then use machine learning classifiers to learn features to classify the texts, such as TextCNN [69] and TextRCNN [70]. Inspired by the text classification tasks, many computational methods for identifying nucleic acid binding proteins based on protein sequences have been proposed. For example, TriPepSVM [71] uses Kmer to encode short amino acid motifs in protein sequences and uses SVM to predict RNA binding proteins. PSFM-DBT [72] and PPCT [73] rely on extracting text features in FastText [74] and design specific formulas to extract sequence features, and use SVM and RF to learn the hidden features to identify nucleic acid binding proteins. Inspired by TextRCNN [70], DeepDRBP-2L [75] employs a two-layer framework to identify nucleic acid binding proteins by using the CNN-biLSTM networks to capture the functional domains and the global sequence information. iDRBP_MMC [76] treats nucleic acid binding protein identification as a multi-label text classification task. It incorporates prior information into the prediction model by using the motif-based CNN and improves the prediction performance of nucleic acid binding protein identification [76]. Please refer to Figure 5 for more details.

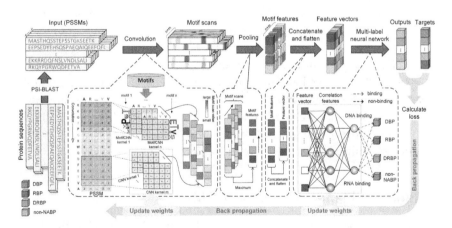

Figure 5. This figure illustrates the network architecture and the working flowchart of iDRBP_MMC. Only those weights of the network in yellow boxes will be updated during training.

4.2.2 *Nucleic acid binding residue prediction*

The nucleic acid binding residues are defined as the amino acids in a protein that interact with the nucleic acids. If the protein sequence is represented as text and the amino acid is represented as word, the nucleic acid binding residues can be considered as the named entities in the text. Therefore, the idea of named entity recognition (NER) [77,78] and other sequence labeling tasks in the natural language process can be applied to the identification of nucleic acid binding residues in proteins. However, most of the current computational methods treat nucleic acid binding residue identification as a classification task. They split protein sequences into short fragments, extract residue features based on short sequence fragments, and train a prediction model in a supervised manner. This is like when a sentence is split into multiple phrases and language entities in the phrase are predicted in a supervised manner. For example, DBS-PSSM [79] employs a neural network to extract evolutionary information from protein sequences to predict DNA binding residues. Pprint [80] predicts RNA-binding residues by combining evolutionary information features and SVM. Some methods predict both DNA-binding residues and RNA-binding residues by using various classification algorithms, such as DRNApred [67] and SVMnuc [81]. Although these methods facilitate development of this very important field, they ignore global and long-distance dependencies among residues in a protein. Inspired by named entity recognition and sequence labeling tasks, NCBRPred predicts both DNA-binding residues and RNA-binding residues in proteins based on the multi-label sequence labeling model [82]. It treats identification of the DNA-binding residues and RNA-binding residues as a multi-label learning task and trains a biGRU-based sequence labeling model. Since the sequence labeling model measures the global and long-distance dependencies among residues and captures the sequential characteristics of the nucleic acid binding residues, NCBRPred improves the prediction performance for nucleic acid binding residue recognition [82]. Please refer to Figure 6 for more details.

5. Biological language models (BLM)

As discussed above, the concepts and techniques derived from the NLP field play increasingly important roles in protein structure and function prediction. Therefore, the platforms, systems or tools that facilitate the

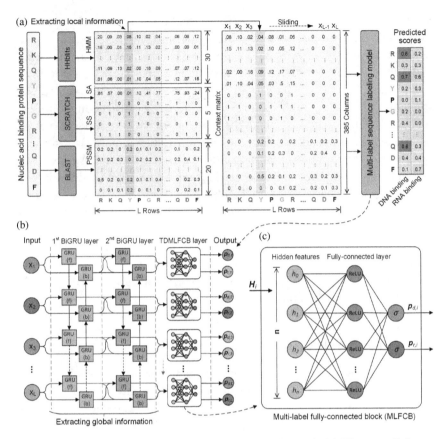

Figure 6. The framework and architecture of NCBRPred. (a) The overall framework of NCBRPred. Both DNA-binding proteins and RNA-binding proteins are fed into NCBRPred for training and test. The sliding window strategy was used to capture the local dependencies among residues in a protein. (b) The network architecture of MSLM. It contains three layers, including two BiGRU layers and a TDMLFCB layer. The two BiGRU layers measure the correlations among residues along the protein in a global fashion so as to capture the long and short-distance dependencies among residues. The TDMLFCB layer predicts DNA-binding residues and RNA-binding residues based on the learned hidden features by the former two BiGRU layers. The red, blue, orange and gray circles in the input layer represent DNA-binding residue, RNA-binding residue, DNA and RNA-binding residue, and non-DNA/RNA-binding residue, respectively. (c) The network architecture of MLFCB. It integrates the predictive results for binding residues via the multi-label learning strategy trained with both DNA and RNA-binding residues, leading to a lower cross-prediction rate.

development of computational methods based on the NLP techniques for the protein sequence analysis are highly desired. In this regard, several platforms have been proposed. For example, Pse-in-One [83] incorporates 28 modes to generate nearly all the possible feature vectors for DNA, RNA and protein sequences. BioSeq-Analysis [65] is a powerful web server, which automatically completes three main steps for constructing a predictor. The user only needs to upload the benchmark data set. BioSeq-Analysis generates an optimized predictor based on the benchmark data set and reports the corresponding performance measures. BioSeq-Analysis was further extended to the residue-level task, and the updated version BioSeq-Analysis2.0 has been established [66]. Based on similarities between natural languages and biological sequences, the BLMs, motivated by language models (LMs) in the field of natural language processing, have been introduced and discussed [2,84]. Moreover, a new platform called BioSeq-BLM [2] has been established. It incorporates 155 different BLMs. Please refer to Figure 7 for more details. The web server of BioSeq-BLM is freely available at http://bliulab.net/BioSeq-BLM/.

6. Summary and recommendations

Protein sequences and natural language sentences are relatively similar. Inspired by these similarities, techniques derived from NLP can be exported to drive development and improve performance of methods for the protein structure and function prediction. However, for most of the tasks in protein sequence analysis, the NLP techniques cannot be directly applied because sequence analysis is arguably more difficult than typical NLP tasks. Therefore, the BioSeq-BLM platform [2] is extremely useful for selecting and optimizing the NLP techniques for the biological sequence analysis by providing 155 BLMs and various machine learning techniques derived from NLP. We anticipate that the NLP techniques will be the key to uncover the meaning of the "book of life".

7. Acknowledgments

This work was supported by the National Natural Science Foundation of China (Nos. 62102030 and 61822306).

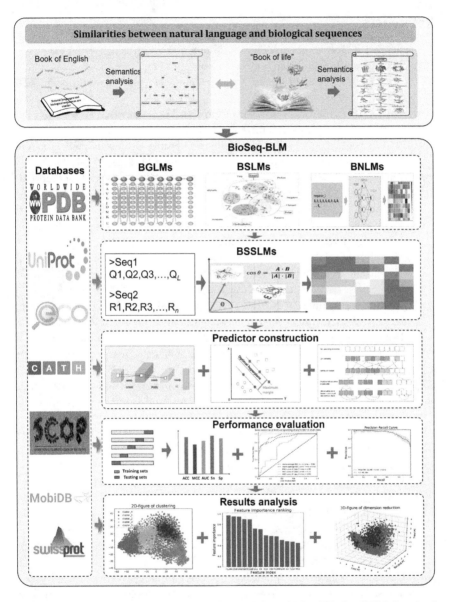

Figure 7.   The main components and their relationships of BioSeq-BLM. Inspired by the similarities between the natural languages and biological sequences, the BioSeq-BLM is constructed. There are four main components in BioSeq-BLM, including Biological Language Models (BLMs), predictor construction, performance evaluation and result analysis.

References

[1] Searls DB. The language of genes. *Nature*, 2002. **420**(6912): 211–217.

[2] Li H, *et al.* BioSeq-BLM: A platform for analyzing DNA, RNA, and protein sequences based on biological language models. *Nucleic Acids Research*, 2021. **49**(22): e129–e129.

[3] UniProt C, UniProt: The universal protein knowledgebase in 2021. *Nucleic Acids Research*, 2021. **49**(D1): D480–D489.

[4] Remmert M, *et al.* HHblits: Lightning-fast iterative protein sequence searching by HMM-HMM alignment. *Nature Methods*, 2011. **9**(2): 173–1735.

[5] Jin X, *et al.* SMI-BLAST: A novel supervised search framework based on PSI-BLAST for protein remote homology detection. *Bioinformatics*, 2021. **37**(7): 913–920.

[6] Jin X, *et al.* PL-search: A profile-link-based search method for protein remote homology detection. *Briefings in Bioinformatics*, 2021. **22**(3).

[7] Shao J, *et al.* ProtRe-CN: Protein remote homology detection by combining classification methods and network methods via learning to rank. *IEEE/ACM Transactions on Computational Biology and Bioinformatics*, 2021.

[8] Ke G, *et al.* Lightgbm: A highly efficient gradient boosting decision tree. *Advances in Neural Information Processing Systems*. 2017. **30**: 3146–3154.

[9] Dong Q, *et al.* A new taxonomy-based protein fold recognition approach based on autocross-covariance transformation. *Bioinformatics*, 2009. **25**(20): 2655–2662.

[10] Liu B, *et al.* A discriminative method for protein remote homology detection and fold recognition combining Top-n-grams and latent semantic analysis. *BMC Bioinformatics*, 2008. **9**(1): 510.

[11] Liu B, *et al.* DeepSVM-fold: Protein fold recognition by combining support vector machines and pairwise sequence similarity scores generated by deep learning networks. *Briefings in Bioinformatics*, 2020. **21**(5): 1733–1741.

[12] Liu B, *et al.* HITS-PR-HHblits: Protein remote homology detection by combining PageRank and Hyperlink-Induced Topic Search. *Briefings in Bioinformatics*, 2020. **21**(1): 298–308.

[13] Liu B and Zhu YL. ProtDec-LTR3.0: Protein remote homology detection by incorporating profile-based features into learning to rank. *IEEE Access*, 2019. **7**: 102499–102507.

[14] Cui X, *et al.* CMsearch: Simultaneous exploration of protein sequence space and structure space improves not only protein homology detection but also protein structure prediction. *Bioinformatics*, 2016. **32**(12): i332–i340.

[15] Wei L and Zou Q. Recent progress in machine learning-based methods for protein fold recognition. *International Journal of Molecular Sciences*, 2016. **17**(12): 2118.

[16] Yan K, *et al.* Protein fold recognition based on sparse representation based classification. *Artificial Intelligence in Medicine*, 2017. **79**: 1–8.

[17]   Pang Y and Liu B. SelfAT-fold: Protein fold recognition based on residue-based and motif-based self-attention networks. *IEEE/ACM Transactions on Computational Biology and Bioinformatics*, 2020.

[18]   Shao J, *et al.* FoldRec-C2C: Protein fold recognition by combining cluster-to-cluster model and protein similarity network. *Briefings in Bioinformatics*, 2021. **22**(3): bbaa144.

[19]   Shao J and Liu B. ProtFold-DFG: Protein fold recognition by combining directed fusion graph and PageRank algorithm. *Briefings in Bioinformatics*, 2021. **22**(3): bbaa192.

[20]   Yan K, *et al.* Protein fold recognition based on multi-view modeling. *Bioinformatics*, 2019. **35**(17): 2982–2990.

[21]   Yan K, *et al.* Protein fold recognition by combining support vector machines and pairwise sequence similarity scores. *IEEE/ACM Transactions on Computational Biology and Bioinformatics*, 2020.

[22]   Yang Y, *et al.* Improving protein fold recognition and template-based modeling by employing probabilistic-based matching between predicted one-dimensional structural properties of query and corresponding native properties of templates. *Bioinformatics*, 2011. **27**(15): 2076–2082.

[23]   Zhu J, *et al.* Improving protein fold recognition by extracting fold-specific features from predicted residue–residue contacts. *Bioinformatics*, 2017. **33**(23): 3749–3757.

[24]   Yan K, *et al.* MLDH-Fold: Protein fold recognition based on multi-view low-rank modeling. *Neurocomputing*, 2021. **421**: 127–139.

[25]   Liu Y, *et al.* A comprehensive review and comparison of existing computational methods for intrinsically disordered protein and region prediction. *Briefings in Bioinformatics*, 2019. **20**(1): 330–346.

[26]   Oldfield CJ, *et al.* Introduction to intrinsically disordered proteins and regions. In *Intrinsically Disordered Proteins*. Salvi N (Ed.), Academic Press, 2019. 1–34.

[27]   Lieutaud P, *et al.* How disordered is my protein and what is its disorder for? A guide through the "dark side" of the protein universe. *Intrinsically Disord Proteins*, 2016. **4**(1): e1259708.

[28]   Uversky VN. Natively unfolded proteins: a point where biology waits for physics. *Protein Science*, 2002. **11**(4): 739–756.

[29]   Dyson HJ and Wright PE. Intrinsically unstructured proteins and their functions. *Nature Reviews Molecular Cell Biology*, 2005. **6**(3): 197–208.

[30]   Iakoucheva LM, *et al.* Intrinsic disorder in cell-signaling and cancer-associated proteins. *Journal of Molecular Biology*, 2002. **323**(3): 573–584.

[31]   Uversky VN, *et al.* Intrinsically disordered proteins in human diseases: Introducing the D2 concept. *Annual Review of Biophysics*, 2008. **37**: 215–246.

[32]   Uversky VN, *et al.* Unfoldomics of human diseases: Linking protein intrinsic disorder with diseases. *BMC Genomics*, 2009. **10**(Suppl 1): S7.

[33] Cheng Y, *et al.* Abundance of intrinsic disorder in protein associated with cardiovascular disease. *Biochemistry*, 2006. **45**(35): 10448–10460.

[34] Konrat R. NMR contributions to structural dynamics studies of intrinsically disordered proteins. *Journal of Magnetic Resonance*, 2014. **241**: 74–85.

[35] Receveur-Brechot V, *et al.* Assessing protein disorder and induced folding. *Proteins*, 2006. **62**(1): 24–45.

[36] Jumper J, *et al.* Highly accurate protein structure prediction with AlphaFold. *Nature*, 2021. **596**(7873): 583–589.

[37] Zhao B and Kurgan L. Surveying over 100 predictors of intrinsic disorder in proteins. *Expert Review of Proteomics*, 2021: 1–11.

[38] Tang YJ, *et al.* IDP-Seq2Seq: Identification of intrinsically disordered regions based on sequence to sequence learning. *Bioinformatics*, 2021. **36**(21): 5177–5186.

[39] Zhu Y, *et al.* CAN-NER: Convolutional attention network for Chinese named entity recognition. *Proceedings of the 2019 Conference of the North American Chapter of the Association for Computational Linguistics: Human Language Technologies*, Minneapolis, Minnesota, Association for Computational Linguistics, 2019. 3384–3393.

[40] Liu Y, *et al.* IDP(-)CRF: Intrinsically disordered protein/region identification based on conditional random fields. *International Journal of Molecular Sciences*, 2018. **19**(9): 2483.

[41] Liu Y, *et al.* Identification of intrinsically disordered proteins and regions by length-dependent predictors based on conditional random fields. *Molecular Therapy Nucleic Acids*, 2019. **17**: 396–404.

[42] Wang S, *et al.* AUCpreD: Proteome-level protein disorder prediction by AUC-maximized deep convolutional neural fields. *Bioinformatics*, 2016. **32**(17): i672–i679.

[43] Hanson J, *et al.* Improving protein disorder prediction by deep bidirectional long short-term memory recurrent neural networks. *Bioinformatics*, 2017. **33**(5): 685–692.

[44] Liu Y, *et al.* RFPR-IDP: Reduce the false positive rates for intrinsically disordered protein and region prediction by incorporating both fully ordered proteins and disordered proteins. *Briefings in Bioinformatics*, 2021. **22**(2): 2000–2011.

[45] Bahdanau D, *et al.* Neural machine translation by jointly learning to align and translate. arXiv, 2014.

[46] Pappas N and Popescu-Belis A. Multilingual hierarchical attention networks for document classification. *The 2016 Conference of the North American Chapter of the Association for Computational Linguistics: Human Language Technologies*, 2016. 1480–1489.

[47] Tang Y, *et al.* DeepIDP-2L: Protein intrinsically disordered region prediction by combining convolutional attention network and hierarchical attention network. *Bioinformatics*, 2022. **38**(5): 1252–1260.

[48] Quaglia F, *et al.* DisProt in 2022: Improved quality and accessibility of protein intrinsic disorder annotation. *Nucleic Acids Research*, 2021. **50**(D1): D480–D487.

[49] Meng F and Kurgan L. DFLpred: High-throughput prediction of disordered flexible linker regions in protein sequences. *Bioinformatics*, 2016. **32**(12): i341–i350.

[50] Peng Z, *et al.* APOD: Accurate sequence-based predictor of disordered flexible linkers. *Bioinformatics*, 2020. **36**(Suppl 2): i754–i761.

[51] Yan J, *et al.* Molecular recognition features (MoRFs) in three domains of life. *Molecular BioSystems*, 2016. **12**(3): 697–710.

[52] Hanson J, *et al.* Identifying molecular recognition features in intrinsically disordered regions of proteins by transfer learning. *Bioinformatics*, 2020. **36**(4): 1107–1113.

[53] Mohan A, *et al.* Analysis of molecular recognition features (MoRFs). *Journal of Molecular Biology*, 2006. **362**(5): 1043–1059.

[54] Tompa P. Intrinsically unstructured proteins. *Trends in Biochemical Sciences*, 2002. **27**(10): 527–533.

[55] Peng Z and Kurgan L. High-throughput prediction of RNA, DNA and protein binding regions mediated by intrinsic disorder. *Nucleic Acids Research*, 2015. **43**(18): e121.

[56] Zhang F, *et al.* DeepDISOBind: Accurate prediction of RNA-, DNA- and protein-binding intrinsically disordered residues with deep multi-task learning. *Briefings in Bioinformatics*, 2021. **23**(1): bbab521.

[57] Katuwawala A, *et al.* DisoLipPred: Accurate prediction of disordered lipid binding residues in protein sequences with deep recurrent networks and transfer learning. *Bioinformatics*, 2021. **38**(1): 115–124.

[58] Meng F and Kurgan L. High-throughput prediction of disordered moonlighting regions in protein sequences. *Proteins*, 2018. **86**(10): 1097–1110.

[59] Barik A, *et al.* DEPICTER: Intrinsic disorder and disorder function prediction server. *Journal of Molecular Biology*, 2020. **432**(11): 3379–3387.

[60] Hu G, *et al.* flDPnn: Accurate intrinsic disorder prediction with putative propensities of disorder functions. *Nature Communications*, 2021. **12**(1): 4438.

[61] Zhang J, *et al.* Comprehensive review and empirical analysis of hallmarks of DNA-, RNA- and protein-binding residues in protein chains. *Briefings in Bioinformatics*, 2019. **20**(4): 1250–1268.

[62] Yan J, *et al.* A comprehensive comparative review of sequence-based predictors of DNA- and RNA-binding residues. *Briefings in Bioinformatics*, 2016. **17**(1): 88–105.

[63] Gerstberger S, *et al.* A census of human RNA-binding proteins. *Nature Reviews Genetics*, 2014. **15**(12): 829–845.

[64] Hudson WH and Ortlund EA. The structure, function and evolution of proteins that bind DNA and RNA. *Nature Reviews Molecular Cell Biology*, 2014. **15**(11): 749–760.

[65] Liu B. BioSeq-Analysis: A platform for DNA, RNA and protein sequence analysis based on machine learning approaches. *Briefings in Bioinformatics*, 2019. **20**(4): 1280–1294.

[66] Liu B, *et al.* BioSeq-Analysis2.0: An updated platform for analyzing DNA, RNA and protein sequences at sequence level and residue level based on machine learning approaches. *Nucleic Acids Research*, 2019. **47**(20): e127.

[67] Yan J and Kurgan L. DRNApred, fast sequence-based method that accurately predicts and discriminates DNA- and RNA-binding residues. *Nucleic Acids Research*, 2017. **45**(10): 84–84.

[68] Berman HM, *et al.* The protein data bank. *Nucleic Acids Research*, 2000. **28**(1): 235–242.

[69] Kim Y. Convolutional neural networks for sentence classification. arXiv, 2014.

[70] Lai S, *et al.* Recurrent convolutional neural networks for text classification. *National Conference on Artificial Intelligence AAAI Press*, 2015.

[71] Annkatrin B, *et al.* TriPepSVM: De novo prediction of RNA-binding proteins based on short amino acid motifs. *Nucleic Acids Research*, 2019. **47**(9): 4406–4417.

[72] Zhang J and Liu B. PSFM-DBT: Identifying DNA-binding proteins by combing position specific frequency matrix and distance-bigram transformation. *International Journal of Molecular Sciences*, 2017. **18**(9): 1856.

[73] Wang N, *et al.* IDRBP-PPCT: Identifying nucleic acid-binding proteins based on position-specific score matrix and position-specific frequency matrix cross transformation. *IEEE/ACM Transactions on Computational Biology and Bioinformatics*, 2021.

[74] Joulin A, *et al.* Bag of tricks for efficient text classification. 2017.

[75] Zhang J, *et al.* DeepDRBP-2L: A new genome annotation predictor for identifying DNA binding proteins and RNA binding proteins using convolutional neural network and long short-term memory. *IEEE/ACM Transactions on Computational Biology & Bioinformatics*, 2021. **18**: 1451–1463.

[76] Lv H, *et al.* iDNA-MS: An integrated computational tool for detecting DNA modification sites in multiple genomes. *iScience*, 2020. **23**(4): 100991.

[77] Xia C, *et al.* Multi-grained named entity recognition. *The 57th Annual Meeting of the Association for Computational Linguistics (ACL)*, 2019.

[78] Lample G, *et al.* Neural architectures for named entity recognition. *Proceedings of the 2016 Conference of the North American Chapter of the Association for Computational Linguistics: Human Language Technologies*, 2016.

[79] Ahmad S and Sarai A. PSSM-based prediction of DNA binding sites in proteins. *BMC Bioinformatics*, 2005. **6**(1): 33.

[80] Kumar M, *et al.* Prediction of RNA binding sites in a protein using SVM and PSSM profile. *Proteins*, 2008. **71**(1): 189–194.

[81] Su H, *et al.* Improving the prediction of protein–nucleic acids binding residues via multiple sequence profiles and the consensus of complementary methods. *Bioinformatics*, 2019. **35**(6): 930–936.

[82] Zhang J, *et al.* NCBRPred: Predicting nucleic acid binding residues in proteins based on multilabel learning. *Briefings in Bioinformatics*, 2021. **22**(5): bbaa397.

[83] Liu B, *et al.* Pse-in-One: A web server for generating various modes of pseudo components of DNA, RNA, and protein sequences. *Nucleic Acids Research*, 2015. **43**(W1): W65–W71.

[84] Elnaggar A, *et al.* ProtTrans: Towards cracking the language of lifes code through self-supervised deep learning and high performance computing. *IEEE Transactions on Pattern Analysis and Machine Intelligence*, 2021.

https://doi.org/10.1142/9789811258589_0004

Chapter 4

NLP-based Encoding Techniques for Prediction of Post-translational Modification Sites and Protein Functions

Suresh Pokharel[*,§], Evgenii Sidorov[*,§], Doina Caragea[†] and Dukka B KC[*,‡]

*Department of Computer Science, Michigan Technological university, Houghton, MI, USA, 49931
†Department of Computer Science, Kansas State University, Manhattan, KS, USA, 66502
‡corresponding author: dbkc@mtu.edu
§equal first authors

With advancements in sequencing and proteomics approaches, computational functional annotation of proteins is becoming increasingly crucial. Among these annotations, prediction of post-translational modification (PTM) sites and prediction of function given a protein sequence are two very important problems. Recently, there have been several breakthroughs in Natural Language Processing (NLP) area. Consequently, we have observed an increase in the application of NLP-based techniques in the field of protein bioinformatics. In this chapter, we review various NLP-based encoding techniques for representation of protein sequences.

Especially, we classify these approaches based on local/sparse encodings, distributed representation encodings, context-independent word embeddings, contextual word embedding and recent language models based pre-trained encodings. We summarize some of the recent approaches that make use of these NLP-based encodings for the prediction of various types of protein PTM sites and protein functions based on Gene Ontology (GO). Finally, we provide an outlook on possible future research directions for the NLP-based approaches for PTM sites and protein function predictions.

1. Introduction

One of key steps used by machine learning-based models in the field of protein bioinformatics is to represent protein sequences using an encoding scheme that assigns numerical values to each amino acid. Machine learning-based approaches have been utilized for some time and there currently have been a plethora of deep-learning based approaches for various bioinformatics prediction problems [1–4]. However, until recently, a lot of these deep-learning models still used encoding schemes that were developed in the pre-deep-learning era e.g., orthogonal encoding or 'one-hot-encoding', substitution matrix, etc. With the advancements of algorithms in the field of Natural Language processing (NLP) [5], we now witness a surge of the NLP-based representation of amino acid sequences.

The protein alphabet consists of 20 commonly occurring amino acids (AAs) and protein sequences are made up of these amino acids. Protein sequences can be illustrated as sentences or text in human language. In this regard, just like in natural language where sequences can be tokenized into letters, words, or other substrings, protein sequences can be similarly tokenized into individual amino acids (1-mer), di-mer, tri-mer, common motifs, domain, etc. Once tokenized, the bag-of-words representation can be used to count unique tokens in the sequence to convert every input sequence into a fixed-size vector. The NLP-based encoding for protein sequences can be classified into three types based on the unit of tokenization: character-level (amino acid level) encoding, word (n-mer) level encoding, and sequence level encoding. While protein sequences share a lot of similarities with natural language, one major difference between the two is that natural language has a well-defined vocabulary (e.g., ~millions of words in English) whereas proteins lack a clear vocabulary [6]. It is also noteworthy here to mention that a lot of bioinformatics works consider amino acids (1-mer) as words rather than characters.

With the recent advances in NLP-based approaches and their application in protein bioinformatics, a few review papers have been dedicated to the application of NLP methods for study of proteins. Ofer *et al.* [6] recently summarized the success and application of NLP methods for the study of proteins. This excellent review discusses various concepts of application of NLP like tokenization etc. to the study of protein sequences. Additionally, Iuchi *et al.* [7] summarized various protein sequence representation techniques for vectorizing biological sequences by providing an overview of NLP-based methods and then describing various dense representation techniques such as Word2Vec [8,9] and others.

The NLP-based encodings differ based on whether the task in hand is a local-level task, e.g., post-translational modification (PTM) sites prediction (interrogating whether a particular amino acid undergoes some modification or not) or a global-level task that predicts an annotation/ property for the whole protein. Especially, for the local-level task, often a window-based peptide sequence centered around the site of interest is the input to the machine learning apparatus, whereas for the global-level task, the input is the complete sequence (Fig. 1). Needless to mention, the size of the window sequence is going to be a parameter but once that size is

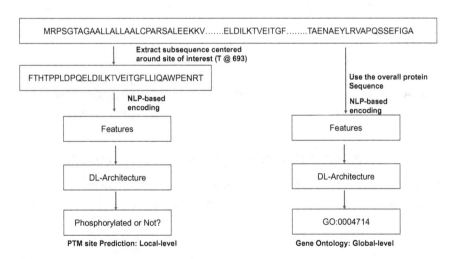

Figure 1.   Schematic of post-translational modification prediction and gene ontology prediction

fixed, the window size for all proteins for that particular problem remains the same. In contrast, the size of protein sequences can vary a lot (e.g., from less than a hundred residues to a few thousand residues). In that regard, as the size of the input vector should remain the same for all inputs for a particular problem, it is often challenging in case of a global-level task to come up with a fixed size input. Often, zero-padding (adding zeros on either end of the protein sequence) or truncation of the sequence has to be performed when one-hot encoding is used for protein representation, as the size of protein sequences are different. Please refer to Lopez-del Rio *et al.* for analysis on the effect of sequence padding [10].

Given this background, we briefly summarize selected NLP methods that are used for the study of proteins by focusing on two types of tasks: for local-level task, we discuss prediction of protein PTM sites and for global-level task, we will focus on annotation of proteins based on GO terms. In the case of PTM sites prediction, the task is related to part of speech tagging or entity relationship identification where we are trying to make predictions over specific residues in the input protein sequence. For the GO terms case, we are interested in predicting the global property (function) of the input protein sequence.

In Section 2 we discuss tokenization, local and distributed representations followed by a brief survey of existing NLP-based encoding techniques for protein sequences. In Section 3, we summarize recent NLP-based encoding approaches for the prediction of various types of PTM sites. In Section 4, we provide a brief summary of various NLP-based encoding approaches for the prediction of GO terms. In addition, we also discuss various pre-trained embeddings (from end-to-end learning) and examine different datasets that are used to generate these pre-trained models. Finally, we conclude by providing an outlook of possible future directions for the NLP-based encoding in the field of protein bioinformatics.

2. NLP-based encoding techniques for protein sequence

We discuss selected NLP-based encoding techniques both for local (sparse) representation as well as distributed representation. We begin with a short description of how to break protein sequences into tokens (sub-units). In Figure 2, we describe a schematic of classification of the NLP-based encoding schemes.

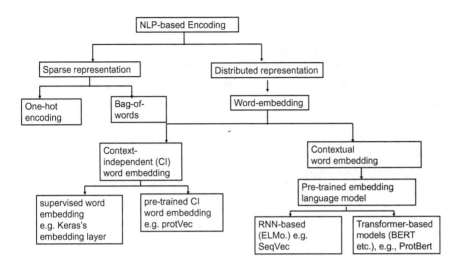

Figure 2. Classification of NLP-based encoding for protein sequences

2.1 *Tokenization*

One of the main roles in analyzing languages is played by the tokenizers. Tokenization (splitting the text into tokens regarded as a useful semantic unit) is a fundamental step in NLP. In human languages, the obvious tokens are words, although they can also be concepts consisting of two or more words. In NLP, tokenization can occur at multiple levels: typical tokens are individual characters, words, and sentences. Also, a common practice is to use subword segmentation, where the algorithm splits the rare words into meaningful subwords [11] and unigram language model [12]. There are also approaches like SentencePiece where the model directly learns subwords from raw sentences [13].

The corresponding tokenization in protein sequences are amino acid (character-level, $k = 1$), k-mer ($k = 2,3,4,..$) (word-level), and protein sequence (sentence-level). Unlike the NLP-based approaches where a set of predefined words exists, word-level tokenization of proteins is a little bit ambiguous: some approaches [6] also consider k-mer ($k = 1$, equivalent to individual amino acid) as word-level tokenization. In addition, subword segmentation is also being introduced for representing protein sequences as in ProtvecX [14].

2.2 Local/sparse representation of protein sequences

Once the protein sequence is tokenized, the next step is to numerically represent the tokens of the protein sequence. There are two types of representations: local (or sparse) representation and global representation. The one-hot encoding of words to vectors is one way to get a local representation. The one-hot encoding (also known as *sparse encoding*) assigns each word (referred to as token in NLP) in the vocabulary an integer N used as the Nth component of a vector with the dimension of the vocabulary size (number of different words). Each component is binary, i.e., either 0 if the word is not present in a sentence/text or 1 if it is. Figure 3 depicts the schematic of tokenization and local and global representation of protein/peptide. The bag-of-words is another local representation of protein sequences where first protein sequence is tokenized and the frequency of each unit/token is recorded.

2.3 Distributed representation of protein sequences

Distributed representation [9] refers to the fact that the representation of any single concept is distributed over many processing units. Thus, unlike a local representation where the unit values in the vectors are 0's and 1's,

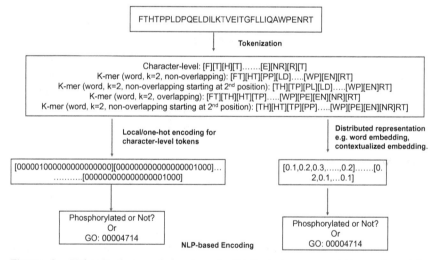

Figure 3. Tokenization and local and distributed representation of protein sequences

the unit values in vectors are continuous values. In other words, concepts are no longer localized in one unit, hence the name "distributed". Distributed representations have a couple of advantages over local representations: i) it is possible to learn new concepts without adding new units (just change the values of the units), and ii) better ability to represent concepts of similarities. Figure 3 gives the schematic representation of local and distributed representations. As shown in the figure, unlike one-hot encoding of amino acid "F" (first residue) where only 1 "unit" (= 6th) has a value of 1 in the vector representation, the corresponding distributed representation of F would have (some) continuous values in all the units. Among various types of distributed representation, word embedding is one of the most popular distributed representations in NLP. Below, we describe various word embeddings for protein sequences.

2.3.1 *Word embedding in proteins*

Word embeddings are a family of algorithms that represent tokens (e.g., words) with fixed-size dense vectors ("embeddings"), such that similar words obtain similar vector representations [15]. In that regard, word embeddings are densely distributed, fixed-length word vectors built using word co-occurrence statistics [15]. The main idea of word embeddings is to take a corpus of text and come up with a distributed vector representation of words that retain information of semantic similarity between these words. Word embedding is an improvement over the one-hot encoding (as well as the bag-of-words) as these words are represented by dense vectors rather than sparse vectors. These word embeddings can be learned from a large corpus of data in an unsupervised manner or can also be learned as part of a training of a deep learning model in a self-supervised-fashion (language model). Depending upon the context, learned word embedding can be context-independent where the embedding of a unit is the same regardless of the context and contextual where embedding of a unit differs based on the context. Moreover, the context-independent word embedding may be classified as supervised word embedding and pre-trained context-independent word embedding. The embedding layer of Keras [16,17] provides a mechanism to learn the word embedding in a supervised context. This embedding layer is initialized with random weights and updates/learns the embedding for all of the words in the training data. In that regard, word embeddings can be broadly classified as pre-trained word embedding and learning embedding from scratch (part of the supervised learning, mainly using Keras's embedding layer).

Word2Vec [8,9] is a popular context-independent pre-trained word embedding. It is an unsupervised shallow, two-layer NN which is trained to learn word embeddings from raw text and comes in two flavors: the continuous-bag-of-words (CBOW) model and the Skip-Gram model. Since the availability of Word2Vec, it has become the de facto standard to replace feature engineering in NLP tasks. The embedding techniques like Word2Vec [8,9], Glove [18], and fastText [19] are also called as static word embeddings as the words have the same representation regardless of the context where they occur/appear. Sometimes, they are also referred as context-independent word embeddings. Due to this drawback, dynamic word embedding or contextual word embedding models were developed. These contextualized embedding models include ELMo [20,21] and CoVe [22] that are RNNs-based, as well as BERT [23] and ALBERT [24] which are transformer-based. ELMo uses representations derived from the hidden states of bidirectional LSTMs. CoVe is a seq2seq model with attention, originally developed for language translation. A CoVe model pre-trained on translation can then be used on other tasks. Here, we describe a few protein sequence representations based on word embedding. Please refer to an excellent review for various types and classification of word embedding [15]. Any pre-trained embeddings can be leveraged to downstream tasks in two ways: i) as feature extractors (not trainable) to offer static pre-trained input features, and ii) a fine-tuning where the embeddings are adjusted on a specific downstream dataset.

2.3.2 *Context-independent word embedding for protein sequence*

Here, we briefly describe ProtVec for the context-independent word embedding of protein sequences.

ProtVec: Word2Vec-based protein sequence representation

ProtVec [14] is an unsupervised data-driven distributed representation for protein sequences. Just like in NLP, a large corpus of protein sequence is required for the training of ProtVec: 546,790 protein sequences obtained from the Swiss-Prot database are used as training dataset. ProtVec regards 3-mers of amino acids as words and thus uses 1,640,370 sequences of

3-mers(grams). Subsequently, Word2Vec [8,9] using the skip-gram model is applied to the dataset, and 100-dimensional protein vectors are calculated for each 3-gram.

Originally, ProtVec was evaluated based on protein family classification and disordered protein prediction accuracies and it achieved high performance in both cases [14]. ProtVec is available at https://github.com/jowoojun/biovec.

2.3.3 Contextual word embedding for protein sequence

PhosContext2vec

PhosContext2vec [25] is a tool (as well as a distributed representation approach) to predict both general and kinase-specific phosphorylation site prediction that uses a distributed representation of residue-level sequence contexts called Context2Vec [26]. Context2Vec uses two strategies to generate contextual features: Word2Vec [8,9] and Doc2vec [27]. Context2Vec encodes residue-level sequence contexts of phosphorylation sites in the form of distributed vectors. It takes into consideration the dependence between phosphorylation sites and their contextual patterns by generating a distributed representation of the local sequence context for each phosphorylation site and then uses this representation to predict both types of phosphorylation sites. In addition to Context2Vec, PhosContext2vec also uses other features like Shannon's entropy.

Here, we briefly summarize the process of creating distributed representation of protein sequence using Context2Vec. After performing an *n-gram (3-gram) split* of the contextual window, a list of biological words is generated for 551,704 sequences from Uniprot. Then, the 3-gram biological words is fed to the Doc2vec to train a vector-generating model. Next, distributed representation of words is generated by two approaches: (i) by summing up the distributed representation of all its biological words, named context2vecadd, and (ii) by feeding the list of biological words to the pre-trained distributed memory network N, named context2vecinference. A separate SVM model is constructed for each of the three phosphorylation site types: S, T, and Y, and each of the five kinase families: AGC/PKC, AGC/PKA, CMGC/CK2, CMGC/CDK, and TK/SRC based on contextual feature vector along with six other residue-level feature groups (Shannon entropy, the relative entropy, protein disordered property, secondary structures, Taylor's overlapping properties, and the average cumulative hydrophobicity). The final feature vector of this phosphorylation site prediction is constructed by stacking all the seven types of feature vectors together, resulting in a 126-dimensional feature vector. To improve the classification performance, PhosContext2vec can be used in combination with other informative residue-level features. The web server of PhosContext2vec is publicly available at http://phoscontext2vec.erc.monash.edu/.

SeqVec: ELMo-based protein sequence representation

Word2Vec [8] embeddings do not take context into account whereas context could be important in some cases. Thus, Heinzinger *et al.* [28] proposed SeqVec representation that is learned using ELMo [20,21] model for protein sequences. SeqVec is a way to represent protein sequences as continuous vectors (embeddings) where the model is trained using the UniRef50 dataset. In NLP, ELMo is trained on unlabeled text-corpora such as Wikipedia to predict the most probable next word in a sentence, given all previous words in this sentence. By learning a probability distribution for sentences, these models autonomously develop a notion for syntax and semantics of language. One of the salient features of an ELMo-based model is that the trained vector representations (embeddings) are contextualized, i.e., the embeddings of a given word depends on its context. This has the advantage that two identical words can have different embeddings, depending on the words surrounding them. In contrast to previous non-contextualized approaches such as Word2Vec [8], this allows one to take the ambiguous meaning of words into account. SeqVec performes better than one-hot encoding, context-independent language models such as ProtVec in various protein prediction tasks [28]. The SeqVec is available at https://github.com/Rostlab/SeqVec.

UDSMProt

UDSMProt [29] is another language model representation extractor using a variant of (Long Short-Term Memory) LSTM pre-trained on unlabeled protein sequences from the Swiss-Prot database. The basic idea in UDSMProt is to apply self-supervised pre-training on a corpus of protein sequences. USDM relies on AWD-LSTM [30] language model, which is a three-layered bi-LSTM that introduces different types of dropout methods to achieve accurate word-level language modeling.

UDSMProt was initially trained on the Swiss-Prot database and then fine-tuned for specific tasks, such as enzyme commission classification, gene ontology prediction, and remote homology detection. Comparison results show that UDSMProt trained with external data performed favorably to the existing methods that were tailored to the task using a position-specific scoring matrix (PSSM) [29].

Unirep

Unirep (Unified Representation) [31] is a representation of protein sequence developed at Church's lab at Harvard based on a training of a recurrent neural network (specifically mLSTM, multiplicative long-short-term-memory) on ~24 million UniRef50 protein sequences. The mLSTM was trained using a 1,900-hidden unit. Technically, the protein sequence is modeled using a hidden state vector, which is recursively updated based on the previous hidden state vector. Essentially, the model is trained to perform the next amino acid prediction. This means that the method learns by scanning a sequence of amino acids and predicting the next one based on the sequence it has seen so far, while training the model revises the way its hidden state is constructed resulting in a better representation of the sequence.

UniRep outputs vectors of 1,900 dimensions. For a detailed explanation on how to retrieve the UniRep embedding, please refer to the GitHub repository at https://github.com/churchlab/UniRep. Additionally, a performant implementation of UniRep is available at https://github.com/ElArkk/jax-unirep.

ProtTrans

Rost Lab developed the ProtTrans [32], a representation of protein sequence using Transformers. As BERT [23] has become the standard model of solving NLP problems, ProtTrans uses BERT for sequence analysis. Essentially in ProtTrans [32], there are two autoregressive models (Transformer-XL [33], XLNet [34]) and four auto-encoder models (BERT, ALBERT, Electra [35], T5 [36]) on data from UniRef [37] and BFD [38], using roughly 2,122 million protein sequences and 572 GB of disk space. Through downstream computational analysis, it has shown that embeddings captured various biophysical properties of the amino acids, structure classes of proteins, domains of life and viruses, and protein functions in conserved motifs [32]. One of the most notable observations in ProtTrans is that the embeddings from the trained ProtT5 model, for the first time, surpassed the state-of-the-art methods in the secondary structure prediction task, without using any evolutionary information.

Additionally, despite BERT being just recently adopted for proteins, it has rapidly gained popularity. Hiranuma *et al.* [39] used ProtBERT embeddings along with several structural features to guide and improve protein structure refinement. Littmann *et al.* [40] investigated the effectiveness of BERT embeddings in Gene Ontology prediction.

2.4 *Variety of databases for generating pre-trained language models*

Pre-trained language models have brought advancements in the NLP field. Essentially, a language model is pre-trained on a large corpus of unlabeled (text) data using an unsupervised learning algorithm to enable the models to learn structures of the language. Recently, word embeddings as well as pre-trained language models have seen a surge in the field of protein bioinformatics. More recently, advances in machine learning have partially shifted the focus from hand-crafted features to automatic representation learning, where a neural network is used to learn features that are useful for the prediction task. Many such neural network methods have been proposed, which use a variety of architectures [41]. In that context, various protein sequence databases like UniRef, Swiss-Prot [42] etc. have been used for pre-training of the neural language models for protein sequences.

In the next sections, we discuss some of the most recent NLP-based encoding approaches for the PTM sites and GO predictions.

3. Methods using NLP-based encoding for PTM prediction: local-level task

PTMs are modifications that occur after the synthesis of the polypeptide chain and are responsible for increasing the diversity of the proteome. Till now, over 450 different types of PTMs have been identified. They are involved in a myriad of cellular processes ranging from maintenance of protein structure to cellular recognition. Elucidation of PTMs and PTM sites are indispensable for understanding cellular biology. Recently, mass spectrometry has become one of the most prevalent experimental techniques to characterize PTMs [43]. Due to the time and cost associated with experimental elucidation of PTMs and PTM sites, development of complementary computational approaches is gaining attention. PTM prediction is a local-level task where the objective is to predict which sites are likely to be PTM sites given a protein sequence. See Figure 1 for schematic of local-level task and global-level task.

Since 1998, we have seen a plethora of computational approaches for prediction of PTM sites. Recent computational methods are increasingly using Deep Learning-based (DL-based) approaches [1]. With the increasing number of computational tools for PTMs, we find several review articles highlighting these computational approaches. Trost and Kusalik [44] summarized machine learning-based approaches for the PTM prediction. We also summarized the most recent DL-based approaches for PTM site prediction [1]. Although this article provides a comprehensive overview of the DL-based approaches for the prediction of PTM sites, it does not cover the use of the NLP-based encoding/embedding in this field. We summarize the recent advances in computational approaches for prediction of PTMs by paying particular attention to the NLP-based encoding techniques (Table 1).

3.1 *Local/sparse representation-based methods for PTM prediction*

We focus on the PTM prediction approaches that use local/sparse representation (especially, one-hot encoding). Some of these approaches may use other types of features in addition to the NLP-based local/sparse representations, but our emphasis is on the NLP-based encoding. Additionally, we do not specifically distinguish between character-level, word-level and sequence-level encoding but annotate it when possible.

MusiteDeep

MusiteDeep [45,46] is a DL-based model for predicting general and kinase-specific phosphorylation sites. It uses a character-level one-hot encoding algorithm to encode 33-residue protein sequence segments centered at the prediction site. Each amino acid is represented by a binarized vector of length 21 including 20 common amino acids and a virtual amino acid (hyphen character '-'). Since the size of the protein fragment is 33, the

Table 1. PTM prediction methods using the NLP-based encoding described in this chapter.

Tool Name	Prediction Task	Encoding Approach	Deep Learning Architecture	Reference
MusiteDeep	General and kinase-specific phosphorylation sites	Character-level one-hot encoding	CNN	[45,46]
DeepNitro	Nitration and nitrosylation site	Character-level one-hot encoding, K-Space spectrum encoding	DNN	[47]
DeepTL-Ubi	Ubiquitination site	Character-level one-hot encoding	(DCCNN)	[48]
DeepUbiquitylation	Ubiquitination site	Character-level one-hot encoding, physio-chemical properties, PSSM	DNN	[49]
DeepUbi	Ubiquitination site	One-hot encoding, IPCP, CKSAAP PseAAC	CNN	[50]
DeepACEt	Acetylation Sites	Character-level one-hot encoding, CKSAAP, BLOSUM62, PSSM	MLP	[51]
DeepPSP	Phosphorylation Sites	Character-level one-hot encoding PSSM	SENet, Bi-LSTM	[52]
DeepPhos	Phosphorylation Sites	Character-level one-hot encoding	1D CNN	[53]
MDCAN-Lys	Succinylation Sites	Character-level one-hot encoding	Convolutional Block Attention Module (CBAM)	[54]
nhKcr	Crotonylation sites	Character-level one-hot encoding, Blosum62	2D CNN	[55]

Name	PTM (Succinylation sites)	Encoding	Model	Ref
DeepSuccinylSite	PTM (Succinylation sites)	Supervised embedding layer, character-level one-hot encoding	CNN	[56]
DTL-DephosSite	Phosphorylation Sites	Supervised embedding layer	Bi-LSTM	[57]
Chlamy-EnPhosSite	Phosphorylation Sites	Supervised embedding layer	2D CNN, LSTM	[58]
LSTMCNNsucc	Succinylation sites	Supervised embedding layer	Bi-LSTM, CNN	[59]
Ubicomb	Ubiquitylation sites (Plant-specific)	Supervised embedding layer + other physio-chemical	1D CNN, LSTM	[60]
LEMP	Mammalian Malonylation Sites	Supervised embedding layer	LSTM, RF	[61]
SohoKo-Kcr	Histone lysine crotonylation (Kcr) Sites	Supervised embedding layer	LSTM, BiLSTM, BiGRU	[62]
DL-Malosite	Malonylation Sites	Supervised embedding layer	CNN, RF	[63]
DeepDigest	Digestibility prediction (trypsin, ArgC, chymotrypsin, GluC, LysC, AspN, LysN, and LysargiNase)	Supervised embedding layer	CNN, LSTM	[64]
PhosContext2vec	General and Kinase-Specific Phosphorylation	One-hot Encoding Word2Vec and ProtVec	SVM	[25]
Deep-KCr	Histone lysine crotonylation (Kcr) Sites	Word2Vec, CKSAAP, EBGW	CNN	[65]
DeepIPs	Phosphorylation sites	Supervised embedding layer, Word2Vec, GloVe and fastText	CNN-LSTM	[66]
BERT-Kcr	Kcr sites	Contextual language model (BERT)	Bi-LSTM	[67]

total length of the feature vector to represent a sequence is 693 (33 × 21). A stacked convolutional neural network with attention mechanism is trained using the encoded features with their corresponding label. MusiteDeep uses the transfer learning, where the weights learned from the general phosphorylation data is applied to kinase-specific initialize phosphorylation prediction tasks. MusiteDeep [45,46] achieved impressive results for general phosphorylation sites and obtained competitive results for kinase-specific phosphorylation prediction. The stand-alone version of this tool is freely available at https://github.com/duolinwang/MusiteDeep_web.

DeepNitr

DeepNitro [47] is a protein nitration and nitrosylation site prediction tool based on deep neural network architecture. To train the model, both positive and negative sites (nitration and nitrosylation) are extracted from 3,113 unique proteins resulting in 1,210 positive and 8,043 negative sites for tyrosine (Y) nitration and 66 positive and 155 negative sites for tryptophan (W) nitration were collected. Similarly, 3,409 positive and 17,453 negative sites are collected for S-nitrosylation that occurs at Cystine (C). For model evaluation purposes, an independent test set of 189 positive and 1,182 negative tyrosine nitration sites and, 485 positive and 4,947 negative sites for S-nitrosylation are extracted apart from the training set. The peptide length of 41 is considered using various experiments to maximize the classification performance.

Besides character-level one-hot encoding, DeepNitro uses three other kinds of conventional feature encoding schemes including physical and chemical properties, k-space spectrum encoding, and PSSM (position-specific scoring matrix) were used. All the features from four different encoding techniques are combined to form a feature vector of length 3,311. The character-level one hot encoding in DeepNitro converts each amino acid to a 21-dimension feature vector. A predefined order of 20 natural amino acid residues and a gap character is set and assigned 1 and 0 according to their position. For the window size of 41, each peptide is represented by 861 (41 × 21) one-hot encoded features.

The computed features are then used to train a six-layer deep neural network. It is observed that the feature combination of PSSM and k-space feature group is the best feature for Tyrosine, Tryptophan, Cysteine nitration or nitrosylation PTM prediction [47]. On an independent test benchmarking, DeepNitro performs quite well [47]. The tool is freely available at http://deepnitro.renlab.org.

DeepTL-Ubi

DeepTL-Ubi [48] is a DL-based approach for predicting ubiquitination sites in the given protein sequences of multiple species. In DeepTL-Ubi a dataset consisting of 62,324, 15,495, 7,012, 2,452, 4,598, 1,173, 848, and 616 positive ubiquitination Lysine (K) sites for *H. sapiens, M. musculus, S. cerevisiae, R. norvegicus, A. nidulans, A. thaliana, T. gondii,* and *O. sativa* respectively was curated from various sources including PhosphoSitePlus, mUbiSida, and PLMD [68].

Subsequently, a peptide of length 31 is extracted around the identified ubiquitinated residue in the sequences. The prepared peptides are encoded using a one-hot encoding technique to generate a feature vector of dimension 31×21 (20 residue + 1 unknown residue). The DL-architecture used in DeepTL-Ubi is a convolutional neural network (CNN). A transfer learning approach is used to transfer the weights learned from a source species (e.g., *Homo Sapiens*) to train a model for a target species (e.g., *Saccharomyces cerevisiae*) by simultaneously minimizing both species transfer (ST) loss and classification loss. The results obtained on an independent dataset show that DeepTL-Ubi improves the performance of species-specific ubiquitination site prediction compared to the existing species-specific and general ubiquitination site prediction tools [48]. DeepTL-Ubi is available at https://github.com/USTC-HIlab/DeepTL-Ubi.

DeepUbiquitylation

DeepUbiquitylation [49] is a DL-based tool for the prediction of ubiquitination sites based on multimodal deep learning architecture. The training data set was constructed using a polypeptide fragment of window size 49 centered around the ubiquitination site from the respective proteins.

In DeepUbiquitylation, character-level one-hot feature encoding technique is used in addition to physio-chemical properties, and PSSM features. The feature vector of dimension 1,029 (i.e., 49×21) generated using a one-hot encoding technique is fed to a four-layer 1D convolutional neural network with three dense layers. Secondly, features from the top 13 physio-chemical properties of length (49 × 13) are fed to a three-layer Deep Neural Network (DNN). Thirdly, PSSM vectors are passed into three layers 1D CNN to obtain deep evolutionary characterization among different sequence positions. Finally, the generative deep representations corresponding to all three feature representations are combined and fed to two-layer neural networks for classification. The model is evaluated on the largest ubiquitination site database called PLMD, and it achieves an accuracy of 66.43%, and MCC of 0.221 [49]. DeepUbiquitylation [49] achieves better results compared to several popular protein ubiquitination site prediction tools. The source code and other resources are freely available at https://github.com/jiagenlee/deepUbiquitylation.

DeepUbi

DeebUbi [50] is a CNN-based ubiquitination site prediction model. In addition to one-hot encoding, DeepUbi uses informative physicochemical properties (IPCP), Compositions of K-spaced amino acid pairs (CKSAAP), and Pseudo Amino Acid Composition (PseAAC). For the one-hot encoding, DeepUbi utilizes 21 bits (20 + 1) where 20 bits is used to represent 20 different amino acids and 1 bit is used to indicate that the window has spanned out of the N-terminal or C-terminal. The training data for DeepUbi is collected from an online data repository called PLMD database [68], which is the biggest dataset for protein lysine (K) modification. After removing redundant sequences using CD-HIT, 53,999 ubiquitination and 50,315 non-ubiquitination sites are extracted from 12,053 different non-redundant proteins. A window size of 31 is used to extract peptide from the protein sequences with lysine (K) residues at the center. In DeepUbi, 651 features generated by one-hot encoding are combined with the features of CKSAAP to make hybrid feature representation that exhibites the best area under the Receiver Operating Characteristic curve (AUC) of 0.9066 and an MCC of 0.7 [50]. The trained model and other resources are freely available at https://github.com/Sunmile/DeepUbi.

DeepACEt

DeepACEt [51] is a Multilayer Perceptron(MLP)-based acetylation prediction tool trained using six different kinds of features: one-hot, BLOSUM62 matrix, a composition of K-space amino acid pairs(CKSAAP), information gain, physicochemical properties(A Index), and a position-specific scoring matrix(PSSM). For the one-hot encoding, DeepACEt uses character-level one-hot encoding technique where each amino acid is converted to a 21-dimension feature vector. A predefined order of 20 natural amino acid residues and a gap character is set and assigned 1 or 0 according to their position. The dataset having 16,107 positive (human lysine-acetylated sites) and 57,443 negative (non lysine-acetylated sites) peptides is prepared using the CPLM database. The under-sampling technique is used to match the size of the positive and negative dataset in order to avoid class imbalance problems. The optimal length of the window is found to be 31 from k-fold validation experiments. A F-Score feature selection technique is used to select the most optimal features. 60% of the total dataset is considered as a training set and the rest is taken as an independent test set. The selected 2,199 features of the training set are fed to a six-layer MLP to train the model. While testing the model on the independent test set, DeepACEt achieves an accuracy of 84.87%, AUC of 0.8407, specificity of 83.46%, the sensitivity of 86.28%, and MCC of 0.6977 [51]. The trained model and other resources of this work are freely available at https://github.com/Sunmile/DeepAcet.

DeepPSP

DeepPSP [52] is a phosphorylation site prediction tool developed by combining local information and global information around potential phosphorylation sites. DeepPSP consists of two different local and global components. The window size of 51 and 2,000 is used for local and global windows respectively. The feature vectors generated by character-level one-hot encoding are fed to Convolution blocks, two squeeze-and-excitation (SENet) blocks and Bi-LSTM blocks. SENet estimates the residue contribution at each position around the potential site. SENet is basically an extension of CNNs by adding channel parameters that improves channel interdependencies at less cost. The rationale for using the global information in DeepPSP is attributed to the fact that interaction between widely separated modification sites may influence functioning of complex cellular information processing. After training and evaluation of the model, it is observed that one-hot encoding is one of the simplest yet more effective encoding techniques as compared to other similar sequence information-based encoding like position-specific scoring matrix (PSSM) and secondary structures (SS) [52]. Also, DeepPSP uses the notion of learning to overcome the problem of not having enough training data for specific-kinase phosphorylation sites. The source codes and other resources of DeepPSP [52] tool are publicly available at https://github.com/gankLei-X/DeepPSP.

DeepPhos

DeepPhos [53] is a DL-based prediction tool for phosphorylation sites. Given a protein sequence as an input, DeepPhos predicts both general and kinase-specific phosphorylation sites. DeepPhos uses character-level one-hot encoding technique to encode the given sequence into vector representation, where each amino acid is encoded to a vector of length 21. The encoded features are fed to multi-layer one-dimensional CNNs. DeepPhos also uses inter- and intra- (Intra-BCL) block concatenation (functional architecture) layers that help to transfer the abstract information of previous layers with different levels to the current layer as the network deepens. Compared to the existing architectures like CNN, LSTM, RNN, and Fully Connected Network (FCN), the architecture used in DeepPhos [53] has better performance in predicting phosphorylation Serine / Threonine (S/T) and Tyrosine (Y). Similarly, it performed relatively well when compared against general site prediction tools like NetPhos3.0, PPSP, Musite, MusiteDeep and also against kinase-specific phosphorylation site prediction jobs when compared to existing kinase-specific predictors like GPS 2.0 and PPSP [53]. The trained model of DeepPhos tool and other resources are available at https://github.com/USTCHIlab/DeepPhos. [Currently Link is broken]

MDCAN-Lys

MDCAN-Lys [54] is a binary classifier tool to predict succinylated sites trained using multi-lane dense convolutional attention networks. In MDCAN-Lys, the sequence information, physicochemical properties, and structural properties of the sequences are extracted using dense convolutional blocks. These blocks are considered as a more effective way of representing the low-level and high-level features of the sequences compared to CNN, LSTM etc.

A training dataset of 5,885 positive and 64,140 negative peptides and a test dataset of 684 positive and 6,709 negative peptides are extracted from Protein Lysine Modification Database (PLMD) [68]. MDCAN-Lys uses one-hot encoding to convert sequence information into vector representation. In addition to the one-hot encoding, five different atchley factors, namely, polarity, codon diversity, secondary structure, molecular volume, and electrostatic charge are used to encode the sequence in features. The class weighting method is used to avoid the possible class imbalance problem due to the different sizes of positive and negative dataset. Finally, a multilane dense convolutional attention network is trained using positive and negative training samples. While comparing with other succinylation site predictors, MDCAN-Lys performs relatively well [54].

nhKcr

nhKCR [55] is a DL-based framework that identifies lysine crotonylation (Kcr) sites. The dataset used in this work consists of 15,605 positive and 75,111 negative samples of window size 29 with the lysine (K) located at the center from unique non-histone proteins collected from HeLa cell, lung cell, A549 and HCT116 cell databases.

In addition to the one-hot encoding, nhKcr uses AAindex and Blosum62. In nhKcr, these three types of encoding schemes are deemed as the RGB channels of a color image and are processed by a two-dimensional convolution neural network. The final model consists of an input layer which takes in 29×29 matrices corresponding to the RGB channels of a color image. While comparing model performances with existing tools, the nhKcr achieved better results based on several performance metrics [55]. The webserver of nhKcr tool is available at http://nhKcr.erc.monash.edu/.

3.2 *PTM site prediction approaches using distributed representation*

We review some of the recent PTM site prediction approaches that utilize distributed representation of protein sequences. As discussed earlier, the distributed representation-based methods can be broadly classified as context-independent word embedding and contextual word embedding approaches. Furthermore, the context-independent word embedding can be classified as supervised word embedding (aka supervised embedding layer) and pre-trained embedding methods.

3.2.1 *Context-independent supervised word embedding-based PTM site-prediction approaches (aka supervised embedding layer)*

We review some of the supervised word embedding-based PTM site-prediction approaches, where the word embedding is learned from scratch (generally initialized with random numbers) within the context of end-to-end deep learning. After the neural network is trained, one achieves word embedding as a side-product. In this type of embedding, initially, a vocabulary is decided. In the case of protein sequences, often the size of vocabulary is taken to be 20, corresponding to the number of the canonical amino acids. After that, a dimension of each word embedding vector is set. Initially, the vectors of the embeddings are randomly initialized and then these word embedded vectors are trained. Practically, Keras's embedding layer is used to obtain/learn this type of embedding.

DeepSuccinylSite

DeepSuccinylSite [56] is a supervised word embedding-based deep learning approach developed in our lab to predict succinylation sites in given protein sequences. It uses two different encoding techniques viz. one-hot encoding, and supervised word embedding implemented using Keras's Embedding layer. Initially, the 21 amino acid type, including a virtual amino acid, are mapped into unique integers. Then, the one-hot encoding is used to convert each amino acid to a binary code consisting of a sequence of zeros and a singular one for the location that encodes the amino acid. A window size of N is represented as a N × 21 input size vector using this encoding. On the other hand, a supervised word embedding is implemented using the Embedding layer of Keras to convert into an embedded encoding vector. The predefined integer representation of each sequence is provided as an input to the embedding layer, which lies at the beginning of the DL architecture. The embedding layer is initialized with random weights, which then learns better vector-based representations with subsequent epochs during training. Each vectorization is an orthogonal representation in another dimension, thus preserving its identity. Hence, making it more dynamic than the static one-hot encoding. In DeepSuccinylSite, the learned word embedding vector for K is: [−0.03372079, 0.01156038, −0.00370798, 0.00726882, −0.00323456, −0.00622324, 0.01516087, 0.02321764, 0.00389882, −0.01039953, −0.02650939, 0.01174229, −0.0204078, −0.06951248, −0.01470334,

−0.03336572, 0.01336034, −0.00045607, 0.01492316, 0.02321628, −0.02551141] in 21-dimensional vector space after training. Since the embedding-based representation performed better than one-hot encoding, DeepSuccinylSite uses an embedding dimension of 21 from window size 33. Combined with a CNN-based DL-architecture the embedding-based DeepSuccinylSite achieves MCC of 0.48 on an independent test set, which is better than one-hot encoding [56]. The model and dataset for DeepSuccinySite are available at https://github.com/dukkakc/DeepSuccinylSite.

DTL-DephosSite

DTL-DephosSite [57] is a dephosphorylation site prediction tool developed in our lab that exploits the fact that phosphorylation and dephosphorylation are closely related to one another at a molecular level. The tool consists of two DL-based prediction components: S/T dephosphorylation site predictor and Y dephosphorylation site predictor. The concept of transfer learning is used to train the model where previously trained weights in similar problems are used to initialize the weights in the current problem. Initially, a model is learned on the phosphorylation sites using a Bi-LSTM (Bidirectional Long Short-Term Memory) and the learned weights are subsequently applied to initialize the weights before training on a dephosphorylation dataset named ComDephos. MusiteDeep's phosphorylation dataset is initially trained on a Bi-LSTM model to extract the pre-trained weights, which are transferred to a new Bi-LSTM architecture to train on the Dephos data containing the S/T residues in the center. The appropriate window length for the problem is found to be 33 after evaluating five-fold cross validation results.

DTL-DephosSite also uses supervised word embedding technique, where each amino acid in the 33 window is assigned to a unique integer and further encoded using the *embedding layer* of the keras library. The embedding layer of keras module is a popular framework for text processing, where each word is assigned a unique integer. In this work, each character is considered as a separate word and vocabulary size is set to 21, including 20 amino acids and one gap character. The results show that the embedding layer with transfer learning-based model, DTL-DephosSite, performed markedly better than those developed using embedding with Bi-LSTM alone [57]. Resources for the DTL-DephosSite tool are available at https://github.com/dukkakc/DTLDephos.

Chlamy-EnPhosSite

Chlamy-EnPhosSite [58] is a DL-based organism-specific phosphorylation site tool developed in our lab that predicts the presence of phosphorylation sites in *Chlamydomonas reinhardtii*, a single cell green-alga of length around 10 micrometers that swims with two flagella. In Chlamy-EnPhosSite, CNNs and a long short-term memory (LSTM) are combined together to make a prediction tool. Chlamy-EnPhosSite also uses the supervised word embedding. For encoding, the 21 amino acid characters including one pseudo-residue are mapped to specific integers ranging from 0 to 21. Then, those integers are fed to the embedding layer of Keras Library that converts the input peptide of length 57 into a vector representation. The word embedding is learned as a byproduct of the training. Finally, the power of stacked 2D convolutional neural networks and stacked long short-term memory were combined to predict the S, S/T, T, Y phosphorylation sites [58] . Also, we propose Chlamy-MwPhosSite based on multiple windows, which concatenates the features extracted by five CNN models trained on five different window sizes; Chlamy-MwPhosSite minimizes the hassle to choose one particular best performing window size. The trained model, test dataset, and other resources are available at https://github.com/dukkakc/Chlamy-EnPhosSite.

LSTMCNNsucc

LSTMCNNsucc [59] is an integrated DL-based model to predict the presence of Succinylation sites in the given protein sequences that combines LSTM and CNN. The training data set with 6,512 positive and 6,512 negative samples, and testing set of 1,479 positive and 1,6457 negative samples in LSTMCNNsucc are extracted from PLMD (Protein Lysine Modifications Database) [68].

LSTMCNNsucc uses an embedding layer as the first layer in the architecture to translate the sequence to a vector representation. Essentially, a peptide of size 31 centered around the site of interest is passed through this embedding layer to learn a supervised word embedding that translates the input to a vector of shape (31,64) resulting in a vector of length 1,984. Those embedded features are then fed to LSTM and 1-D CNN networks and the results from both networks were concatenated together and sent to a fully connected layer. The overall performance of the LSTMCNNsucc on the independent test set is found to be slightly better compared to existing succinylation prediction tools [59]. The model is available as a web server at http://8.129.111.5/.

UbiComb

Ubicomb [60] uses a hybrid DL-based approach by utilizing CNN and LSTM for identifying plant-specific ubiquitination sites. It uses embedding encoding and physicochemical properties-based encoding techniques for vector representation. Initially, each amino acid is considered as a word and a dictionary of residues is created by an integer encoding the amino acid and pseudo-amino acid. The supervised embedding encoding approach uses an embedding layer where the initial embedding weight matrix is initialized randomly and these weights are learned during training. The embedding layer is followed by LSTM in one module whereas in the second module, the latent five-dimensional feature of physicochemical properties are extracted using 1D convolutional layers followed by a max-pooling layer. The outputs of these modules are concatenated and passed as an input to dense layers for deeper feature extraction. The embedding encoding is similar to index-based integers encoding ranging from 1 to 22, whereas, for the physicochemical properties-based encoding, a small five-dimensional latent vector which corresponds to meaningful major principal components: hydrophobicity, size, number of degenerated triplet codons, preference of amino acid residues in a beta-strand, and frequency of occurrence of amino acid residues in a beta-strand were generated from 237 physicochemical properties for each residue of amino acid in 31 window fragment. This work considers local window fragments of 31, where ubiquitination site (K) residue is at the middle of the window fragment. The training data for UbiComb consists of 7,000 protein fragments constructed from plant subset data containing 3,500 positive and 3,500 negative fragments, which were selected randomly. From this, 5,500 were used for training and 1,500 sequences were used for independent testing. UbiComb demonstrated a better generalization capability than other compared ubiquitination predictors [60]. The webserver of the UbiComb model can be found at http://nsclbio.jbnu.ac.kr/tools/UbiComb/.

LEMP

LEMP [61] is a LSTM-based ensemble mammalian malonylation site predictor which integrates an LSTM-based classifier using a supervised word embedding encoding technique with a random forest (RF) classifier using an enhanced amino acid content (EAAC). The LSTM classifier uses five layers: the input layer that takes 31 residue peptide fragments, embedding layer that converts amino acid into a five-dimension word vector, recurrent layer, fully connected layer, and output layer. The word embedding technique in the LSTM-based classifier is found to be a better encoding technique as compared to other traditional feature extraction algorithms including one-hot encoding [61]. A dataset with 288 positive sites (K-malonylated) and 88,636 (non K-malonylated) negative sites are collected from 2,127 unique sequences. A window length of size 31 is chosen using a ten-fold cross validation where the model was trained over different window lengths (15, 19, 23, 27, 31, and 35) and the window with the largest area under the receiver operating characteristic (ROC) curve (AUC) is selected.

Since the size of the training dataset should be large enough for deep learning models, this work also experiments with the effectiveness of informative features with traditional machine learning algorithms. The RF-based classifier uses EAAC. The final model named LEMP (LSTM-based ensemble malonylation predictor) is constructed by combining the LSTM model with word embedding and Random Forest model with a novel encoding of enhanced amino acid content (EAAC) using a logistic regression. LEMP performed relatively well for the prediction of mammalian malonylation sites [61]. LEMP is available at http://www.bioinfogo.org/lemp.

SohoKo-Kcr

SohoKo-Kcr [62] is a deep learning and RNN-based Bidirectional Gated Recurrent Unit (BiGRU)-based model for prediction of protein lysine crotonylation sites. The training set consists of 9,964 positive and 9,964 negative crotonylation sites. SohoKo-Kcr uses supervised embedded encoding where each peptide is treated as a sentence and each amino acid as words (k-mer, k = 1). Each amino acid residue is then converted into an integer encoding (0–19) to represent 20 amino acids. The peptide input is set to 31. Then, these integers encoded inputs are passed through an embedding layer that embeds the input to a vector space of 256 dimensions. The embedding weight matrix is initialized randomly and is learned during the training. The output of embedding is passed to BiGRU with 256 memory units and then are subsequently flattened. In addition to k = 1, the authors also tested various values of k-mers (k = 2,3) and observed that k = 1, meaning treating each amino acid as words, produces better results [62]. Sohoko-Kcr can be accessed at https://sohoko-research-9uu23. ondigitalocean.app.

DL-Malosite

DL-Malosite [63] is a DL-based approach for prediction of malonylation sites developed in our lab. DL-MaloSite encodes amino acid sequence utilizing a supervised embedding technique using an embedding layer. The dataset for DL-Malsosite consists of 3,978 positive training sites and 988 positive test sites curated from experimentally verified malonylation sites from mice and humans. The same number of negative data is randomly generated from the sequences in order to avoid the class imbalance problem.

The DL-MaloSite uses CNN as the architecture, where each sequence fragment of length 29 is encoded to integers from 0 to 20 by following the predefined mapping function. Finally, those integers are encoded with the window size 29 and fed into the two layered convolutional neural network to train DL-MaloSite model. In order to address the fact that deep learning models need large amounts of data to train the model properly, the transfer learning is used to transfer the initial weights from DeepPhos. Finally, DL-MaloSite models are tested on an independent test set. The results were compared with SPRINT-Mal where DL-MaloSite achieves better results in terms of MCC score [63]. Data and source code of this project is available at https://github.com/dukkakc/DL-MaloSite-and-RF-MaloSite.

DeepDigest: Prediction of proteolytic digestion

DeepDigest [64] is a digestibility (another type of PTM) prediction tool for predicting eight different proteases (trypsin, ArgC, chymotrypsin, GluC, LysC, AspN, LysN, and LysargiNase) developed by integrating CNN and LSTM network. The convolutional layers are used to extract local sequence features called low-level features and an LSTM layer incorporates the inter amino acid dependencies. The different large datasets covering the eight different proteases are collected for four organisms *including E. coli*, yeast, mouse, and human. A window size of 31 is considered while preparing peptides to train the model.

Unlike handcrafted features used in other machine learning-based tools, DeepDigest uses a supervised word embedding technique where each amino acid in the peptide is mapped into a 21-dimensional feature by an embedding layer. Each positive and negative peptide is converted into a vector of length 693(33 × 21). The embedding weights are initialized randomly and learned throughout the training process. The constructed feature vectors from the embedding layer are fed to two convolution layers followed by an LSTM layer. The proposed model is evaluated using 10-fold cross-validation and independent testing dataset. Compared with the results of traditional machine learning algorithms (SVM, RF, LR), DeepDigest shows improved performance for all the eight proteases on a variety of datasets in terms of AUCs, F1 scores, and Matthews correlation coefficients (MCCs) [64]. The DeepDigest tool and test data can be downloaded from http://fugroup.amss.ac.cn/software/DeepDigest/DeepDigest.html.

3.2.2 Context-independent pre-trained word embedding-based approaches for PTM prediction

Deep-Kcr

Deep-KCR [65] is a DL-based approach for predicting histone lysine crotonylation (Kcr) sites. A window size of 31 residues with lysine (K) located at the center is extracted from the 3,734 unique histone proteins in the HeLa cell. A total of 9,964 positive and 9,964 negative samples are selected using data undersampling technique. Deep-Kcr uses six types of encoding schemes to convert protein sequences into feature vectors including Word2Vec [8]. Word2Vec is a natural language processing (NLP)-based vectorizing tool that takes text corpus as an input and converts that input to corresponding feature vectors called neural word embeddings. The CBOW model of Word2Vec is used in Deep-Kcr as it uses context words in a neighboring window to predict the word. The dimension of the Word2Vec-based feature vector in this model is set to 200. Finally, all the extracted features are combined to form a 3,184-dimensional feature set. This feature set is then passed into convolutional neural networks (CNN) for classification purposes. The model trained on the fusion feature set achieves better performance (AUC = 0.8671) compared with the models trained on the individual original feature set [65].

The feature importance analysis reveals that the features contained in the optimal feature set are 40% Word2Vec. Furthermore, authors use a popular Information Gain feature selection method to fuse feature set and obtain AUC of 0.8846. The prediction performance of different feature encoding schemes, different classifiers and while comparing with existing tools Deep-Kcr [65] shows robustness and generalization ability of the model. The Deep-Kcr can be accessed from the webserver at http://lin-group.cn/server/Deep-Kcr/.

3.2.3 Methods using both context independent supervised word embedding + context independent pre-trained model

DeepIPs

DeepIPs [66] is a DL-based model that identifies phosphorylation sites in host cells infected with SARS-CoV-2. The dataset used in this work consists of 5,387 positive samples and 5,387 negative samples of S/T sites, 102 positive samples and 102 negative samples of Y sites curated from verified phosphorylation sites of human A549 cells infected with SARS-CoV-2.

The model is based on Convolutional Neural Network — Long Short-Term Memory (CNN-LSTM). DeepIPs uses word embedding method of natural language processing to convert the protein sequences into vector representation. DeepIPS uses two different types of encoding techniques including Supervised Embedding Layer (SEL) and unsupervised embedding based on pre-trained word embedding methods (Word2Vec, GloVe [18], and fastText [19]). The supervised embedding is achieved through Keras's embedding layer that consists of a fully connected neural network that converts given positive integers into dense vectors of predefined size. The weights are initialized randomly and continuously updated using a back-propagation approach. Additionally, DeepIPS uses various unsupervised embedding: Word2Vec, GloVe and fastText. In DeepIPS, Word2Vec is used to train a distributed representation embedding for protein sequences. A protein peptide of fixed-length k is considered as a 'word'. The Continuous Bag of Words (CBOW) model of Word2Vec is adopted over the Skip-Gram model. Additionally, the GloVe is used to represent sequences into vectors. Initially, the frequencies of the words that occur together are stored in a matrix called word-word co-occurrence matrix. Then, the weights are learned based on that matrix.

For the DL-architecture, a hybrid deep learning architecture that consists of CNN followed by a LSTM layer is utilized. For S/T, the supervised embedding layer produced slightly better results and for Y, GloVe produced slightly better results compared to other implemented encoding schemes [66]. Hence, these models are chosen as final models for S/T and Y respectively in DeepIPS. The model is available as a web-server at http://lin-group.cn/server/Deep-Kcr/DeepIPs.

3.2.4 *Methods using pre-trained contextual embedding language model (BERT-based)*

BERT-Kcr

BERT-Kcr [67] is a lysine crotonylation site prediction model given a protein sequence. BERT-Kcr uses the concept of transfer learning method with pre-trained BERT (bidirectional encoder representations from transformers) model. In this work, the BERT model is implemented considering each amino acid as a word and a sequence as a document. The features prepared from the pre-trained model are then supplied to Bidirectional — Long Short-Term Memory (BiLSTM) network to train the final BERT-Kcr model.

A total of 9,964 positive samples and the same number of negative samples are collected from HeLa cells. Each sample contains 31 amino acids with the lysine in the middle. Finally, the model is tested on an independent test set (30% of the total data) where the result outperforms the Deep-Kcr [65] model. To observe the effectiveness of the proposed model, the results were also compared with other popular pre-trained models: ELMo [20,21] and fastText, where the BERT [67] model achieved better results. The BERT-Kcr model is available on http://zhulab.org.cn/BERT-Kcr_models/.

4. Methods using NLP-based encoding for GO-based protein function prediction: Global-level task

As an illustrative example of the global-level task, we consider Gene Ontology prediction. Prediction of protein function is a crucial part of genome annotation. Computational protein function prediction serves as a complementary approach to experimental function annotation. Although there is some variability in the vocabulary used for defining protein function, Gene Ontology (GO) [69] is arguably the most comprehensive and widely used functional classification. GO is composed of three ontologies: molecular functional ontology (MFO), biological process ontology (BPO), and cellular component ontology (CCO) [70]. MFO describes the elemental activities of a gene product at the molecular level (i.e., binding and catalysis); BPO captures the beginning and end, pertinent to the functioning of integrated living units: cells, tissues, organs, and organisms; CCO describes the parts of cells and their extracellular environments. Automated function prediction (AFP) in terms of GO is a challenging problem and Critical Assessment of Function Annotation (CAFA) [71,70] has been contributing enormously in helping spur the development of various computational approaches for function prediction. CAFA is an experiment designed to provide a large-scale assessment of computation methods for predicting protein function where different computational approaches are evaluated by their ability to predict GO terms in the catergories of Molecular Function (MF), Biological Processe (BP) and Cellular Component (CC) [71,70]. So far, four CAFAs (CAFA1–CAFA4) have been organized. In addition, the Function special Interest group [72] also is contributing singnificanlty in the endeavor by organinzing meetings. KC is also one of the members of the group.

In that regard, there are a few excellent reviews summarizing recent approaches for Gene Function prediction. Zhao highlighted various machine learning techniques for protein function prediction based on GO [73]. Bonetta and Valentino summarize the recent machine learning techniques for GO prediction highlighting the various ML/DL approaches as well as different types of features [41]. To our knowledge, we are the first to survey recent GO prediction approaches that use the NLP-based protein sequence encoding. We provide a summary of these methods in Table 2.

4.1 Local/sparse representation-based methods GO-based protein function prediction

We briefly review selected local/sparse representation-based methods for the GO prediction.

DeepGOPlus

DeepGOPlus [76] is a new version of DeepGO [75] that solves some limitations of DeepGo related to sequence length and number of predicted classes. DeepGo could not predict the function of nearly 10% of proteins

Table 2. GO function prediction methods using the NLP-based encoding described in the chapter.

Tool Name	Encoding Approach	Deep Learning Architecture	Reference
SDN2GO	Word-level encoding, Word2Vec, n-grams (trigrams), pre-trained embedding	CNN	[74]
DeepGO	Word-level encoding, n-grams (trigrams), supervised embedding layer	CNN	[75]
DeepGOPlus	Character-level one-hot encoding,	CNN	[76]
DeepAdd	Word-level encoding, n-grams (trigrams), supervised embedding layer	CNN	[77]
NetGO 2.0	Embedding layer, text, domain and network	BiLSTM, logistic regression	[78]
GOLabeler	Sentence-level encoding, trigram frequencies	Logistic regression	[79]
DeepText2GO	Sentence-level encoding: InterPro	Logistic regression	[80]
DeepFRI	Character-level encoding, contextual word embedding based on pre-trained LSTM language model	CNN	[81]
goPredSim	Sentence-level encoding, pre-trained embeddings (SeqVec and ProtBert)	Embedding similarity-based transfer	[40]
ProLanGO	Word-level encoding	LSTM	[82]
GCN for Structure and Function	Contextual word embedding from pre-trained ELMo language model	Graph Convolution Network	[83]
DeepSeq	Supervised embedding layer	CNN	[84]

as these proteins are larger than 1,002 amino acids and could only predict 2,000 functions out of more than 45,000 functions in GO. In contrast to DeepGO, DeepGOPlus uses local/sparse representation of amino acids (character-level encoding) based on the one-hot encoding. The input sequence is converted to a vector of size 21×2000 where 21 is the type of amino acids and 2,000 is the length of the protein sequence. The model works for sequences of length up to 2,000 resides covering almost 99% of existing proteins in the UniProt database. Later on, these local representations are fed onto the CNN to produce 8,192 dimensional vectors representing motifs in the target protein. Subsequently, passing them to fully connected classification layer which can output more than 5,000 classes to produce a prediction of GO terms. The source code is available at https://github.com/bio-ontology-research-group/deepgoplus and the tool is available online at https://deepgo.cbrc.kaust.edu.sa/deepgo/.

GOLabeler

GOLabeler [79] is an ensemble approach that uses five-component classifiers trained from different features, including GO term frequency, sequence alignment, amino acid trigram, domains and motifs, and biophysical properties using Learning to Rank (LTR) to integrate these multiple classifiers. An amino acid trigram is the NLP-based approach in GOLabeler and is basically the frequency of the types of three consecutive amino acids which has a dimension of 8,000 ($20 \times 20 \times 20$) features. In that regard, GOLabeler is a bag-of-words based approach and achieves fairly good results compared to other function prediction methods.

GOLabeler implement a consensus of five models: homology-based transfer (BLAST-KNN), Naïve Bayes approach, and three sequence-based models. The latter are logistic regression models. The LR-mer module converts protein sequence to a frequency vector of length of 8,000 (20^3 is a number of 3-mers over the alphabet of 20 amino acids) then the logistic regression predicts probabilities for each GO term. LR-InterPro is a multiclass logistic regression model. Instead of frequency tables, InterPro encodes each protein into a 33 879 binary vector by looking up motifs and domains in biological databases. LR-ProFET presents a protein as a vector of handcrafted features of the length of 1,170. A final vector encodes various biophysical, statistical, and structural information.

4.2 Distributed representation-based methods GO-based protein function prediction

Recently, a few distributed representation-based methods for the GO prediction are published. We briefly review these methods. As discussed above, the distributed representation approaches can be classified as approaches using context-independent and contextual distributed representation. Furthermore, contextual approaches are broadly classified as supervised context-independent approaches and pre-trained contextual approaches.

4.2.1 Context-independent supervised word embedding-based GO prediction approaches word (aka supervised embedding layer)

DeepSeq

DeepSeq [84] is a context-independent supervised word embedding-based approach that extracts (or learns) features as part of its end-to-end learning. DeepSeq uses an amino acid sequence of protein as input for predicting functions for *Homo Sapiens* using Gene Ontology. The model takes amino acid sequences up to 2,000 residues. Each amino acid (character-level encoding) is represented as a 23-dimensional vector (22 amino acids and one more character for 0-padding). The DL-architecture in DeepSeq is a convolutional neural network with some modifications. The network architecture consists of the embedding layer, convolutional layers, dropout layers as well as global average pooling, linear transformation, batch normalization, and activation layers. One salient feature of the CNN architecture in DeepSeq compared to traditional CNN is that instead of simple flattening, a global pooling operation is utilized to reduce the number of parameters. The output from the pooling layer which is a feature vector is then passed to the output layer which uses ADAM classifier and binary cross entropy loss function. DeepSeq achieves an overall validation accuracy of 86.72% on prediction of *Homo Sapiens* molecular functions [84]. The source code is published at https://github.com/recluze/deepseq.

DeepGO

DeepGO [75] is a method for predicting protein functions from protein sequence and known interactions. The input to DeepGo is the amino acid sequence of a protein. DeepGO uses a trigram embedding layer to represent protein sequences. A trigram of AA from the sequence is generated which is of length 8,000 ($20 \times 20 \times 20$). Instead of representing the trigrams using one-hot encoding vectors of sparse nature, in DeepGO, a distributed representation of trigrams is used. Essentially, embedding is used where the vectors are randomly initialized and then learn the representation during training. Using this vocabulary, a sequence of length 1,002 is represented by a vector of 1,000 indices. Sequences longer than 1,002 are ignored and shorter are padded with zeros. Given a vocabulary of 8,000 trigrams, each of a trigram encoded initially by its index is mapped to 128-dimensional space, resulting in 1,000 by 128 dense matrix representation for a protein sequence. A CNN is selected as an architecture model combined all together with protein-protein interactions network embeddings. A hierarchical classifier is in place for predicting GO terms. In addition to protein sequence, DeepGO [75] also uses protein-protein interaction network features. The source code is published at https://github.com/bio-ontology-research-group/deepgo. The online server is available at http://deepgo.bio2vec.net.

NetGO2.0

NetGO2.0 [78] is an improved version of NetGO and consists of seven component methods: Naive, BLAST-KNN, LR-3mer, LR-InterPro, NetKNN, LR-Text and Seq-RNN. From these components LR-3mer, LR-Text and Seq-RNN are the NLP-based encoding schemes. LR-3mer is a logistic regression of the frequency of amino acid trigram. The LR-Text is a component of DeepText2GO method. In LR-text, the abstract of the article for the UniprotID is combined to form a document and for this document, its sparse TF-IDF is combined with dense semantic representation generated by Doc2Vec. For each GO term, a Logistic regression model is trained. In addition, NetGO2.0 also uses Seq-RNN to extract deep representations of protein sequences. Essentially, in Seq-RNN, each first amino acid (character-level encoding) is represented by embedding and then this embedding is input into a BiLSTM. Finally, a max pooling layer generates the overall representation of the protein sequence. The benchmark dataset is available at https://drive.google.com/drive/folders/1wSS-R335UcNMToMskx3dE4XcTaLvCAOc.

4.2.2 *Context-independent pre-trained word embedding-based approaches for the GO prediction*

In this section, we describe a few context-independent pre-trained word embedding-based approaches for the GO prediction.

SDN2GO

SDN2GO [74] is a context-independent pre-trained word embedding-based approach for the GO prediction. SDN2GO predicts GO labels based on protein sequences, protein domain content and protein-protein interaction networks. SND2GO works for proteins less than 1,500 amino acids and uses representation of proteins sequences of ProtVec [14] (for the details of ProtVec please refer to Section 2.3.2) of BioVec. Following ProtVec, SDN2GO uses a sliding window of length 3 with a step of 1 producing 3-grams sequences with a length of 1,498 for each protein of length 1,500.

Subsequently, distributed representations of 3-grams are obtained through ProtVec-100d-3grams look-up table from the BioVec library. Thus, each trigram is mapped into a 100-D vector. Essentially, each protein is represented by 1,498 × 100 vector matrix, which is used as an input to the subsequent Deep Learning model. In addition to representation from protein sequence, SDN2GO also uses integrated deep learning architecture: the sequence model utilizes 1-D CNN, the PPI uses Neural Network and the domain model utilizes 1-D CNN. Finally, the output features of the three sub-models are combined to produce the GO output. The source code and dataset are available at https://github.com/Charrick/SDN2GO.

DeepAdd

DeepAdd [77] is another context-independent pre-trained word embedding approach for the GO prediction that uses NLP model to generate word embedding of protein sequence. The main workflow of DeepAdd uses Word2Vec [8], a word-based encoding, to represent query protein sequence. The protein sequence of length less than or equal to 1,000 residues is broken into 3-mers. The k-mers dictionary is later mapped into 4-dimensional vector space using a continuous bag of words model. DeepAdd uses CBOW model of Word2Vec. The output of Word2Vec is the input to the two CNN models that follow. In addition to the Word2Vec features, DeepAdd uses two additional features: Protein-Protein Interaction (PPI) network features and Sequence Similarity Profile features (SSP).

Subsequently, two CNN models with multiple convolution blocks that map the encoded feature to these two-feature vectors representation are created. Compared to DeepGo [75], DeepAdd yields the best results across the three GO categories in most of the metrics on the CAFA3 dataset [77].

4.2.3 *Contextual word embedding-based protein GO prediction*

In this section, we review selected contextual word embedding-based GO prediction approaches. The main characteristics of these methods is that each amino acid (token) gets a representation based on context. Thus, the same amino (e.g., 'A') appearing in different locations in the sequence is encoded differently.

DeepFRI

DeepFRI [81] is a contextual word embedding-based GO prediction approach that uses Graph Convolutional by leveraging a large number of protein sequences and protein structures including the modeled structures through SWISS-MODEL [85]. In essence, DeepFRI combines sequence features extracted from a protein language model and protein structures. In DeepFRI, for those protein sequences for which PDB structures are not available, the structures are obtained using SWISS-MODEL. This approach is expected to become even better with the availability of the AlphaFold2 [38] structures. DeepFRI, which stands for Deep Functional Residue Identification, uses a two-stage architecture that takes a protein structure and a sequence representation from a pre-trained language LSTM model based on ~10 million Pfam sequences as input to extract residue-level features for the sequence. The LSTM-LM is trained on 10 million protein domain sequences from Pfam to predict an amino acid residue in the context of its position in a protein sequence. In addition, DeepFRI is trained on protein structures mainly from contact maps using the Graph Convolutional Networks. DeepFRI [81] performs quite well compared to other approaches like DeepGo [75]. The source code and dataset are published at https://github.com/flatironinstitute/DeepFRI. The webserver of DeepFRI is available at https://beta.deepfri.flatironinstitute.org/

GCN for Structure and Function

Villegas-Morcillo *et al.* [83] developed a contextual unsupervised protein embedding-based approach to predict molecular function of a given protein sequence. The authors extracted both sequence-based and structure-based features. The sequence-based features for a protein of sequence length is extracted using a ELMo-based [20,21] model which outputs a feature vector of 1,024 dimension for each amino acid in the sequence. From these features, the protein-level features are extracted by averaging each feature over the L-amino acids where L is the size of the protein. In order to compare ELMo with other features, one-hot encoding as well as k-mer counts (with k = 3, 4, 5) are also extracted.

For structural information, the protein distance map is considered. Essentially, a 398-dimensional protein level feature vector from distance map is extracted from DeepFold [86]. Various classifiers like kNN, logistic regression and multilayer perceptron, and several CNNs are trained on these representations. The unsupervised ELMo embeddings of protein sequences perform much better than other representations, including one-hot encoding and others in MF, BP, and CC prediction [83]. The source code and dataset are available at https://github.com/stamakro/GCN-for-Structure-and-Function.

goPredSim

GoPredSim [40] predicts GO terms through annotation transfer, based on proximity of proteins in the embeddings from the language models [40]: mainly SeqVec embedding [28] and ProtBert [32] rather than in the sequence space. These embeddings originate from deep learned language models (LMs) for protein sequences (SeqVec) learned on 33 million protein sequences. Based on the premise of the usefulness of embedding space proximity in NLP, SeqVec learns to predict the next amino acid given the entire previous sequence on unlabeled data.

The authors rely on two pre-trained embeddings: SeqVec [28] and ProtBert [32]. The SeqVec is a stack of bidirectional LTSMs with an objective to predict the next amino acid residue in a sequence. BERT is a transformer model with the masked language objective. They both produce a vector of fixed size (1,024) to represent a protein in a fixed dimensional space. Given this fact, the authors introduce Euclidean and cosine similarity distances to measure similarity between two embeddings then transfer GO terms from the annotated to the unknown protein. In order to be able to represent various size proteins sequences by a fixed-length vector, initially, a matrix of size $L \times 1{,}024$ for a protein with L residues and then, the resulting vector of size 1,024 for each protein is obtained by averaging over the length dimension (also called global average pooling).

Unlike previous approaches where the GO annotations are transferred from a labeled protein with the highest percentage pairwise sequence identity (PIDE), this approach transfers annotations from the labeled protein to another protein with the smallest distance in embedding space. Based on a benchmarking of all 3,328 CAFA3 targets, this approach is as good as other top-performers in CAFA3 [87]. In addition, the authors compared embeddings derived from two different language models SeqVec (LSTM-based) and ProtBert (transformer-based) and found similar performance for both embeddings. The source code along with data and embeddings are available at https://github.com/Rostlab/goPredSim and the tool is available online at https://embed.protein.properties/.

4.2.4 Other approaches

In this section, we briefly summarize a couple of approaches that use machine translation models and text-based information for the GO prediction.

ProLanGO

Neural machine translation models, such as Google Translate, allow text to be translated between arbitrary pairs of languages using ANNs. ProLanGO [82] is a novel method to convert protein function problems into language translation problems. Essentially, ProLanGo solves the protein function prediction problem by first proposing protein sequence language 'ProLan' and the function language 'GoLan' and then building a neural machine translation model for GO term prediction based on a recurrent neural network to translate 'ProLan' to 'GoLan'.

Essentially, in this model, protein sequences and GO term annotations are treated as languages and a mapping function is learnt by an ANN to translate between the semantics of particular protein sequences and their equivalent GO term semantics. The model is an RNN, composed of LSTM units.

The words in protein sequences are constructed from all k-mers of length 3, 4 or 5 that occur in UniProt > 1,000 times. The GO term language is constructed by assigning each GO term to a unique four-letter code word in base 26. The four-letter code is the index of the term from a depth-first search of the GO DAG. For example, there are 28,768 terms in the biological process ontology, so the root node of the ontology, GO:0008150, is the 28,768th term to be visited in the depth-1st search, which corresponds to

BQKZ in four-letter code. ProLanGo model performed quite well on the CAFA3 dataset [87].

DeepText2GO

DeepText2GO [80] is a GO prediction method that uses text information in addition to protein sequence. The study is motivated by an exploration of both evidence and sequence-based methods. The novelty of the method is in the introduction of published papers abstract on the target protein with its sequence information. Given a target protein, first the associated PubMed identifiers (PMID) based on the protein's Uniprot ID and the sequence for the protein are retrieved. Then, corresponding abstracts are obtained from the PMIDs and combined to form a single document. Then, three NLP-based representations viz. Term Frequency-Inverse Document Frequency (TFIDF) [88], Document to Vector (Doc2Vec) [27] and concatenation of D2V and TFIDF representations (D2V-TFIDF) are extracted. Essentially, these three representations belong to the group of deep semantic representations.

Additionally, InterPro modules [89] encode a sequence as a binary vector. It represents domain boundaries and conserved regions that are coined in various NCBI databases. Thus, each protein sequence is obtained through the InterPro lookup algorithm that walks across the sequence and identifies structural features. Finally, the text-based approach and sequence-based approach is integrated by a consensus approach to form the final model in DeepText2GO. A consensus is defined for each GO term separately in a probabilistic manner, thus simple homology-based transfer is supported by the scientific evidence found in literature and structural properties of the protein.

5. Conclusion and discussion

NLP is one of the hottest research areas in AI and ML, spurring substantial amount of new research in this field. Similarities between natural language and protein sequences inspired usage of various NLP-based encoding approaches in both protein PTM prediction and protein GO term prediction. Additionally, just like in the NLP, various contextual protein sequence representations learned from different databases using language models were developed. The key difference among these language models originates from two main factors: i) the language model architecture and ii) the database used for training these language models. Especially, ProtBert- [32] and ProtT5- [32] based approaches have shown great promise in various protein bioinformatics tasks [90]. It is likely that we will see some approaches combining the language model-based encoding with multiple sequence alignment as well as other protein sequence- and structure-based features.

Although language models derived embeddings of protein sequences have shown a lot of promise, one of the bottlenecks for applying protein language model-based embeddings is the massive hardware requirements for training these types of models and unavailability of these pre-trained models. Additionally, as protein language models show promising results/ potential in generating descriptive representations of protein sequences, we are also likely to see more tools/pipelines for generating pre-trained language models for protein sequences. Bio_embeddings is one such example [91]. Developed at the Rost lab, bio_embeddings is likely to spur growth of use of protein language model-based representations of protein sequences. The availability of these large pre-trained models should promote use of various pre-trained protein language models for the PTM sites and GO term predictions. Recently, Min *et al.* [92] compared various protein language models (UniRep [31] , SeqVec [28], etc.) for the prediction of heat shock proteins and observed that Transformer-based model (ProtBERT [32]) has the best performance for predicting heat shock protein. They also observed that sparse representation (based on one-hot encoding) has the lowest F1-score. As these pre-trained embeddings language models become available, they can be leveraged to downstream tasks in two ways: i) as feature extractors (not trainable) to offer static pre-trained input features, and ii) a fine-tuning where the embeddings are adjusted on a specific downstream dataset. Although, we have seen some approaches using these language models as feature extracts, we have yet to see a fine-tuning approach trained on a specific downstream dataset.

Additionally, as there are no natural words in protein sequence space, there are some challenges in tokenizing proteins sequences. In NLP, generally three types of tokenizers are used on Transformers: Byte-Pair encoding [93], wordPiece [94], and SentencePiece [13]. A possible future research direction is to devise these types of tokenizers for protein sequence. Another future development in protein language models could be to design models that consider unique properties of protein. Although there are various similarities between natural language and protein sequences, there are also various differences and the new language models should exploit these differences [95]. As our ability to develop these new protein language models is impacted by high computational costs, it is critical to perform studies on the size of the dataset that these language models require as well as the number of parameters these models should consider [95]. Although fine-tuning the pre-trained models can help with a specific task, the fine-tuning process still takes high computing cost. In that regard, the other possible direction for these protein language models is to build purpose-built protein language models.

References

[1]   Pakhrin SC, *et al.* Deep learning-based advances in protein post-translational modification site and protein cleavage prediction. *Methods in Molecular Biology Series*, 2022. **2499**: 285–322.

[2]   Min S, *et al.* Deep learning in bioinformatics. *Briefings in Bioinformatics*, 2017. **18**(5): 851–869.

[3]   Pakhrin SC, *et al.* Deep learning-based advances in protein structure prediction. *The International Journal of Molecular Sciences*, 2021. **22**(11): 5553.

[4]   Tang B, *et al.* Recent advances of deep learning in bioinformatics and computational biology. *Frontiers in Genetics*, 2019. **10**: 214.

[5]   Torfi A, *et al.* Natural language processing advancements by deep learning: A survey. arXiv, 2020.

[6]   Ofer D, *et al.* The language of proteins: NLP, machine learning & protein sequences. *Computational and Structural Biotechnology Journal*, 2021. 19: 1750–1758.

[7]   Iuchi H, *et al.* Representation learning applications in biological sequence analysis. *Computational and Structural Biotechnology Journal*, 2021. 19: 3198–3208.

[8]   Mikolov T, *et al.* Efficient estimation of word representations in vector space. arXiv, 2013.

[9]   Mikolov T, *et al.* Distributed representations of words and phrases and their compositionality. *Advances in Neural Information Processing Systems*, 2013. **2**: 3111–3119.

[10] Lopez-del Rio A, *et al.* Effect of sequence padding on the performance of deep learning models in archaeal protein functional prediction. *Scientific Reports*, 2020. **10**(1): 1–14.

[11] Sennrich R, *et al.* Neural machine translation of rare words with subword units. arXiv, 2015.

[12] Kudo T. Subword regularization: Improving neural network translation models with multiple subword candidates. arXiv, 2018.

[13] Kudo T and Richardson J. Sentencepiece: A simple and language independent subword tokenizer and detokenizer for neural text processing. arXiv, 2018.

[14] Asgari E, *et al.* Probabilistic variable-length segmentation of protein sequences for discriminative motif discovery (DiMotif) and sequence embedding (ProtVecX). *Scientific Reports*, 2019. **9**(1): 1–16.

[15] Almeida F and Xexéo G. Word embeddings: A survey. arXiv, 2019.

[16] Gulli A and Pal S. Word Embeddings. In *Deep Learning with Keras*. Packt Publishing Ltd, 2017. 140–174.

[17] Ketkar N. Introduction to keras. In *Deep learning with Python*. Apress, Berkeley, CA, 2017. 97–111.

[18] Pennington J, *et al.* Glove: Global vectors for word representation. *Proceedings of the 2014 Conference on Empirical Methods in Natural Language Processing (EMNLP)*, Doha, Qatar, Association for Computational Linguistics. 2014. 1532–1543.

[19] Bojanowski P, *et al.* Enriching word vectors with subword information. *Transactions of the Association for Computational Linguistics*, 2017. **5**: 135–146.

[20] Peters M, *et al.* Deep contextualized word representations. arXiv, 2018.

[21] Sarzynska-Wawer J, *et al.* Detecting formal thought disorder by deep contextualized word representations. *Psychiatry Research*, 2021. **304**: 114135.

[22] McCann B, *et al.* Learned in translation: Contextualized word vectors. In *Proceedings of the 31st International Conference on Neural Information Processing Systems (NIPS'17)*. Red Hook, NY, USA. Curran Associates Inc., 2017. **30**: 6297–6308.

[23] Devlin J, *et al.* Bert: Pre-training of deep bidirectional transformers for language understanding. arXiv, 2018.

[24] Lan Z, *et al.* Albert: A lite bert for self-supervised learning of language representations. arXiv, 2019.

[25] Xu Y, *et al.* PhosContext2vec: A distributed representation of residue-level sequence contexts and its application to general and kinase-specific phosphorylation site prediction. *Scientific Reports*, 2018. **8**(1): 1–14.

[26] Melamud O, *et al.* context2vec: Learning generic context embedding with bidirectional LSTM. *Proceedings of the 20th SIGNLL Conference on Computational Natural Language Learning*, Berlin, Germany, Association for computational Linguistics, 2016. 51–61.

[27] Lau JH and Baldwin T. An empirical evaluation of doc2vec with practical insights into document embedding generation. arXiv, 2016.

[28] Heinzinger M, *et al.* Modeling aspects of the language of life through transfer-learning protein sequences. *BMC Bioinformatics*, 2019. **20**(1): 1–17.

[29] Strodthoff N, *et al.* UDSMProt: Universal deep sequence models for protein classification. *Bioinformatics*, 2020. **36**(8): 2401–2409.

[30] Merity S, *et al.* Regularizing and optimizing LSTM language models. arXiv, 2017.

[31] Alley EC, *et al.* Unified rational protein engineering with sequence-based deep representation learning. *Nature Methods*, 2019. **16**(12): 1315–1322.

[32] Elnaggar A, *et al.* ProtTrans: Towards cracking the language of Life's code through self-supervised deep learning and high performance computing. arXiv, 2020.

[33] Dai Z, *et al.* Transformer-xl: Attentive language models beyond a fixed-length context. arXiv, 2019.

[34] Yang Z, *et al.* Xlnet: Generalized autoregressive pretraining for language understanding. *Advances in Neural Information Processing Systems*, 2019. **32**(517): 5753–5763.

[35] Clark K, *et al.* Electra: Pre-training text encoders as discriminators rather than generators. arXiv, 2020.

[36] Raffel C, *et al.* Exploring the limits of transfer learning with a unified text-to-text transformer. arXiv, 2019.

[37] Suzek BE, *et al.* UniRef: Comprehensive and non-redundant UniProt reference clusters. *Bioinformatics*, 2007. **23**(10): 1282–1288.

[38] Jumper J, *et al.* Highly accurate protein structure prediction with AlphaFold. *Nature*, 2021. **596**(7873): 583–589.

[39] Hiranuma N, *et al.* Improved protein structure refinement guided by deep learning based accuracy estimation. *Nature Communications*, 2021. **12**(1): 1–11.

[40] Littmann M, *et al.* Embeddings from deep learning transfer GO annotations beyond homology. *Scientific Reports*, 2021. **11**(1): 1–14.

[41] Bonetta R and Valentino G. Machine learning techniques for protein function prediction. *Proteins: Structure, Function, and Bioinformatics*, 2020. **88**(3): 397–413.

[42] Boeckmann B, *et al.* The SWISS-PROT protein knowledgebase and its supplement TrEMBL in 2003. *Nucleic Acids Research*, 2003. **31**(1): 365–370.

[43] Zhao Y and Jensen ON. Modification-specific proteomics: Strategies for characterization of post-translational modifications using enrichment techniques. *Proteomics*, 2009. **9**(20): 4632–4641.

[44] Trost B and Kusalik A. Computational prediction of eukaryotic phosphorylation sites. *Bioinformatics*, 2011. **27**(21): 2927–2935.

[45] Wang D, *et al.* MusiteDeep: A deep-learning based webserver for protein post-translational modification site prediction and visualization. *Nucleic Acids Research*, 2020. **48**(W1): W140–W146.

[46] Wang D, *et al.* MusiteDeep: A deep-learning framework for general and kinase-specific phosphorylation site prediction. *Bioinformatics*, 2017. **33**(24): 3909–3916.

[47] Xie Y, *et al.* DeepNitro: Prediction of protein nitration and nitrosylation sites by deep learning. *Genomics Proteomics Bioinformatics*, 2018. **16**(4): 294–306.

[48] Liu Y, *et al.* DeepTL-Ubi: A novel deep transfer learning method for effectively predicting ubiquitination sites of multiple species. *Methods*, 2021. **192**: 103–111.

[49] He F, *et al.* Large-scale prediction of protein ubiquitination sites using a multimodal deep architecture. *BMC Systems Biology*, 2018. **12**(Suppl 6): 109.

[50] Fu H, *et al.* DeepUbi: A deep learning framework for prediction of ubiquitination sites in proteins. *BMC Bioinformatics*, 2019. **20**(1): 86.

[51] Wu M, *et al.* A deep learning method to more accurately recall known lysine acetylation sites. *BMC Bioinformatics*, 2019. **20**(1): 49.

[52] Guo L, *et al.* DeepPSP: A global-local information-based deep neural network for the prediction of protein phosphorylation sites. *The Journal of Proteome Research*, 2021. **20**(1): 346–356.

[53] Luo F, *et al.* DeepPhos: Prediction of protein phosphorylation sites with deep learning. *Bioinformatics*, 2019. **35**(16): 2766–2773.

[54] Wang H, *et al.* MDCAN-Lys: A model for predicting succinylation sites based on multilane dense convolutional attention network. *Biomolecules*, 2021. **11**(6): 872.

[55] Chen YZ, *et al.* nhKcr: A new bioinformatics tool for predicting crotonylation sites on human nonhistone proteins based on deep learning. *Briefings in Bioinformatics*. **22**(6): bbab146.

[56] Thapa N, *et al.* DeepSuccinylSite: A deep learning based approach for protein succinylation site prediction. *BMC Bioinformatics*, **21**(Suppl 3): 63.

[57] Chaudhari M, *et al.* DTL-DephosSite: Deep transfer learning based approach to predict dephosphorylation sites. *Frontiers in Cell and Developmental Biology*, 2021. **9**: 662983.

[58] Thapa N, *et al.* A deep learning based approach for prediction of Chlamydomonas reinhardtii phosphorylation sites. *Scientific Reports*, 2021. **11**(1): 12550.

[59] Huang G, *et al.* LSTMCNNsucc: A bidirectional LSTM and CNN-based deep learning method for predicting lysine succinylation sites. *BioMed Research International*, 2021. **2021**: 9923112.

[60] Siraj A, *et al.* UbiComb: A hybrid deep learning model for predicting plant-specific protein ubiquitylation sites. *Genes (Basel)*, 2021. **12**(5): 717.

[61] Chen Z, *et al.* Integration of a deep learning classifier with a random forest approach for predicting malonylation sites. *Genomics, Proteomics & Bioinformatics*, 2018. **16**(6): 451–459.

[62] Tng SS, *et al.* Improved prediction model of protein lysine crotonylation sites using bidirectional recurrent neural networks. *Journal of Proteome Research*, 2021. **21**(1): 265–273.

[63] Al-Barakati H, *et al.* RF-MaloSite and DL-Malosite: Methods based on random forest and deep learning to identify malonylation sites. *Computational and Structural Biotechnology Journal*, 2020. **18**: 852–860.

[64] Yang J, *et al.* DeepDigest: Prediction of protein proteolytic digestion with deep learning. *Analytical Chemistry*, 2021. **93**(15): 6094–6103.

[65] Lv H, *et al.* Deep-Kcr: Accurate detection of lysine crotonylation sites using deep learning method. *Briefings in Bioinformatics*, 2021. **22**(4): bbaa255.

[66] Lv H, *et al.* DeepIPs: Comprehensive assessment and computational identification of phosphorylation sites of SARS-CoV-2 infection using a deep learning-based approach. *Briefings in Bioinformatics*, 2021. **22**(6): bbab244.

[67] Qiao Y, *et al.* BERT-Kcr: Prediction of lysine crotonylation sites by a transfer learning method with pre-trained BERT models. *Bioinformatics*, 2021. **38**(3): 648–654.

[68] Xu H, *et al.* PLMD: An updated data resource of protein lysine modifications. *The Journal of Genetics and Genomics*, 2017. **44**(5): 243–250.

[69] Ashburner M, *et al.* Gene ontology: Tool for the unification of biology. *Nature Genetics*, 2000. **25**(1): 25–29.

[70] Dessimoz C, *et al.* CAFA and the open world of protein function predictions. *Trends in Genetics*, 2013. **29**(11): 609–610.

[71] Radivojac P, *et al.* A large-scale evaluation of computational protein function prediction. *Nature Methods*, 2013. **10**(3): 221–227.

[72] Wass MN, *et al.* The automated function prediction SIG looks back at 2013 and prepares for 2014. *Bioinformatics*, 2014. **30**(14): 2091–2092.

[73] Zhao Y, *et al.* A literature review of gene function prediction by modeling gene ontology. *Frontiers in genetics*, 2020. **11**: 400.

[74] Cai Y, *et al.* SDN2GO: An integrated deep learning model for protein function prediction. *Frontiers in Bioengineering and Biotechnology*, 2020. **8**: 391.

[75] Kulmanov M, *et al.* DeepGO: Predicting protein functions from sequence and interactions using a deep ontology-aware classifier. *Bioinformatics*, 2018. **34**(4): 660–668.

[76] Kulmanov M and Hoehndorf R DeepGOPlus: Improved protein function prediction from sequence. *Bioinformatics*, 2021. **36**(2): 422–429.

[77] Du Z, *et al.* DeepAdd: Protein function prediction from k-mer embedding and additional features. *Computational Biology and Chemistry*, 2020. **89**: 107379.

[78] Yao S, *et al.* NetGO 2.0: Improving large-scale protein function prediction with massive sequence, text, domain, family and network information. *Nucleic Acids Research*, 2021. **49**(W1): W469–W475.

[79] You R, *et al.* GOLabeler: Improving sequence-based large-scale protein function prediction by learning to rank. *Bioinformatics*, 2018. **34**(14): 2465–2473.

[80] You R, *et al.* DeepText2GO: Improving large-scale protein function prediction with deep semantic text representation. *Methods*, 2018. **145**: 82–90.

[81] Gligorijevic V, *et al.* Structure-based protein function prediction using graph convolutional networks. *Nature Communications*, 2021. **12**(1): 3168.

[82] Cao R, *et al.* ProLanGO: Protein function prediction using neural machine translation based on a recurrent neural network. *Molecules*, 2017. **22**(10): 1732.

[83] Villegas-Morcillo A, *et al.* Unsupervised protein embeddings outperform hand-crafted sequence and structure features at predicting molecular function. *Bioinformatics*, 2021. **37**(2): 162–170.

[84] Nauman M, *et al.* Beyond homology transfer: Deep learning for automated annotation of proteins. *Journal of Grid Computing*, 2019. **17**(2): 225–237.

[85] Kiefer F, *et al.* The SWISS-MODEL repository and associated resources. *Nucleic Acids Research*, 2009. **37**(Suppl 1): D387–D392.

[86] Liu Y, *et al.* Learning structural motif representations for efficient protein structure search. *Bioinformatics*, 2018. **34**(17): i773–i780.

[87] Zhou N, *et al.* The CAFA challenge reports improved protein function prediction and new functional annotations for hundreds of genes through experimental screens. *Genome Biology*, 2019. **20**(1): 244.

[88] Ramos J. Using tf-idf to determine word relevance in document queries. *Proceedings of the First Instructional Conference on Machine Learning*, 2003. **1**:29–48.

[89] Mitchell A, *et al.* The InterPro protein families database: The classification resource after 15 years. *Nucleic Acids Research*, 2015. **43**(D1): D213–D221.

[90] Stärk H, *et al.* Light attention predicts protein location from the language of life. *Bioinformatics Advances*, 2021. **1**(1): vbab035.

[91] Dallago C, *et al.* Learned embeddings from deep learning to visualize and predict protein sets. *Current Protocols*, 2021. **1**(5): e113.

[92] Min S, *et al.* Protein transfer learning improves identification of heat shock protein families. *Plos One*, 2021. **16**(5): e0251865.

[93] Shibata Y, *et al.* Byte Pair encoding: A text compression scheme that accelerates pattern matching. 1999.

[94] Wu Y, *et al.* Google's neural machine translation system: Bridging the gap between human and machine translation. arXiv, 2016.

[95] Bepler T and Berger B. Learning the protein language: Evolution, structure, and function. *Cell Systems*, 2021. **12**(6): 654–669. e653.

https://doi.org/10.1142/9789811258589_0005

Chapter 5

Feature-Engineering from Protein Sequences to Predict Interaction Sites Using Machine Learning

Dana Mary Varghese, Ajay Arya and Shandar Ahmad*

SciWhyLab, School of Computational and Integrative Sciences, Jawaharlal Nehru University, New Delhi-110067, India
*corresponding author: shandar@jnu.ac.in

Prediction of interaction sites at a residue level from protein sequence or structure is critical to our understanding of and ability to modulate cellular functions. Since the three-dimensional structures are available for only a limited number of proteins, researchers have made substantial efforts to develop computational methods for predicting interacting residues directly from protein sequence. These methods often rely on machine learning algorithms that require input features which are directly computed and contextually derived from the input sequence, and which utilize sequence-based predictions. Feature engineering from the sequence and labeling of the binding residues from structure to create training and validation data are essential steps to develop the sequence-based predictors using machine learning. We review various sequence-derived features that are used in the literature, provide rationale for their use, and summarize methods that compute them. We focus primarily on the prediction of the interacting residues for three common types of partners, namely proteins, DNA and RNA. However, the methods that we present here can be easily transferred to similar predictive problems that target other similar aspects of protein function and structure.

1. Introduction

All vital functions in the cell necessitate interaction of molecules, among which, interactions of proteins in response to small molecules, such as carbohydrates and metals, and larger ones ranging from peptides, proteins, various types of RNA and genomic DNA, are essential for living systems. Since these interactions take place in a three-dimensional space, the natural settings to study them are based on the structures of these molecules on the one hand and the larger cellular context on the other. However, only a limited number of structures is available. Moreover, the X-ray crystallography and/or NMR-based collection of accurate structures of all the known proteins under all biological conditions is not feasible while this information is important to elucidate interacting regions on proteins, irrespective whether they are specific or non-specific to their target molecules. Due to the limited throughput of the experimental methods, there has been a substantial interest in the scientific community in predicting interaction sites on proteins from the readily available sequence or sequence-derived information. Nonetheless, the sequence-based prediction methods require an accurately labeled interaction data, largely taken from the experimentally solved structures of complexes of proteins. Problem of predicting interacting amino acid residues in an input protein sequence can thus be summed up in three steps:

1. Data labeling: Annotation of interaction sites from protein structures solved in complex with their target molecules such as other proteins, RNA and DNA. This information could be extracted using other methods such as next generation sequencing (NGS) and protein-binding microarray (PBM).
2. Data featurization: Representation of amino acid sequence of proteins at the level of segments that are expected to contribute to interactions. These locally annotated features are often supplemented by features that quantify relevant characteristics of the whole protein sequence that may introduce useful protein-specific biases.

3. Model selection and training: Once the problem has been computationally defined in terms of (1) and (2), a suitable statistical or machine learning (ML) model is selected to carry out training/testing/deployment steps, much like a standard ML problem.

A large number of ML methods to predict protein interactions from sequence has been developed with an ever-improving prediction performance [1–6]. However, the primary focus of the published works has been on step (3) of this problem. Such methods have been extensively reviewed and analyzed in other studies [7–12]. Yet, not much attention has been paid to steps (1) and (2) and a systematic overview of these aspects is expected to accelerate research in this subfield of protein bioinformatics. We summarize key information concerning the data labeling and featurization for the prediction of protein interactions at the sequence level, and we review the published literature addressing binding site prediction problems.

2. Data labeling for the prediction of interaction sites

The primary source for deriving the annotations of interaction sites in proteins are the three-dimensional structures solved in complex with corresponding target molecules. To annotate the interacting residues, high resolution structures of protein complexes are employed. Interaction labeling or binding site annotation on these proteins at the amino acid level can be performed by explicitly identifying chemical nature of interactions e.g., hydrogen or covalent bond, or by its approximate variant in which any atom of a residue coming closer than a threshold to its ligand are treated as interacting. A number of softwares, such as HBPlus, BioLip, ProteinPlus, TCBRB and NPIDB [13–18], automate the process of annotating interactions from considerations of the actual bonding patterns. However, the latter more permissive definition of interactions has prevailed in the literature, especially in the context of the development of predictive methods. This technique, while not rigorous in a biological sense, provides a more consistent way to handle data, is easily automated, and takes care of interactions that are difficult to describe directly. Furthermore, the loss of information when using this approach is arguably small in the context of the potential loss in the predictive performance of the resulting predictors, because the sequence-based predictions are limited in their scope since they do not account for the actual cellular context or possible structural variations introduced by biochemical and thermodynamic conditions.

Thus, in practice, an amino acid is labeled as interacting if any of its atoms comes closer than a 3.5–5.0 Å (most popularly 3.5 Å for protein-nucleic acids and 4.5 Å for protein-protein interactions) of the atoms of the ligand in the complex (Table 1). Another popular approach is to use a specific atom of protein, such as its Cα-atom, as a reference and label them interacting at longer distances, such as 12 Å, from a specific atom of its ligand, such as the Cα-atom of a partner protein when considering the protein-protein interactions [19]. A systematic comparison of these binding site annotation methods and implications to their predictability is not yet available in the public domain and should be considered in the future. However, we argue that it is intuitive that the differences resulting from applying different definitions are likely relatively small and should have limited effect on the predictive performance of the resulting ML models.

One issue that arises in annotating interactions using distance (geometrical considerations) is that the number of annotated interactions depends on the nature and completeness of available three-dimensional structures. The currently available three-dimensional structures are limited in number and may be biased towards smaller structures which are easier to solve. To quantify this issue, we performed a quick analysis of protein lengths in Swiss-Prot, Protein Data Bank (PDB), which showed that the average length of protein sequences in PDB is 266.2 amino acids compared to the corresponding average in Swiss-Prot sequences without structures being 497.4 amino acids in March 2022 (detailed data not being provided) [20–23]. The emergence of structural genomics programs has revealed the structure of numerous new proteins, the functions of which remain unknown [24–26]. As of March 2022, there are 566,996 sequence entries in Swiss-Prot, but the total number of structures in PDB was only 187,844. One might be misled to think that 33% of sequences have a solved 3-D structure. However, PDB entries have a high degree of redundancy with multiple structures of the same protein at different resolutions, showing the structure of only one domain or terminal region. Closer inspection of uniquely mapped Swiss-Prot sequences showed that only 31,099 of the 566,997 (5.4%) mapped to at least one PDB structure entry as per the DBREF record of PDB. In fact, many of these mapped structures represent only a part of the full-length protein sequence. In addition UniProt contains a massive data set of 230 million translated protein coding sequences, called TrEMBL, derived from EMBL nucleotide sequences without manual curation, most of which do not have a corresponding structure in PDB [27]. Thus, the need to rapidly annotate structures and interactions from

Table 1. List of representative methods utilizing various feature groups and types for the prediction of binding sites.

Feature Group	Feature Type	Method (Ligand)	Machine Learning Model	Binding Residue Definition	Webserver
Direct protein sequence features	Sparse/one-hot encoding	PPIPP [86] (Protein)	ANN	≤6 Å	http://sciwhylab.jnu.ac.in/servers/ppipp/
		SPRINT [87] (Protein)	SVM	≤3.5 Å	http://sparks-lab.org/
		MCDPPI [88] (Protein)	SVM	NA	http://csse.szu.edu.cn/staff/youzh/MCDPPI.zip
		NAPS [4] (DNA, RNA)	C4.5 decision tree algorithm	NA	http://proteomics.bioengr.uic.edu/NAPS
		DP-Bind [89] (DNA)	SVM, KLR, PLR	≤4.5 Å	http://lcg.rit.albany.edu/dp-bind
		SRCPred [51] (RNA)	ANN	≤3.5 Å	http://sciwhylab.jnu.ac.in/servers/srcpred
		DeepDISOBind [90] (DNA, RNA, Protein)	DNN	NA	https://www.csuligroup.com/DeepDISOBind/
	Amino acid index	DNABind [91] (DNA)	SVM	≤4.5 Å	http://mleg.cse.sc.edu/DNABind/
	Physiochemical property-based encoding	PPIPre [92] (Protein)	LibD3C (Ensemble Method)	NA	http://lab.malab.cn/soft/PPIPre/PPIPre.html
		NIP-SS, NIP-RW[8] (Protein)	DNN	NA	http://mlda.swu.edu.cn/codes.php?name=NIP
		DRNApred [5] (DNA, RNA)	LR	≤3.5 Å	http://biomine.cs.vcu.edu/servers/DRNApred/

(*Continued*)

Table 1. (*Continued*)

Feature Group	Feature Type	Method (Ligand)	Machine Learning Model	Binding Residue Definition	Webserver
		DP-BINDER [93] (DNA)	RF, SVM	NA	http://lilab.ecust.edu.cn/NABind
		NAbind [94] (RNA)	RF	≤3.5 Å, ≤5 Å	
		BindN-RF [95] (DNA)	RF	≤3.5 Å	http://bioinfo.ggc.org/bindn-rf/
		SCRIBER [96] (Protein)	LR	<0.5 Å+Van der Waal's radii of two atoms	http://biomine.cs.vcu.edu/servers/SCRIBER/
		DNAgenie [97] (DNA)	LR, KNN, RF, SVM, NB	<0.5 Å+Van der Waal's radii of two atoms	http://biomine.cs.vcu.edu/servers/DNAgenie/
		PROBselect [98] (Protein)	Consensus of predictors (PBRs)	<0.5 Å+Van der Waal's radii of two atoms	http://bioinformatics.csu.edu.cn/PROBselect/home/index
Global protein sequence features		WSRC [99] (Protein)	WSRC	NA	
		iPPBS-Opt [100] (Protein)	KNN	ASA (Ri|P)-ASA(Ri|PP)> 1 Å	http://www.jci-bioinfo.cn/iPPBS-Opt
		SNBRfinder [3] (DNA, RNA)	SVM, HMM	≤4.5 Å	http://ibi.hzau.edu.cn/SNBRFinder
		iDBPs [101] (DNA)	RF	NA	http://idbps.tau.ac.il/
		DR_bind [102] (DNA)	Rank Clustering	≤3.5 Å, ≤4.0 Å	http://dnasite.limlab.ibms.sinica.edu.tw/

	Method	Algorithm	Criteria	URL
	PRIdictor [103] (RNA)	SVM	Hydrogen bond <3.7 Å Hydrophobic groups < 5.4 Å Waters bridge <5 Å	http://bclab.inha.ac.kr/pridictor
Derived sequence features	NABind [94] (RNA)	RF	≤3.5 Å, ≤5 Å	http://lilab.ecust.edu.cn/NABind
PSSM	RVKDE [104] (Protein)	RVKDE	NA	
	PPIevo [105] (Protein)	RF	NA	http://lbb.ut.ac.ir/Download/LBBsoft/PPIevo/
	Ens-PPI [106] (Protein)	Rotation Forest (RF)	NA	
	CNN-FSRF [107] (Protein)	CNN, Rotation Forest	NA	
	BindN-RF [95] (DNA)	RF	≤3.5 Å	http://bioinfo.ggc.org/bindn-rf/
	DBS-PSSM [108] (DNA)	NN	≤3.5 Å	http://www.netasa.org/dbs-pssm/
	DNABinder [109] (DNA)	SVM	NA	http://www.imtech.res.in/raghava/dnabinder/
	NAPS [4] (DNA,RNA)	C4.5 decision tree algorithm	NA	http://proteomics.bioengr.uic.edu/NAPS
PSSM and entropy	SNBRFinder [3] (DNA, RNA)	SVM, HMM	≤4.5 Å	http://ibi.hzau.edu.cn/SNBRFinder
PSSM and HMM profile	NucBind [110] (RNA)	SVM, COACH	≤3.5 Å, ≤5 Å	http://yanglab.nankai.edu.cn/NucBind

(Continued)

Table 1. (*Continued*)

Feature Group	Feature Type	Method (Ligand)	Machine Learning Model	Binding Residue Definition	Webserver
Predicted features	Accessible surface area	PSIVER [111] (Protein)	NB	Relative solvent accessibility <5%	http://tardis.nibio.go.jp/PSIVER/
		RVKDE [104] (Protein)	RVKDE	NA	
	Relative solvent accessibility	SPPIDER [112] (Protein)	LDA, SVM, ANN	Change in Relative solvent accessibility >4%, change in exposed surface area >5 Å	http://sppider.cchmc.org/
		PiRaNhA [113] (RNA)	SVM	≤3.9 Å	http://www.bioinformatics.sussex.ac.uk/PIRANHA
	Secondary structure	DisoRDPbind [114] (DNA, RNA, Protein)	LR	NA	http://biomine.ece.ualberta.ca/DisoRDPbind/
		DeepDISOBind [90] (DNA, RNA, Protein)	DNN	NA	https://www.csuligroup.com/DeepDISOBind/
	Secondary structure and solvent accessibility	DNABind [91] (DNA)	SVM	≤4.5 Å	
		SNBRFinder [3] (DNA,RNA)	SVM, HMM	≤4.5 Å	http://ibi.hzau.edu.cn/SNBRFinder
		PETs [115] (Protein)	ET	≤4.5 Å	https://github.com/BinXia/PETs

Method	Algorithm	Threshold	URL
DISIS [116] (DNA)	NN, SVM	≤6 Å	http://cubic.bioc.columbia.edu/services/disis
ProNA2020 [1] (DNA, RNA, Protein)	SVM, ANN, ProtVec	≤6 Å	https://github.com/Rostlab/ProNA2020.git
PRINTR [117] (RNA)	SVM	Hydrogen bond ≤3.9 Å, stacking ≤5 Å, hydrophobic ≤5 Å, Van der Waals interaction ≤ Van der Waals radii of two atoms+0.8 Å	http://210.42.106.80/printr/
RBRpred [118] (RNA)	SVM, WildSpan	≤6 Å	
SCRIBER [96] (Protein)	LR	<0.5 Å+Van der Waal's radii of two atoms	http://biomine.cs.vcu.edu/servers/SCRIBER/
DNAgenie [97] (DNA)	LR, KNN, RF, SVM, NB	<0.5 Å+Van der Waal's radii of two atoms	http://biomine.cs.vcu.edu/servers/DNAgenie/
PROBselect [98] (Protein)	Consensus of predictors (PBRs)	<0.5 Å+Van der Waal's radii of two atoms	http://bioinformatics.csu.edu.cn/PROBselect/home/index

SVM: Support Vector Machine, ANN: Artificial Neural Network, KLR: Kernel Logistic Regression, PLR: Penalized Logistic Regression, DNN: Deep Neural Network, LR: Logistic Regression, RF: Random Forest, KNN: K-Nearest Neighbours, HMM: Hidden Markov Model, NV: Naïve Bayes, CNN: Convolution Neural Network, NN: Neural Network, LDA: Linear Discriminant Analysis, ET: Extremely Randomised Trees , WSRC: weighted sparse representation based classifier, RVKDE: relaxed variable kernel density estimator sequences is a critical step in the functional analysis of living systems. A way to increase the number of annotations of interactions is to use the transfer of annotations from other structurally similar proteins. This is done with the help of alignment that identifies aligned/corresponding residues [24–26]. However, this comes with a risk of error since structural similarity alone does not guarantee identical functions, as there are structurally similar proteins with different functions.

While the currently available amount of structure-derived interaction data is sufficient to design, train and test sequence-based prediction methods, there is an interest in modelling interactions annotated from other experimental techniques. For example, protein-protein interactions have been experimentally determined using several techniques including yeast two hybrid (Y2H), phage display (PD), tandem affinity purification, co-immunoprecipitation (Co-IP), glutathione S-transferase pull-down (GST pull-down), cellular co-localization, far-western blotting, virus overlay protein binding assay (VOPBA), bimolecular fluorescence complementation (BiFC), surface plasmon resonance (SPR), and fluorescence resonance energy transfer (FRET) [28,29]. Interactions between proteins and nucleic acids can be been identified using a variety of experimental approaches, including pull-down assays [30,31], yeast one-hybrid system(Y1H) [32], electrophoretic mobility shift assays(EMSA) [33], Southwestern blotting, reporter gene, Co-IP, peptide nucleic acid (PNA)-assisted identification of RNA binding proteins (PAIR), chromatin immunoprecipitation (ChIP), MicroChIP, Fast ChIP, ChIP-chip, ChIP-seq, and ChIP-exo [34–37]. Due to the rapid improvement of next-generation sequencing technology, ChIP-seq has become the most widely used technique [38]. The newly improved ChIP-exo technology improves spatial resolution to the single nucleotide level. Over the years, research has evolved from analyzing single molecules and their specific relationships to broadly parallel, simultaneous investigations of several molecules and their interactions.

3. Featurization of protein sequences

Table 1 provides a comprehensive list of web-based methods that predict interaction residues from a sequence, together with the features that these methods employed, ML methods that they use and the location of the corresponding web-servers. Historically, the earliest featurization of protein sequences for ML models was employed for predicting secondary structures and solvent accessibility of proteins [39,40]. The featurization in the context of the prediction of the nucleic acid interactions of proteins dates back to 2004, when Ahmad *et al.* made early attempts to use evolutionary and sequence information to predict binding residues with an artificial neural network (ANN)[41]. In the following, we list a few methods for encoding primary or derived information from protein sequences, focusing on methods that target prediction of interaction sites (defined at the single residue level) on proteins. We can group sequence-derived information into three categories. The first is the direct representation of amino acids in the sequence. The second attempts to represent the amino acids in a particular context, such as evolutionary patterns. Finally, the third group uses sequence-based predictions of certain structural properties, which are then used as inputs to predict interactions. Kurgan *et al.* identified that the relative solvent accessibility (RSA), evolutionary conservation and propensity of amino acids (AAs) can characterize different types of interactions (RNA versus DNA versus protein), the residues of the protein might be involved in [42]. This research also found that multi-ligand binding residues are more conserved than single-ligand binding residues [42]. Following, we describe these three groups of features.

4. Direct protein sequence features

4.1 *One-hot/sparse encoding*

By far, the most popular representation of protein sequences is the sparse or one-hot encoding. Based on this approach, each amino acid in the input protein sequence is represented by a 20-dimensional binary vector where all positions, except that associated to a given amino acid, are represented by 0, while 1 is used for the position that represents this amino acid type. Despite being popular, this method has its limitations. It treats all amino acids as independent and distinct, whereas in reality, certain amino acids share some structural and biophysical properties, i.e., some amino acids are more similar to each other than other amino acids. While the "true" dimensionality of amino acid space is unclear, several attempts have been made to investigate alternatives to sparse encoding. These methods consider known general properties of amino acids, and in particular residue positions [43,44]. This is because an amino acid in a given protein is not in a free state and general biophysical properties of amino acids may not be fully relevant when they are peptide-bonded to their sequence neighbors.

Nonetheless, the native amino acid properties at sequence positions may provide an alternative to the one-hot/sparse encoding, which is better informed by the physical properties of the atomic environment in given positions. Some of the non-sparse representations of amino acids in proteins are described below.

4.2 *Amino acid indices*

The AAindex database contains numerical indices that quantify different physicochemical and biological characteristics of amino acids and amino acid pairs [45]. AAindex is currently divided into three sections: AAindex1 for the amino acid index with 20 numerical values, AAindex2 for the amino acid substitution matrix, and AAindex3 for statistical protein contact potentials [46]. AAindex currently contains data about 566 experimentally determined chemical properties of amino acids.

4.3 *Physicochemical property-based encoding*

Amino acid can be encoded according to different residue types based on their physicochemical properties like charge, polarity and hydrophobicity. In some sense, this is a special case of AAindex-based features. However, if the properties, such as hydrophobicity, are derived from proteins instead of free amino acids, they may be better representative of the protein environment. For example, hydrophobicity is a key factor that affects the packaging of amino acids during protein folding. Hydrophobic amino acids are often located within the globular proteins and are not present on the exposed surface where the interactions occur [47,48]. These properties may be helpful to develop more accurate predictive models that factor in the physical nature of interactions, while operating in a lower dimensional space, as compared to the one-hot encoding. This might reduce overfitting, although this aspect has not yet been properly examined in the literature.

4.4 *Global protein sequence features*

Local amino acid properties (i.e., properties computed within a small sliding window in a protein sequence) are the primary inputs for many of sequence-based predictors of interactions. However, the context of the whole protein expressed by its length/size, protein-level amino acid composition, and presence of sequences patterns could also be useful for the prediction of interactions. For example, longer proteins may fold back to cover a protein segment, which would otherwise be available for an interaction. Also, there may be a bias in interaction sites shared by proteins with similar amino acid composition. For example, Ahmad *et al.* found that charged residues, such as Arginine and Lysine, are significantly overrepresented in the DNA binding sites, while Serine and Glycine are more abundant in the DNA binding proteins in terms of their overall content [41]. Thus, adding the global/protein-level features have been shown to improve predictive performance of the sequence-based predictors of interactions [49–51]. These features primarily quantify the sequence length and the 20-dimensional amino acid composition of the sequence. They need a normalization step to bring them to the same scale as the other direct features that are computed in a local window.

4.5 *Window size*

Sliding sequence window is a useful tool to perform the encoding when predicting binding sites. This approach is used to provide local (within the confines of the window) context for the prediction of the residue in the middle of the window. Window size is used as an important parameter when performing the sequence encoding. Using the N-neighbor window leads to using 2*N+1 amino acids [49–51], since the neighboring residues are extended in both directions from the central residue. In some works, explicit patterns of co-occurrence of binding sites within a window have been analyzed and five to six residue windows were shown to be best [41]. However, it is not unusual to use long windows with as many as 11 neighbors (window size of 23 residues) [5,52,53]. There is no clear evidence that the predictive performance is enhanced when using such long windows. Based on our experience and published work, large windows may be suitable only if dealing with large amounts of training data. However, if specific interactions with fewer proteins of interest are to be used for a specialized model, a smaller window size of three to four neighbors should remain effective to avoid overfitting.

5. Derived sequence features

Arguably, the most widely used derived features are those that place protein sequence in the evolutionary context and represent patterns observed in the resulting evolutionary profiles instead of the source sequences. In a simplest form, evolutionary information can be summarized by conservation scores or information context for individual amino acids. In a more complex form, the frequencies of occurrence for each position in the sequence are processed statistically and their 20-dimensionally log odd values are used in the form of the position specific scoring matrix (PSSM), also called an evolutionary profile. PSSMs for protein sequences are often created by running the position specific iterative basic local alignment search tool (PSI BLAST) against a non-redundant database [54]. The PSSM matrix essentially indicates the likelihood of each residue being conserved at a particular place/position in the sequence. In an early use for the prediction of binding sites, Ahmad *et al.* (2004) and Jun *et al.* (2019) predict DNA binding sites in proteins using only PSSM. Additionally, PSSM can also be created using HHblits [55], HMMER3 [56] and other algorithms [57].

The state to step ratio score (SSR) is another example of a derived sequence property. Originally proposed by Wang *et al.* in 2008 [58], SSR measures evolutionary conservation. This method uses the multiple sequence alignment produced by PSI-BLAST to create trees, which are used to compute variation patterns for each residue starting from the root of the tree.

The importance of conservation stems from the fact that it is linked to the underlying protein function. Multiple sequence alignments, which are produced by the above algorithms (PSI-BLAST, HHblits and others), are used to quantify conservation. To construct the multiple sequence alignment, the query protein is typically aligned against a (large) non-redundant database of protein sequences. The sequence conservation score is then calculated using Shannon entropy, von Neumann entropy, and relative entropy [59]. Webservers and tools, such as Scorecons [60], can be used to generate these conservation profiles. There are other algorithms like Wildspan [61] used in ProteRNA [62], and MaxHom [63] used in DISIS [116], which can be used to produce the conservation values.

6. Predicted features

Interactions take place in a context of three-dimensional structures and therefore it is useful to utilize structural information when developing the sequence-based predictive models. In cases where protein structures are already solved, structure and homology-based methods as well as docking may be effective to predict interactions [64,65]. However, in the absence of protein structures, residue-level predicted structural properties can be used. Among the relevant one-dimensional structural properties of proteins (i.e., structural properties that can be encoded along the protein sequence), arguably the most widely investigated are secondary structure and solvent accessibility. Both these characteristics are relevant to binding sites [41,66–68]. There are three basic categories of secondary structures (beta-strand, alpha-helix, and coils), which are further categorized into eight subcategories. There are three kinds of helices (3–10 helix, alpha helix, and pi helix), two strand types (beta strand and beta bridge), and three forms of coils (beta-turn, high curvature loop, and irregular). Programs such as PSIPRED [69], and RaptorX [70] are popular choices to predict secondary structures from sequences utilizing evolutionary conservation. Many other methods for the prediction of secondary structure are available [71–75].

The amino acids with higher exposed surface area in a protein structure are more likely to interact with partner molecules than the residues with lower area. The regions on the protein surface that are more likely to be exposed can be calculated using DSSP [39] and NACCESS [76] when structure is available. The solvent exposure values can also be predicted from sequences using tools such as NETASA [77], SAAN [78], PROFphd [79], RVP-net [80] and RaptorX [70], when the structures are unavailable.

7. Summary and conclusions

A large number of ML methods for the sequence-based prediction of interactions have been published in literature [81–85]. Features that are extracted from the input sequences and used by these models are broadly categorized into three groups: direct encoding, derived features, and predicted features. We define, describe and summarize these groups in the context of the predictors that use them in Table 1. Undoubtedly, the most popular methods for featurization of protein sequences are the one-hot encoding from single sequences and PSSM that is produced from the evolutionary profiles that are in turn generated from the sequences. Many studies which we outline above also employ one-dimensional structure features that are predicted from sequences. However, since these derived features may also have been trained on the very sequences for which interactions are to be predicted, independent evaluation of their contribution could be complicated. In terms of predictive performances of individual feature sets, PSSM-based features have been found to be the most powerful in most models and have been utilized in many ML implementations.

Some unanswered questions that emerge from our review of the literature relate to the utility of using specific ML methods with different types of the input feature sets. Furthermore, a brute force approach of pooling together all the features during training versus a more focused approach may need additional investigation. Finally, a theoretical limit of predictability of interactions from any of these sequence features would be an interesting question to address, considering that the biological and structural/oligomeric contexts of interactions are not considered by the sequence-based model. These theoretical limits of predictability may be influenced by the feature sets used and hence, simultaneous research on better feature engineering in parallel with advancing ML techniques remains an area needing further attention in this field.

References

[1]   Qiu J, *et al.* ProNA2020 predicts protein–DNA, protein–RNA, and protein–protein binding proteins and residues from sequence. *Journal of Molecular Biology*, 2020. **432**(7): 2428–2443.

[2]   Chen C, *et al.* LightGBM-PPI: Predicting protein-protein interactions through LightGBM with multi-information fusion. *Chemometrics and Intelligent Laboratory Systems*, 2019. **191**: 54–64.

[3]   Yang X, *et al.* SNBRFinder: A sequence-based hybrid algorithm for enhanced prediction of nucleic acid-binding residues. *PloS One*, 2015. **10**(7): e0133260.

[4]   Carson MB, *et al.* NAPS: A residue-level nucleic acid-binding prediction server. *Nucleic Acids Research*, 2010. **38**(Suppl 2): W431–W435.

[5]   Yan J and Kurgan L. DRNApred, fast sequence-based method that accurately predicts and discriminates DNA-and RNA-binding residues. *Nucleic Acids Research*, 2017. **45**(10): e84–e84.

[6]   Mishra A, *et al.* StackDPPred: A stacking based prediction of DNA-binding protein from sequence. *Bioinformatics*, 2019. **35**(3): 433–441.

[7]   Yan J, *et al.* A comprehensive comparative review of sequence-based predictors of DNA-and RNA-binding residues. *Briefings in Bioinformatics*, 2016. **17**(1): 88–105.

[8]   Zhang L, *et al.* Predicting protein-protein interactions using high-quality non-interacting pairs. *BMC Bioinformatics*, 2018. **19**(19): 105–124.

[9]   Ding X-M, *et al.* Computational prediction of DNA-protein interactions: A review. *Current Computer-Aided Drug Design*, 2010. **6**(3): 197–206.

[10] Emamjomeh A, *et al.* DNA–protein interaction: Identification, prediction and data analysis. *Molecular Biology Reports*, 2019. **46**(3): 3571–3596.

[11] Cirillo D, *et al.* Predictions of protein–RNA interactions. *Wiley Interdisciplinary Reviews: Computational Molecular Science*, 2013. **3**(2): 161–175.

[12] Esmaielbeiki, R, *et al.* Progress and challenges in predicting protein interfaces. *Briefings in Bioinformatics*, 2016. **17**(1): 117–131.

[13] McDonald IK and Thornton JM. Satisfying hydrogen bonding potential in proteins. *Journal of Molecular Biology*, 1994. **238**(5): 777–793.

[14] Hu J and Yan C. A tool for calculating binding-site residues on proteins from PDB structures. *BMC Structural Biology*, 2009. **9**(1): 1–6.

[15] Schöning-Stierand K, *et al.* Proteins plus: Interactive analysis of protein–ligand binding interfaces. *Nucleic Acids Research*, 2020. **48**(W1): W48–W53.

[16] Salentin S, *et al.* PLIP: Fully automated protein–ligand interaction profiler. *Nucleic Acids Research*, 2015. **43**(W1): W443–W447.

[17] Yang J, *et al.* BioLiP: A semi-manually curated database for biologically relevant ligand–protein interactions. *Nucleic Acids Research*, 2012. **41**(D1): D1096–D1103.

[18] Kirsanov DD, *et al.* NPIDB: Nucleic acid — protein interaction database. *Nucleic Acids Research*, 2012. **41**(D1): D517–D523.

[19] Ofran Y and Rost B. Analysing six types of protein–protein interfaces. *Journal of Molecular Biology*, 2003. **325**(2): 377–387.

[20] Berman HM, *et al.* Announcing the worldwide protein data bank. *Nature Structural Biology*, 2003. **10**(12): 980.

[21] Berman HM, *et al.* The worldwide protein data bank (wwPDB): Ensuring a single, uniform archive of PDB data. *Nucleic Acids Research* (Database issue), 2007. **35**(D301): 3.

[22] Consortium w. Protein data bank: The single global archive for 3D macromolecular structure data. *Nucleic Acids Research*, 2019. **47**: D520–D528.

[23] E B, *et al.* UniProtKB/Swiss-Prot. *Methods in Molecular Biology*, 2007. **406**: 89–112.

[24] Baker D and Sali A. Protein structure prediction and structural genomics. *Science*, 2001. **294**(5540): 93–96.

[25] Yokoyama S, *et al.* Structural genomics projects in Japan. *Nature Structural Biology*, 2000. **7**(11): 943–945.

[26] Burley SK. An overview of structural genomics. *Nature Structural Biology*, 2000. **7**(11): 932–934.

[27] Boeckmann B, *et al.* The Swiss-Prot protein knowledgebase and its supplement TrEMBL in 2003. *Nucleic Acids Research*, 2003. **31**: 365–370.

[28] Rao VS, *et al.* Protein-protein interaction detection: Methods and analysis. *International Journal of Proteomics*, 2014. **2014**: 147648.

[29]   Zhou M, *et al.* Current experimental methods for characterizing protein-protein interactions. *ChemMedChem*, 2016. **11**(8): 738–756.

[30]   Jutras BL, *et al.* Identification of novel DNA-binding proteins using DNA-affinity chromatography/pull down. *Current Protocols in Microbiology*, 2012. **24**(1): 1F. 1.1–1F. 1.13.

[31]   Wang IX, *et al.* Human proteins that interact with RNA/DNA hybrids. *Genome Research*, 2018. **28**(9): 1405–1414.

[32]   Gaudinier, A, *et al.* Identification of protein–DNA interactions using enhanced yeast one-hybrid assays and a semiautomated approach. In *Plant Genomics*, Busch W (ed.), Humana Press, New York, NY, 2017. 187–215.

[33]   Hellman LM and Fried MG. Electrophoretic mobility shift assay (EMSA) for detecting protein–nucleic acid interactions. *Nature Protocols*, 2007. **2**(8): 1849–1861.

[34]   O'Neill LP and Turner BM. Immunoprecipitation of chromatin. In *RNA Polymerase and Associated Factors, Part B*, Adhya S (ed.), Academic Press, 1996.

[35]   Carey MF, *et al.* Chromatin immunoprecipitation (chip). *Cold Spring Harbor Protocols*, 2009. **2009**(9): pdb. prot5279.

[36]   Collas P. The current state of chromatin immunoprecipitation. *Molecular Biotechnology*, 2010. **45**(1): 87–100.

[37]   Rhee HS and Pugh BF. ChIP-exo method for identifying genomic location of DNA-binding proteins with near-single-nucleotide accuracy. *Current Protocols in Molecular Biology*, 2012. **100**(1): 21.24. 1–21.24. 14.

[38]   Ma T, *et al.* Genome wide approaches to identify protein-DNA interactions. *Current Medicinal Chemistry*, 2019. **26**(42): 7641–7654.

[39]   Kabsch W and Sander C. Dictionary of protein secondary structure: Pattern recognition of hydrogen-bonded and geometrical features. *Biopolymers: Original Research on Biomolecules*, 1983. **22**(12): 2577–2637.

[40]   Kyte J and Doolittle RF. A simple method for displaying the hydropathic character of a protein. *Journal of Molecular Biology*, 1982. **157**(1): 105–132.

[41]   Ahmad S, *et al.* Analysis and prediction of DNA-binding proteins and their binding residues based on composition, sequence and structural information. *Bioinformatics*, 2004. **20**(4): 477–486.

[42]   Zhang J, *et al.* Comprehensive review and empirical analysis of hallmarks of DNA-, RNA-and protein-binding residues in protein chains. *Briefings in Bioinformatics*, 2019. **20**(4): 1250–1268.

[43]   Mei H, *et al.* A new set of amino acid descriptors and its application in peptide QSARs. *Peptide Science: Original Research on Biomolecules*, 2005. **80**(6): 775–786.

[44]   Maetschke S, *et al.* BLOMAP: An encoding of amino acids which improves signal peptide cleavage site prediction. *Proceedings of the 3rd Asia-Pacific Bioinformatics Conference*, Chen YPP (ed.), World Scientific, 2005. 141–150.

[45]   Kawashima S, *et al.* AAindex: Amino acid index database, progress report 2008. *Nucleic Acids Research*, 2007. **36**(Suppl 1): D202–D205.

[46]   Kawashima S and Kanehisa M. AAindex: Amino acid index database. *Nucleic Acids Research*, 2000. **28**(1): 374–374.

[47]   Rose GD, *et al.* Hydrophobicity of amino acid residues in globular proteins. *Science*, 1985. **229**(4716): 834–838.

[48]   Lins L and Brasseur R. The hydrophobic effect in protein folding. *The FASEB Journal*, 1995. **9**(7): 535–540.

[49]   Walia RR, *et al.* Protein-RNA interface residue prediction using machine learning: An assessment of the state of the art. *BMC Bioinformatics*, 2012. **13**(1): 1–20.

[50]   Li S, *et al.* Quantifying sequence and structural features of protein–RNA interactions. *Nucleic Acids Research*, 2014. **42**(15): 10086–10098.

[51]   Fernandez M, *et al.* Prediction of dinucleotide-specific RNA-binding sites in proteins. *BMC Bioinformatics*, 2011. **12**(Suppl 13): S5.

[52]   Tahir M and Hayat M. Machine learning based identification of protein–protein interactions using derived features of physiochemical properties and evolutionary profiles. *Artificial Intelligence in Medicine*, 2017. **78**: 61–71.

[53]   Kumar M, *et al.* Prediction of RNA binding sites in a protein using SVM and PSSM profile. *Proteins: Structure, Function, and Bioinformatics*, 2008. **71**(1): 189–194.

[54]   Altschul SF, *et al.* Gapped BLAST and PSI-BLAST: A new generation of protein database search programs. *Nucleic Acids Research*, 1997. **25**(17): 3389–3402.

[55]   Remmert M, *et al.* HHblits: Lightning-fast iterative protein sequence searching by HMM-HMM alignment. *Nature Methods*, 2012. **9**(2): 173–175.

[56]   Eddy SR. A new generation of homology search tools based on probabilistic inference. In *Genome Informatics 2009: Genome Informatics Series*, Morishita S, *et al.* (eds.), World Scientific, 2009. 205–211.

[57]   Hu G and Kurgan L. Sequence similarity searching. *Current Protocols in Protein Science*, 2019. **95**(1): e71.

[58]   Si J, *et al.* An overview of the prediction of protein DNA-binding sites. *International Journal of Molecular Sciences*, 2015. **16**(3): 5194–5215.

[59]   Capra JA and Singh M. Predicting functionally important residues from sequence conservation. *Bioinformatics*, 2007. **23**(15): 1875–1882.

[60]   Valdar WS. Scoring residue conservation. *Proteins: Structure, Function, and Bioinformatics*, 2002. **48**(2): 227–241.

[61]   Hsu C-M, *et al.* Efficient discovery of structural motifs from protein sequences with combination of flexible intra-and inter-block gap constraints. *Advances in Knowledge Discovery and Data Mining*, Ng WK, *et al.* (eds.), Springer, Berlin, Heidelberg, 2006. 530–539.

[62]   Huang Y-F, *et al.* Predicting RNA-binding residues from evolutionary information and sequence conservation. *BMC Genomics*, 2010. **11**(Suppl 4): S2.

[63]   Schneider R and Sander C. The HSSP database of protein structure-sequence alignments. *Nucleic Acids Research*, 1996. **24**(1): 201–205.

[64] Ramakrishnan G, *et al.* Homology-based prediction of potential protein-protein interactions between human erythrocytes and plasmodium falciparum. *Bioinformatics and Biology Insights*, 2015. **9**: BBI. S31880.

[65] Murakami Y and Mizuguchi K. Homology-based prediction of interactions between proteins using averaged one-dependence estimators. *BMC Bioinformatics*, 2014. **15**(1): 1–11.

[66] Ahmad S, *et al.* CCRXP: Exploring clusters of conserved residues in protein structures. *Nucleic Acids Research*, 2010. **38**(Suppl 2): W398–W401.

[67] Chen R, *et al.* Redesigning secondary structure to invert coenzyme specificity in isopropylmalate dehydrogenase. *Proceedings of the National Academy of Sciences of the United States of America*, 1996. **93**(22): 12171–12176.

[68] Keskin O, *et al.* Predicting protein–protein interactions from the molecular to the proteome level. *Chemical Reviews*, 2016. **116**(8): 4884–4909.

[69] McGuffin LJ, *et al.* The PSIPRED protein structure prediction server. *Bioinformatics*, 2000. **16**(4): 404–405.

[70] Wang S, *et al.* RaptorX-Property: A web server for protein structure property prediction. *Nucleic Acids Research*, 2016. **44**(W1): W430–W435.

[71] Oldfield CJ, *et al.* Computational prediction of secondary and supersecondary structures from protein sequences. *Methods in Molecular Biology*, 2019. **1958**: 73–100.

[72] Jiang Q, *et al.* Protein secondary structure prediction: A survey of the state of the art. *Journal of Molecular Graphics & Modelling*, 2017. **76**:379–402.

[73] Meng F and Kurgan L. Computational prediction of protein secondary structure from sequence. *Current Protocols in Protein Science*, 2016. **86**: 2.3.1–2.3.10.

[74] Zhang H, *et al.* Critical assessment of high-throughput standalone methods for secondary structure prediction. *Briefings in Bioinformatics*, 2011. **12**(6): 672–88.

[75] Kashani-Amin E, *et al.* A systematic review on popularity, application and characteristics of protein secondary structure prediction tools. *Current Drug Discovery Technologies*, 2019. **16**(2): 159–172.

[76] Hubbard S and Thornton J. *NACCESS. Department of Biochemistry and Molecular Biology.* University College London, 1993.

[77] Ahmad S and Gromiha MM. NETASA: Neural network based prediction of solvent accessibility. *Bioinformatics*, 2002. **18**(6): 819–824.

[78] Joo K, *et al.* Sann: Solvent accessibility prediction of proteins by nearest neighbor method. *Proteins: Structure, Function, and Bioinformatics*, 2012. **80**(7): 1791–1797.

[79] Rost B and Sander C. Conservation and prediction of solvent accessibility in protein families. *Proteins: Structure, Function, and Bioinformatics*, 1994. **20**(3): 216–226.

[80] Ahmad S, *et al.* RVP-net: Online prediction of real valued accessible surface area of proteins from single sequences. *Bioinformatics*, 2003. **19**(14): 1849–1851.

[81] Zhang J and Kurgan L. Review and comparative assessment of sequence-based predictors of protein-binding residues. *Briefings in Bioinformatics*, 2018. **19**(5): 821–837.

[82] Roche D, *et al.* Proteins and their interacting partners: An introduction to protein-ligand binding site prediction methods. *International Journal of Molecular Sciences*, 2015. **16**(12): 29829–29842.

[83] Walia RR, *et al.* Protein-RNA interface residue prediction using machine learning: An assessment of the state of the art. *BMC Bioinformatics*, 2012. **13**: 89.

[84] Yan J, *et al.* A comprehensive comparative review of sequence-based predictors of DNA- and RNA-binding residues. *Briefings in Bioinformatics*, 2016. **17**(1): 88–105.

[85] Zhao H, *et al.* Prediction of RNA binding proteins comes of age from low resolution to high resolution. *Molecular BioSystems*, 2013. **9**(10): 2417–2425.

[86] Ahmad S and Mizuguchi K. Partner-aware prediction of interacting residues in protein-protein complexes from sequence data. *PloS One*, 2011. **6**(12): e29104.

[87] Taherzadeh G, *et al.* Sequence-based prediction of protein–peptide binding sites using support vector machine. *Journal of Computational Chemistry*, 2016. **37**(13): 1223–1229.

[88] You Z-H, *et al.* Prediction of protein-protein interactions from amino acid sequences using a novel multi-scale continuous and discontinuous feature set. *BMC Bioinformatics*, 2014. **15**(Suppl 15): S9.

[89] Hwang S, *et al.* DP-Bind: A web server for sequence-based prediction of DNA-binding residues in DNA-binding proteins. *Bioinformatics*, 2007. **23**(5): 634–636.

[90] Zhang F, *et al.* DeepDISOBind: Accurate prediction of RNA-, DNA- and protein-binding intrinsically disordered residues with deep multi-task learning. *Briefings in Bioinformatics*, 2022. **23**(1): bbab521.

[91] Liu R and Hu J. DNABind: A hybrid algorithm for structure-based prediction of DNA-binding residues by combining machine learning-and template-based approaches. *Proteins: Structure, Function, and Bioinformatics*, 2013. **81**(11): 1885–1899.

[92] Wei L, *et al.* Improved prediction of protein–protein interactions using novel negative samples, features, and an ensemble classifier. *Artificial Intelligence in Medicine*, 2017. **83**: 67–74.

[93] Ali F, *et al.* DP-BINDER: Machine learning model for prediction of DNA-binding proteins by fusing evolutionary and physicochemical information. *Journal of Computer-Aided Molecular Design*, 2019. **33**(7): 645–658.

[94] Sun M, *et al.* Accurate prediction of RNA-binding protein residues with two discriminative structural descriptors. *BMC Bioinformatics*, 2016. **17**(1): 1–14.

[95] Wang L, *et al.* Prediction of DNA-binding residues from protein sequence information using random forests. *BMC Genomics*, 2009. **10**(1): 1–9.

[96] Zhang J and L. Kurgan, SCRIBER: Accurate and partner type-specific prediction of protein-binding residues from proteins sequences. *Bioinformatics*, 2019. **35**(14): i343–i353.

[97] Zhang J, *et al.* DNAgenie: Accurate prediction of DNA-type-specific binding residues in protein sequences. *Briefings in Bioinformatics*, 2021. **22**(6).

[98] Zhang F, *et al.* PROBselect: Accurate prediction of protein-binding residues from proteins sequences via dynamic predictor selection. *Bioinformatics*, 2020. **36**(Suppl 2): i735–i744.

[99] Huang Y.-A, *et al.* Sequence-based prediction of protein-protein interactions using weighted sparse representation model combined with global encoding. *BMC Bioinformatics*, 2016. **17**(1): 1–11.

[100] Jia J, *et al.* iPPBS-Opt: A sequence-based ensemble classifier for identifying protein-protein binding sites by optimizing imbalanced training datasets. *Molecules*, 2016. **21**(1): 95.

[101] Nimrod G, *et al.* iDBPs: A web server for the identification of DNA binding proteins. *Bioinformatics*, 2010. **26**(5): 692–693.

[102] Chen YC, *et al.* DR_bind: A web server for predicting DNA-binding residues from the protein structure based on electrostatics, evolution and geometry. *Nucleic Acids Research*, 2012. **40**(W1): W249–W256.

[103] Tuvshinjargal N, *et al.* PRIdictor: Protein–RNA interaction predictor. *Biosystems*, 2016. **139**: 17–22.

[104] Chang DT-H, *et al.* Predicting the protein-protein interactions using primary structures with predicted protein surface. *BMC Bioinformatics*, 2010. **11**(1): 1–10.

[105] Zahiri J, *et al.* PPIevo: Protein–protein interaction prediction from PSSM based evolutionary information. *Genomics*, 2013. **102**(4): 237–242.

[106] Gao Z-G, *et al.* Ens-PPI: A novel ensemble classifier for predicting the interactions of proteins using autocovariance transformation from PSSM. *BioMed Research International*, 2016. **2016**: 4563524.

[107] Wang L, *et al.* Predicting protein-protein interactions from matrix-based protein sequence using convolution neural network and feature-selective rotation forest. *Scientific Reports*, 2019. **9**(1): 1–12.

[108] Ahmad S and Sarai A. PSSM-based prediction of DNA binding sites in proteins. *BMC Bioinformatics*, 2005. **6**(1): 1–6.

[109] Kumar M, *et al.* Identification of DNA-binding proteins using support vector machines and evolutionary profiles. *BMC Bioinformatics*, 2007. **8**(1): 1–10.

[110] Su H, *et al.* Improving the prediction of protein–nucleic acids binding residues via multiple sequence profiles and the consensus of complementary methods. *Bioinformatics*, 2019. **35**(6): 930–936.

[111] Murakami Y and Mizuguchi K. Applying the Naïve Bayes classifier with kernel density estimation to the prediction of protein–protein interaction sites. *Bioinformatics*, 2010. **26**(15): 1841–1848.

[112] Porollo A and Meller J. Prediction-based fingerprints of protein–protein interactions. *Proteins: Structure, Function, and Bioinformatics*, 2007. **66**(3): 630–645.

[113] Murakami Y, *et al.* PiRaNhA: A server for the computational prediction of RNA-binding residues in protein sequences. *Nucleic Acids Research*, 2010. **38**(Suppl 2): W412–W416.

[114] Peng Z and Kurgan L. High-throughput prediction of RNA, DNA and protein binding regions mediated by intrinsic disorder. *Nucleic Acids Research*, 2015. **43**(18): e121–e121.

[115] Xia B, *et al.* PETs: A stable and accurate predictor of protein-protein interacting sites based on extremely-randomized trees. *IEEE Transactions on Nanobioscience*, 2015. **14**(8): 882–893.

[116] Ofran Y, *et al.* Prediction of DNA-binding residues from sequence. *Bioinformatics*, 2007. **23**(13): i347–i353.

[117] Wang Y, *et al.* PRINTR: Prediction of RNA binding sites in proteins using SVM and profiles. *Amino Acids*, 2008. **35**(2): 295–302.

[118] Zhang T, *et al.* Analysis and prediction of RNA-binding residues using sequence, evolutionary conservation, and predicted secondary structure and solvent accessibility. *Current Protein and Peptide Science*, 2010. **11**(7): 609–628.

Part III

Predictors of Protein Structure
and Function

https://doi.org/10.1142/9789811258589_0006

Chapter 6

Machine Learning Methods for Predicting Protein Contacts

Shuaa M. A. Alharbi* and Liam J. McGuffin*,†

*School of Biological Science, University of Reading,
Reading, UK, RG6 6EX
†corresponding author: l.j.mcguffin@reading.ac.uk

In bioinformatics, the prediction of protein structures and their complexes is a prominent research focus. Predicting the contacts between protein residues has been one of the major obstacles, since folding of proteins into their 3D structures is dictated by the interacting amino acids in their native states. Due to the relevance of contact prediction for the improvement of protein structure modeling techniques, several methodologies and innovative approaches have been used to improve contact prediction accuracy. In this chapter, a brief history of contact prediction methods and their incremental improvements are covered. The significance of contact predictions is discussed, as well as their developing influence and usefulness in the field of protein structure prediction today. Following that, we describe the advancement of classical to modern methods for predicting protein contacts (and distance) maps based on the machine learning algorithms, as well as give examples from existing servers and methodologies. The machine learning approaches that have been used in contact prediction methods include Hidden Markov models, Support Vector machines, Random Forest algorithms, and Classifiers

155

based on the Naïve Bayes principle. Furthermore, residual convolutional neural networks, recurrent neural networks, and end-to-end learning models, which are the current state-of-the-art deep neural network models in use today, have been employed over the years to enhance the accuracy and utility of contact prediction methods.

1. Introduction

Research on the folding of protein sequences into their tertiary structures has received much attention for decades because of its importance in biomedical sciences, biotechnology, and other fields in the life sciences. Understanding how protein sequences fold into three-dimensional (3D) structures is fundamental to helping us to solve biological problems based on the sequence-structure-function paradigm. Researchers can use the knowledge of protein 3D structures to better understand the pathways of biological systems and disease mechanisms. Experimental approaches, such as X-ray crystallography and nuclear magnetic resonance (NMR) spectroscopy, have been developed to determine protein structures [1–3]. However, these methods have drawbacks, as they are expensive, and the effort required to resolve structures can be time consuming, as some structures may take many years to solve. Conversely, obtaining DNA and protein sequences is comparatively very rapid and inexpensive. Therefore, as a result of this disparity, there is a notable gap between the number of protein sequence entries in protein databases (219,174,961 sequences in UniProt: https://www.ebi.ac.uk/uniprot/TrEMBLstats) and the number of protein structures that have been determined by experimental methods (181,535 structures in PDB bank: https://www.rcsb.org/stats/growth/growth-released-structures) [4–6]. This means that most protein sequences have unknown structures and functions, and it would require extensive time and effort to resolve these all experimentally. Fortunately, computational methods have been developed, which provide a rapid and cost-efficient way to elucidate structures compared to experimental structure determination procedures. These methods can work quickly and reasonably accurately to predict 3D models for protein sequences with unknown structures.

The quality of protein structure models is crucial if they are to be used in biomedical fields such as drug discovery. Therefore, protein structure prediction methods have been developed iteratively over the years in order to increase the accuracy of 3D models that they produce [7–9]. This development involves incorporating various protein features from their sequences and structures, such as the inter-residue contact maps, into protein structure prediction pipelines [10–12]. When protein sequences fold, the amino acid residues interact to form 3D structures by creating non-covalent bonds between their atoms [13]. Thus, residue interaction predictions, or contact maps, can provide valuable information describing the tertiary structure, which can be exploited to reconstruct 3D models, leading to enhanced quality (Fig. 1)[14,15]. This information is derived from predicting pairwise contacts in a protein sequence, and is employed by many researchers to predict protein folding, for example, by restricting the conformational space of *ab initio* modelling [16–19]. Moreover, protein contact prediction methods have been integrated in model quality estimation servers to detect both the local (per-residue) and global errors in models [17,20,21]. Furthermore, in refinement processes, contact prediction has been used as part of a "gradual restraint strategy" [22]. For transmembrane proteins, predicting contacts between the transmembrane alpha-helices helps to elucidate the protein fold, which can, in turn, help to predict functions [23].

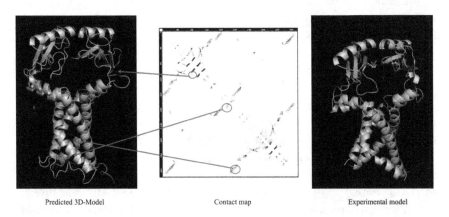

Predicted 3D-Model   Contact map   Experimental model

Figure 1. Illustration of role contact maps. Role contact maps can play in the improvement of 3D protein modeling. A contact map of siderophore reductase FoxB (PDB ID: 7awb) predicted by RaptorX-Contact. An experimental structure of this protein was determined by X-ray diffraction. In the contact map, residues pairs predicted to be in contact are highlighted by red circles. Protein models have been rendered using PyMOL(pymol.org)[19,27–31,46].

Due to their potential usefulness for predicting protein folding, methods for the prediction of contacts between residues have been in development since the early 1990s. Covell and Jernigan used the lattice model to represent amino acid residue contacts for restricting a conformational space of globular proteins to predict "all chain conformations" [6,24]. This approach was useful for predicting a small group of protein structures. In 1996, contact prediction was introduced as a part of the *ab initio* category for secondary and tertiary structure prediction in the second round of the Critical Assessment of Techniques for Protein Structure Prediction (CASP2). In this experiment round, contact prediction methods were developed using the principle of correlated mutation of coevolutionary residues [25,26].

2. Residue contact definitions

Contacting residue pairs in protein structures can be identified by calculating the distance between carbon atoms of the amino acid residues at a specific threshold. The distance threshold between carbon atoms of residues pair in a protein structure has different values depending on the goal of the contact prediction. For predicting helix-helix interactions in transmembrane proteins, contacts between residues are defined as distances less than 5.5 Å between two heavy atoms of the side chain or backbone. An alternative definition considers the contact distance threshold to be less than 8 Å between C-beta ($C\beta$) atoms of side chains [17,32]. For modeling the 3D structures of proteins, contacts have been defined between C-beta ($C\beta$) atoms (or between C-alpha ($C\alpha$) atoms in the case of Gly) using different distance cut-offs of between 7 and 11 Å [32–35]. However, in the contact prediction evaluation process of the Critical Assessment of Techniques for Protein Structure Prediction (CASP) experiments, the formal definition is that residue pairs are in contact if the distance between their $C\beta$ atoms ($C\alpha$ in Gly) is less than 8 Å [17,26,36–38]. All these threshold values are in a range that allows non-covalent interactions to be measured as protein sequences folded up into 3D shapes [4].

2.1 *Contact maps*

To represent contacts between residues computationally, "contact maps" have been devised as two-dimensional (2D) matrices (N × N), where N is the length of the protein sequence. The contacting residue pairs are set to 1 if the distance between their atoms is less than or equal to a given cut-off value; otherwise, 0 indicates the residue pairs that are non-contacting. The distance between the same residue position is also set to 0 and represents the diagonal line in the contact matrix. Thus, a contact between each residue can be represented as a dot and the x- and y- axes represent residue positions along the sequence length (Fig. 2) [3,4,39].

Furthermore, the type of contact is particularly important in determining protein structures. To classify contact types, the number of residues between two residue pairs that are predicted to be in contact determines the type of contact. In other words, if there are more than 24 separate predicted residue pairs, their contact is classified as being long-range; if there are more than 12 but less than 23 residues, the predicted contact is classified as medium-range; and if there are more than 6 residues but less than 12, in this case, predicted contacts are classified as short-range [17,26,36–38]. The long-range contacts contribute to improving the quality of 3D models as

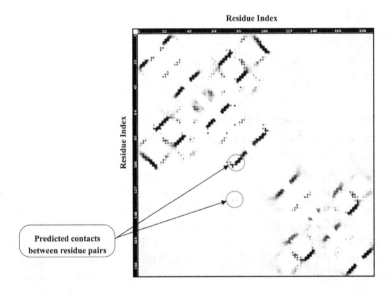

Figure 2. A protein contact map representation. This contact map is for the TCR-017 ectodomain protein (PDB ID:7EA6). The diagonal line corresponds to the residue contacting with itself (set to 0). Contacts between residue pairs are represented as black dots. The contact map was predicted by RaptorX-Contact [19,27–31].

they assist in positioning the secondary structures at the right distance. Therefore, this type of contact can be used as a restraint for conformational spaces in predicting the structures *ab initio* [17,35,39,40]. According to CASPs evaluation system, each residue pair predicted to be in contact can be assigned by calculating the probability score. The length of the target domain (L) with the greatest probability value is used to divide each contact range into subsets (L/5, L/2, L, FL, where FL indicates all predicted contacts in these sets). In this chapter, we will refer to the accuracy of contact prediction by machine learning approaches as L/5 long-range contacts for template-free or free-modelling (FM) targets [37,38].

3. Machine learning algorithms in contact prediction methods

Contact prediction accuracy should be sufficient to capture the most correct contacts that could be used to bring 3D models as close as possible to the native structures of proteins. Therefore, increasing the accuracy of predicted residue–residue contacts became a core challenge in the field of structural bioinformatics. Many approaches and algorithms have been used to extract accurate contact predictions between residue pairs in protein sequences. Using the evolutionary theory of protein folding, correlated mutation-based methods were developed based on the hypothesis that residue pairs in protein sequences are more likely to have correlated mutations to maintain the stability of protein structures. In other words, if one residue mutated, then the corresponding interacting residue/s will also be mutated in a coevolutionary process to stabilize the protein structure, and these residues could be identified in multiple sequence alignments (MSAs) [41]. Therefore, to extract coevolutionary information, multiple sequence alignment methods are used to identify homologous proteins using various rapid algorithms [6,11,17]. However, simple statistical methods are not sufficient to identify contact information between residues because of their inability to extract a precise mutation correlation between pairs of protein residues. In addition, traditional methods for contact prediction have been dependent on the availability of homologous sequences in protein databases and the accuracy relied on the number of aligned sequences [6,42,43]. Therefore, many researchers have sought to exploit the advantages of machine learning in order to improve the accuracy of contact prediction methods.

Machine learning approaches are computational algorithms used to adapt a fitted model for detecting meaningful patterns within data. In the contact prediction field, they learn how to identify contact networks among residues through their properties from protein sequence and structural data. Machine learning methods are trained from protein structures by creating their contact maps based on known coordinates. Protein sequence features are then fed into algorithm models, such as support vector machines, neural networks, and random forests, which are then trained to predict the contact maps (Fig. 3). Many of the machine learning-based contact prediction methods are freely accessible; some have web interfaces and others are provided as downloadable binaries and/or open-source code (Table 1). The output of the machine learning models typically consists of lists of scores (or p-values) for pairs of contacting residues, which inform users how likely is that each residue pair is in contact. These models can combine large sets of protein features and learn from them,

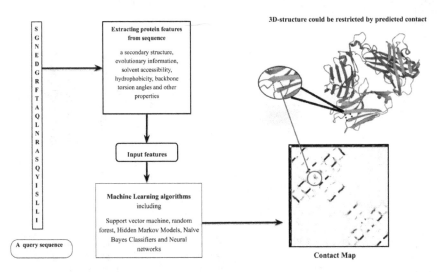

Figure 3. General schematic of contact prediction procedure. Starting with extracting protein features from a query sequence. These features are considered as input data that is fed into machine learning algorithms to predict a contact between each residue pair. The output is a contact map of query sequence which can be used to aid 3D-structure prediction. The example query protein is TCR-017 ectodomain PDB ID: 7EA6. The contact map was predicted by RaptorX-Contact and the 3D structure visualized by Mol*Viewer [19,27–31,46].

Table 1. The available contact prediction methods based on machine learning algorithms

Methods	Brief description	URL or web interface	Citation
DEEPCON	Deep learning-based method using covariance and sequences features as in DNCON2 and DeepCov and integrating these features into four models of fully residual convolutional neural networks with dropout layers and dilated convolution layers.	https://github.com/ba-lab/DEEPCON	[47]
DeepConPred2	The second version of DeepConPred, developed based on three models, the first and second model are deep belief networks, and the third model is a ResNet.	http://structpred.life.tsinghua.edu.cn/DeepConPred2.html	[48]
SPOT-Contact	Deep learning-based method designed based on Recurrent neural networks with LSTM cells and input features predicted from SPIDER3, CMMpred and DCA.	https://sparks-lab.org/server/spot-contact/	[49]
SVMcon	Method developed based on a support vector machine with a large set of features.	http://sysbio.rnet.missouri.edu/multicom_toolbox/SVMcon%201.0.html	[50]
DNCON2	Deep learning-based method improved through predicting protein features from PSIPRED, SCRATCH, CCMpred, FreeContact and PSICOV, which were fed into two-level CNN, where the first level had five CNNs and the second one has one CNN.	https://github.com/multicom-toolbox/DNCON2	[51]
RaptorX-Contact	Deep learning-based method developed by designing two ResNets models for integrating 1D and 2D protein features.	http://raptorx.uchicago.edu/ContactMap/	[28]

Name	Description	URL	Ref.
ResPRE	A method developed by integrating a precision matrix into fully residual convolutional neural networks.	https://zhanggroup.org/ResPRE/ https://github.com/leeyang/ResPRE	[52]
MapPred	Developed by combining two methods DeepMSA and DeepMeta into a dilated residual neural network model.	https://yanglab.nankai.edu.cn/MapPred/	[53]
DeepMetaPSICOV	Developed based on a deep, fully convolutional residual neural network with a set of features predicted from PSICOV, MetaPSICOV, PSICOV, CCMpred and FreeContact.	https://github.com/psipred/DeepMetaPSICOV	[54]
TripletRes	Deep learning-based method developed by integrating three coevolutionary matrices into a residual neural network model.	https://zhanggroup.org/TripletRes/	[87]
NeBcon	Developed by designing naïve Bayes classifier (NBC) to combine eight contact prediction methods, then the NBC output with other features were fed into a neural network model.	https://zhanggroup.org/NeBcon/	[42]
SVMSEQ	Machine learning-based method developed to predict contact maps based on SVM software.	https://zhanggroup.org/SVMSEQ/	[41]
DeepDist	Developed to predict real-value inter-residue distances based on four models of residual convolutional networks (ResNet).	https://github.com/multicom-toolbox/deepdist	[56]
DeepECA	Developed based on an end-to-end learning neural network to predict contact maps directly from MSAs.	https://github.com/tomiilab/DeepECA	[57]

which makes them less dependent on the depth of multiple sequence alignments, and thereby reduces the prediction accuracy when fewer homologous sequences can be identified [41,42,44,45].

3.1 *Hidden Markov Models*

A Hidden Markov Model (HMM) is a statistical model that estimates hidden events using observable events. HMMs comprise of one category of machine learning algorithms that have been used broadly in structural bioinformatics. In the protein structure prediction field, HMMs have been applied for fold recognition pipelines and have been used to enhance performance since CASP2 [58]. FragHMMent is a HMM-based residue-residue contact prediction tool [59]. The HMMs have been applied to detect local protein neighbourhoods that include all inter-residue contacts at different ranges (short-, medium- and long-range) [59]. To this purpose, Björkholm *et al.* [59] used local descriptors of protein structures to identify local neighbourhoods of amino acids. The local structural descriptors comprise of all residues in the neighbourhood's area of desired residue pairs. These descriptors in turn were used to construct multiple backbone segments arranged close together. Hence, The HMMs were trained by combining sequence signals in structurally similar neighbourhoods, with two protein features derived from secondary structure and evolutionary information to create a predicted contact map. It is worth mentioning that the identification of long-range contacts is particularly difficult for *ab initio* structure prediction. Interestingly, FragHMMent has proven to be particularly accurate for proteins with novel folds and is mostly fold-independent, and thus may be useful in this most difficult application field [17,59].

3.2 *Support Vector Machines*

A Support Vector Machine (SVM) is a classification algorithm that maps high-dimensional input as vectors into nonlinear and linear models to solve binary classification problems [50,60]. The performance of machine learning in contact prediction depends on the feature sets and model designs used. Feature sets can be embedded into SVM models in order to classify residues that are either in contact or non-contact in protein structures, and more sufficient input features can be used to program such models to explore the contact patterns between residues [61]. As previously mentioned, contact prediction accuracy is often associated with the quality of the multiple sequence alignment analysis. Furthermore, it may be dependent on the secondary structure prediction accuracy and the frequency of β-sheets [50]. A key advantage of SVMs is that they integrate linear and nonlinear methods: they can be used to design nonlinear models by representing input data nonlinearly into feature space while simultaneously classifying input dots in feature space utilising linear methods [50]. Cheng and Baldi [50] exploited this benefit when creating their contact prediction method SVMcon: they used a SVM model with a large set of protein features, including secondary structure, mutual information, solvent accessibility, and the global and local features of amino acid residues [50,62]. Another contact prediction method is SVMSEQ, which employs a SVM with two windows to predict protein residue contacts. The first window comprises of local window features, including three protein features: position-specific scoring matrices (PSSMs), secondary structure predictions and solvent accessibility predictions. The second window comprises of in-between segment feature sets involving sequence separations, which are the number of residues separating an interesting residue pair, the secondary structure content, the distribution of residues between residue pairs predicted to be in contact and the local properties of five residues distributed evenly in the middle of a desired residue pair [41]. SVM-based methods improved the accuracy of contact prediction for template-free and template-based modelling targets by approximately 25–40% [59]. They have also been integrated with other methods into other server pipelines, such as the Yang-Server and Zhang_Contact server which were ranked as top-performing methods in CASP12 [37].

3.3 *Random Forest Algorithms*

Random Forest (RF) is a model constructed by merging several decision tree algorithms to obtain a final decision based on the most "votes". RFs are often used because they can solve a variety of problems at once, making them particularly suited to dealing with large high-dimensional datasets and identifying noisy input information. They can also be used to build classification models rapidly [63,64]. A standard RF encompasses a set of classification models ("trees"), each of which creates a classifier and "votes" for one of the two classes (positive or negative) [64]. Once it has been designed to consider predicting residue-residue contact as a classification problem, a RF model can be trained to identify residue pairs as being in contact (positive) or non-contacting (negative).

RF models have been used and incorporated with other algorithms to predict residue-residue contact maps. When using a sufficient dataset to derive protein properties, a RF model can extract accurate contact information from known protein structures. The first RF-based method for contact prediction was ProC_S3, developed by Li *et al.* [63]. The RF model they used constructed 500 classification trees for the training and prediction stages and was trained on a large dataset including 1,490 protein structures and feature sets, which considered "the average of [the] maximum accessible surface areas and isoelectric points of the amino acids in two local windows (four features) [and the] f-mean of the between segment (20 features and [the seven] features of the central residue of the segment" [63]. Since ProC_S3 was based on a RF algorithm, it acquired selected features which could determine the relevance of protein features to residue contacts [63]. To investigate the usefulness of this feature selection, the RF-based method TMhhcp was designed to predict contacts in alpha-helical transmembrane proteins based on all their features and the selected feature set [32]. All features constructed from the evolutionary profiles of the residue pairs included the TM helix numbers, residue distance in the sequence, relative distance of two residues in between two helices, residue conservation scores and correlated mutation scores calculated by covariance algorithms, resulting in 408 feature vectors. From this construction, 10 feature subsets have been selected by using the correlation-based feature selection (CFS) [32]. The selected features experiment was conducted to identify a range of distinguishing features which have a greater individual capacity to predict the class (contact) but minimal intercorrelation [32]. Two models, named TMhhcp1 and TMhhcp2, were built based on training data with all the features, and two others, named TMhhcp_cfs1 and TMhhcp_cfs2, were built with selected features. In the latter two models, three protein features were found to produce particularly accurate contact predictions: the residue separation in the main sequence, the relative distance between two residues in helix-helix interaction and the correlated mutation score [32]. Other algorithms incorporating RF models included PhyCMAP, combining an RF model and integer linear program, which predicts contact maps by integrating evolutionary and physical constraints [65,66]. In general, RF-based methods have demonstrated reliable improvement with regards to the accuracy of contact prediction.

3.4 Naïve Bayes Classifiers

A Naïve Bayes Classifier (NBC) is a simple probability classifier based on the assumption that each feature value has an independent effect on a particular class. NBCs improve the accuracy of contact prediction in a complementary way for proteins which lack homologous sequences. He et al.'s [42] NeBcon, is a meta-server for contact prediction, which combines a Bayes classifier and a neural network to predict an accurate contact map by exploiting coevolutionary features and machine learning-based contact methods [42,67]. An NBC was used in this server to compute the contact probability scores of eight contact prediction methods; it was also given a set of posterior probability values for predicted contacts. Subsequently, the output of the NBC along with six structural features extracted from the target protein sequence was fed into a neural network model to predict the final contact map [42]. The improvement of performance of NeBcon was attributed to the integration of the complementary coevolutionary information from eight methods into the NBC model, and the structural features, using neural networks [42]. He et al. [42] demonstrated that the combination of machine learning-based methods with coevolution methods into NBC model improved the accuracy of contact prediction from hard targets, which tends to have low accuracy predicted contacts by coevolution methods.

3.5 Neural Networks

Neural Network (NN) is an artificial neural network composed of a number of computing units known as neurons. These units are linked together by connections, each of which has a weight attached to it [68]. NNs have had a considerable impact on the advancement of machine learning and on the accuracy of contact prediction methods. One of the first applications of neural networks to the problem of contact prediction was when Fariselli and Casadio [69] used them to extract the relationship between contact maps and the chemical interaction between protein residues. The neural network had a high level of adaptability through the combination of different input features such as secondary structure prediction, chemicophysical properties of residues and evolutionary features extracted from multiple sequence alignment (MSA) in its first layer, leading to adequate learning to increase its prediction power [69,70]. Shackelford and Karplus demonstrated that neural networks could play a vital role in improving contact prediction accuracy by integrating several protein features and training on large data sets [41,70]. This confirmed the observation by Fariselli *et al.* [71] who showed that the performance of neural networks in predicting a contact map improved when the amount of input data was raised [71]. In prior neural network models, input features, including protein sequences and predicted secondary structures, mutational information from multiple sequence alignments, and hydrophobicity scores, were investigated for their importance on improving contact prediction accuracy through the design of different NN models, which included different input data [69–72]. Fariselli and Casadio [69] had initially demonstrated that protein features can improve the accuracy of contact prediction if they are combined using neural networks. They subsequently showed that evolutionary information from structure-sequence alignments can provide accurate predicted contacts for proteins with less than 170 residues, while the sequence context, which are five potential couplings for each residue into parallel and antiparallel pairings encoded into three-amino-acid window, plays a role in the accuracy of contact prediction for proteins with sizes more than 170 [69,73]. The accuracy of contact prediction is computed by dividing the correctly predicted contacts by the total predicted contacts. Each protein's accuracy is evaluated separately before being averaged throughout the whole protein dataset [71]. By integrating a variety of protein features, contact prediction accuracy achieved a more reliable value (21% of average accuracy in CASP3) [71], however alternative neural network models have since been developed to further improve the accuracy. For example, Xue *et al.* [45] developed SPINE-2D by designing a deeper NN model with two hidden layers to extract information from residue solvent accessibility and backbone torsion angle features, resulting in increasing average contact accuracy at 26% in CASP8 [45]. Although this neural network improved upon previous neural network-based methods, its accuracy did not achieve a sufficient level to be used for confidently modeling tertiary structures. Therefore, researchers were encouraged to employ deeper neural network models including residual convolutional neural networks, recurrent neural networks, and end-to-end learning models, which will be discussed in the next section.

3.5.1 *Deep neural networks*

Deep neural networks are complex architectures of neural networks designed to obtain extensive knowledge from high-level data. Deep models of neural networks differ from shallow models in terms of architectural construction. While shallow architectures are constructed from two layers (input and output) and a small number of hidden layers, deep models are designed from "deep stacks" of classical neural networks with large number of hidden layers [45,69,74]. Deep learning-based methods often perform better when the depth of neural network layers is increased, enabling them to extract accurate information from very large datasets with numerous input features [17]. Since 2008, deep neural networks have been employed in contact prediction methods, and have led to improvements in their accuracy. NNcon was an early deep learning-based contact prediction method, designed with a 2D recursive neural network for predicting tertiary and secondary structure contacts (β-sheet). In CASP8, NNcon was ranked as one of the top-performing methods [17,75]. Later in 2012, Lena *et al.* designed "a 3D of stack of neural networks" that could extract contact information, where each stack consisted of three neural network layers (one input, one hidden, and one output) [76]. This method improved the accuracy of contact prediction by nearly 30% (from 28% to 35%), which indicated that deep NNs could learn more efficiently than shallow NN models [17,76]. Due to the rapid development in Graphics Processing Units (GPUs), researchers have been focusing on improving deep neural network models by exploiting the increasing GPU card capabilities, which have allowed efficient training on big data sets with complex NN architectures. Another approach was developed by Eickholt and Cheng [77] who designed DNcon, which uses several deep models from restricted Bolzmann machines, combined with the boosting ensemble method. Restricted Bolzmann machines are neural networks with two layers: visible and hidden, with symmetric weights connecting the nodes of both layers. Several restricted Bolzmann machines were combined to construct deep networks (DNs). Many layers were added to each DN in the boosted ensemble model, and DNs were then trained in a stepwise, semi-supervised manner [77,78]. They attributed the improved prediction performance of DNcon to its use of a feature set with a deep architecture [17,77,78]. Many alternative methods have been developed, which employ various deep model architectures with different input data sets, and have shown improvements in contact prediction performance over successive CASPs experiments [19,43,79].

A different approach has also been taken in recent years after multiple sequence alignment analysis tools were developed. These tools help to enhance alignment approaches for extracting correlated mutation features, which is what contact prediction methods have often depended on. The new feature extraction approach uses mutual information predicted from coevolution methods as input data for deep neural networks [80]. The success of this approach was demonstrated when Jones *et al.* [81] designed MetaPSICOV by integrating the PSICOV, FreeContact, and CCMpred methods with a two-stage neural network model [74,81].The CASP11 evaluations revealed that MetaPSICOV outperformed all contact prediction methods, and the accuracy of contact prediction exceeded 30% [80]. Researchers were inspired by MetaPSICOV and developed their own servers through the inclusion of coevolutionary features derived from MSAs and structural properties in a variety of deep neural networks. Consequently, a milestone was achieved when the accuracy of contact prediction in CASP12 reached 47% [37].

3.5.1.1 Residual convolutional neural networks

Convolutional neural network (CNN) is a deep neural network comprised of varied layers (or filters), in which each layer has neurons with larger local receptive fields than the previous one. The input matrix of the CNN model is divided into submatrices, and each submatrix is filtered by each local receptive field to encode the local map. The output consists of multiple local maps, and this operation is called convolution. Following that, the pooling operation, which consists of pooling submatrix values from the convolution output into single values, results in size minimization. Eventually, the classification stage operates in the last layers of CNN and transforms the output probabilities to a range between 0 and 1, with the sum equal to 1 [39].

CNN models often have the problem of overfitting because each neuron in each layer is fully connected with all neurons of the previous layer, which ultimately reduces the accuracy of these models. To fix this problem, skip connection is applied by designing two residual blocks between layers, creating a residual neural network (ResNet). ResNets have been used in protein contact predictions because of the consistent spatial regularity of amino acid residues on the protein sequence. A ResNet can apply the same local filters over all residue positions by requiring a limited number of weights to be adjusted in relation to the input layer and the next layer's dimensionality. This leads to improved computational implementation and output accuracy [10,39]. Therefore, protein sequence alignments can be analysed by ResNets to predict contact maps with far higher accuracy above 40% [37].

As with previous machine learning approaches, a ResNet needs a large set of features to extract accurate contact patterns between protein residues. However, designing the best architecture for certain features is the key to improving contact prediction accuracy and therefore method performance [83]. Wang *et al.* [19] developed RaptorX-Contact with a deep model consisting of two residual convolutional neural network modules. The first module was designed to be one-dimensional (1D) to learn from 1D protein features, including sequence profile, predicted secondary structure, and solvent accessibility. The second module was built in 2D representation to learn from 2D pair-wise properties. To extract contact patterns, the output of the 1D module was converted to a 2D matrix and fed into a 2D module simultaneously with the pair-wise features. In the last step, the probability values of contact prediction are computed by integrating the output of the second module into logistic regression [19,84]. This designed model has the ability to capture contact existence between residues from the complex protein features, increasing contact prediction accuracy (to 47% in CASP12) substantially. RaptorX-Contact was independently benchmarked for the first time in CASP12 and was among the top performing contact prediction tools [19,29].

Residual convolutional neural networks have been employed in various contact prediction servers with various input protein features and architectures, and most are designed to analyse the MSAs resulting from searches of massive sequence databases. In CASP13, Kandathil *et al.* [54] developed DeepMetaPSICOV, a method that improved the accuracy of protein contact prediction by combining multiple protein properties with a deep convolutional residual network and using MSAs from a large sequence dataset [54].

Another contact-map predictor, ResPRE, was developed by combining coevolution-derived precision matrices that improved the analysis of MSA using deep residual convolutional neural networks [12,39,52]. Along with other methods, ResPRE was integrated into the meta-predictor method called NeBcon (described in previous section). In this predictor, the confidence values of predicted contacts from these methods were fed into a nave Bayes classifier (NBC). The output of NBC was integrated with different sequence data in 350 units of a hidden layer connected to a neural network to refine the contact prediction model [12]. The combination of

ResNets with other machine learning methods and the use of large sequence datasets to derive the input MSAs has helped to greatly improve the predictive performance, so much so that the performance of deep learning-based methods raised the accuracy of contact prediction to 70% in CASP13 [38].

Deep residual convolutional networks achieved success with other aspects of residue-residue contact prediction and the development of alignment techniques. Recent studies demonstrated that residue-residue contacts predicted as binary classification provide restricted information. On the other hand, predicting the actual distance between residues produces more precise information, and ResNet models that are trained in the universal network of inter-residual distances allow for the capture of higher-order residue relationship [79,86,88]. Thus, the cutting edge of contact prediction methods now expand to predict distances in their pipelines. For example, DeepPotential was developed by modifying the deep ResNet by adding 10 residual blocks as 1D and 2D representations for predicting inter-residue contacts within different range of distances. Additionally, to predict all inter-residue interactions, another fully ResNet was fed by the outputs of 1D and 2D ResNet and trained by cross-entropy loss. The inter-residue distances considered in this method are side chain contact, backbone contact, torsional angle, and hydrogen-bond interaction at different distance thresholds, ranging from 2 to 10, 13, 16, and 20 Å [86,88].

TripletRes is ranked as a top-performing method in the most recent CASP experiment [79]. In this method, Zhang's group first developed an alignment strategy to improve the quality of MSA inputs for coevolutionary information extraction. The strategy was to construct a deep MSA using several rounds of HHblits, then extract "covariance features (COV), precision matrix features (PRE), and a coupling parameter matrix approximated by pseudolikelihood maximization (PLM)". These features were combined into the ResNets model with four sets of residual blocks and trained by loss function examining a discrete map of the distance information between each residue pair [10,87]. Li *et al.* [87] demonstrated that one of the success factors of TripletRes was incorporating deep neural networks with the three sets of coevolutionary features, which enabled the capture of more accurate contacts [87]. This indicates that ResNets has made an invaluable contribution to the accuracy of contact prediction methods, whether or not they were used in conjunction with the prediction of absolute distances.

3.5.1.2 Recurrent neural network

Recurrent neural network (RNN) is an advanced architecture of neural networks in which nodes are connected in a recurrent pattern to process sequential data [89]. RNNs have been employed to predict protein secondary structures and they have been designed to extract these structural features from multiple sequence alignment data. Di Iena *et al.* [76] used a RNN to predict coarse contact and orientation of secondary structures, while a further deep neural network architecture was then used to generate final more refined contact predictions [17,76,91]. For contact map prediction, 2D, bidirectional, recurrent long short-term memory (2D-BRLSTM) networks, have been employed with residual convolution neural networks for the SPOT-Contact method. The SPOT-Contact method was designed to combine 2D recurrent neural networks (2D-RNNs) with long short-term memory (LSTM) cells. LSTM cells can learn the complicated context of long-range contacts between residues for the whole protein sequence, while 2D-RNNs can generate an accurate model because of their capacity to identify misleading data in all input variables [39,49]. SPOT-Contact is ranked as one of the top-performing methods in CASP13 according to the independent blind evaluation of contact prediction methods [53]. The improvement in mean accuracy of neural network-based contact prediction methods during the CASP experiments can be seen in Figure 4.

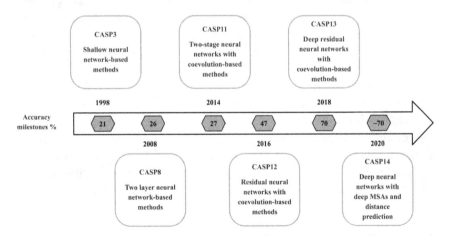

Figure 4. Timeline for the development of neural network-based methods and their average accuracy based on the CASP evaluation procedures. The accuracy of contact prediction is based on L/5 long-range contacts for FM targets.

3.5.1.3 End-to-end learning models

The previously discussed deep neural networks are built with multiple layers and achieved success in detecting coevolution-based features from MSAs, but they gain no information concerning the natural relationship between sequence and structure for proteins if no sequence homologs can be detected. To fix this issue, another class of methods was developed, which use end-to-end differentiable deep learning models based on an explanatory structure-to-sequence maps [92]. In such methods, end-to-end differentiability indicates the ability to use a single approach to optimise a sophisticated multi-stage pipeline from input to output without relying explicitly on coevolutionary information, in which the whole prediction process is represented by a single deep neural network [39,92]. Thus, end-to-end approaches have been employed to enhance contact prediction accuracy in the absence of deep MSAs. The method DeepECA was developed by Fukuda and Tomii to predict contact maps from both deep and shallow MSAs directly in a single neural network. With the availability of homologous sequences, correlated information can be extracted using a covariance matrix, then the coevolution values from this matrix can be used as input for the deep neural network model. The model used was a 1×1 convolutional neural network using end-to-end learning to weight each sequence of an MSA, which helped to eliminate the noisy information from the abundant sequences. The weighting process in the end-to-end model was used to optimize the quality of the MSA analysis and provided the most relevant homologous sequences to the target protein sequence. This method showed an improvement in the contact prediction accuracy, even though predictions were made directly from an MSA alone without any other encoded features. In the case of shallow MSAs, the accuracy of contact prediction can be increased by adding other protein features with correlation information in the CNN model extended to a multi-task model [57,93]. It is worth noting that employing an end-to-end model to improve the procedure of extracting pure and accurate mutation information not only enhances the predictive power of contact predictions for tertiary structures, it may also be used for the currently unresolved problem of predicting protein-protein interactions [94].

4. Conclusions

Protein structure prediction methods have been enhanced by integrating different features predicted directly from protein sequences, including predicted residue-residue contacts. The incorporation of residue contact prediction has led to the enhancement of protein prediction accuracy in various fields, including tertiary structure prediction for 3D protein modelling, 3D model quality estimation and model refinement. Further enhancements to these methods occurred in close association with the considerable improvement of contact prediction methods based on machine learning algorithms. The development and application of machine learning approaches, particularly neural networks, have led to improvements in contact prediction accuracy due to their capability to recognise likely contacting residues from MSAs. The purpose of using Hidden Markov models in protein contact prediction servers was to identify contacting residues within their local neighbourhood, whereas the support vector models were used to detect interaction between residues in 3D space through the linear and nonlinear methods. Random forest algorithms have been used in contact prediction due to their strengths in managing immense high-dimensional datasets and detecting signals from chaotic input data. Naive Bayes classifiers use multi-dimensional protein sequence features in a complementary fashion in order to predict correct contact patterns. In terms of deep neural networks, the design of new architectures greatly influenced the development of contact prediction methods leading to increased performance. Furthermore, the performance of deep learning-based contact prediction methods has also been influenced by advances in MSA methods which search large sequence databases, and the move towards inter-residue distance prediction.

The successes of machine learning-based methods for predicting contacting residues in tertiary structures may help to encourage developers to build upon this research to solve further protein structure prediction problems. Perhaps the next greatest challenge for this field is found in the accurate prediction of protein-protein interactions and quaternary structure modeling. This challenge could be tackled more effectively if developers turned their attention to the problem of accurately predicting the contacting residues between the chains in protein complexes.

5. Acknowledgments

This research was supported in part by the Biotechnology and Biological Sciences Research Council (BBSRC) [BB/T018496/1 to L.J.M] and the Saudi Arabian Government [to S.M.A.A.].

References

[1] Breda A, *et al.* Chapter A06. Protein structure, modeling and applications. In *Bioinformatics in Tropical Disease Research: A Practical and Case-Study Approach*, Gruber A, *et al.* (eds.), Bethesda (MD): National Center for Biotechnology Information, US, 2008. 137–170.

[2] Rangwala H and Karypis G. Introduction to protein structure. In *Introduction to Protein Structure Prediction: Methods and Algorithms*. Rangwala H & Karypis G (eds.), John Wiley & Sons, Inc, 2010. 1–13.

[3] Suh D, *et al.* Recent applications of deep learning methods on evolution- and contact-based protein structure prediction. *International Journal of Molecular Sciences*, 2021. **22**(11): 6032.

[4] Emerson IA and Amala A. Protein contact maps: A binary depiction of protein 3D structures. *Physica A*, 2017. **465**: 782–791.

[5] Li Z, *et al.* Protein contact map prediction based on ResNet and DenseNet. *BioMed Research International*, 2020. **2020**: 1–12.

[6] Pearce R and Zhang Y. Toward the solution of the protein structure prediction problem. *Journal of Biological Chemistry*, 2021b. **297**(1): 100870.

[7] El-Rashidy N, *et al.* Comprehensive survey of using machine learning in the Covid-19 pandemic. *Diagnostics*, 2021. **11**(7): 1155.

[8] Heo L and Feig M. Modeling of severe acute respiratory syndrome coronavirus 2 (SARS-CoV-2) proteins by machine learning and physics-based refinement. *BioRxiv, preprint*, 2020.

[9] Kryshtafovych A, *et al.* Modeling SARS-CoV-2 proteins in the CASP-commons experiment. *Proteins: Structure, Function and Bioinformatics*, 2021. **89**(12): 1987–1996

[10] Pakhrin SC, *et al.* Deep learning-based advances in protein structure prediction. *International Journal of Molecular Sciences*, 2021. **22**: 5553.

[11] Yang J, *et al.* Improved protein structure prediction using predicted interresidue orientations. *Proceedings of the National Academy of Sciences — PNAS*, 2020. **117**(3): 1496–1503.

[12] Zheng W, *et al.* Deep-learning contact-map guided protein structure prediction in CASP13. *Proteins: Structure, Function and Bioinformatics*, 2019. **87**(12): 1149–1164.

[13] McMurry JE, *et al.* Amino acids and proteins. In *Fundamentals of general, organic, and biological chemistry (7th edition)*, McMurry J (ed.), Pearson Education Ltd. 2014. 596–637.

[14] Hou J, *et al.* Protein tertiary structure modeling driven by deep learning and contact distance prediction in CASP13. *Proteins: Structure, Function and Bioinformatics*, 2019. **87**(12): 1165–1178.

[15] Konopka BM and Ciombor M. Automated procedure for contact-map-based protein structure reconstruction. *The Journal of Membrane Biology*, 2014. **247**(5): 409–420.

[16] Adiyaman R and McGuffin LJ. Methods for the refinement of protein structure 3D models. *International Journal of Molecular Sciences*, 2019. **20**(9): 2301.

[17] Jing X, *et al.* Protein inter-residue contacts prediction: Methods, performances and applications. *Current Bioinformatics*, 2019. **14**(3): 178–189.

[18] Lundström, J., *et al.* Pcons: A neural-network-based consensus predictor that improves fold recognition. *Protein Science*, 2001. **10**(11): 2354–2362.

[19] Wang S, *et al.* Analysis of deep learning methods for blind protein contact prediction in CASP12. *Proteins: Structure, Function and Bioinformatics*, 2018. **86**(S1): 67–77.

[20] Cheng J, *et al.* Estimation of model accuracy in CASP13. *Proteins: Structure, Function and Bioinformatics*, 2019. **87**(12): 1361–1377.

[21] McGuffin LJ, *et al.* ModFOLD8: Accurate global and local quality estimates for 3D protein models. *Nucleic Acids Research*, 2021. **49**(W1): W425–W430.

[22] Adiyaman R and McGuffin LJ. ReFOLD3: Refinement of 3D protein models with gradual restraints based on predicted local quality and residue contacts. *Nucleic Acids Research*, 2021. **49**(W1): W589–W596.

[23] Fang C, *et al.* IMPContact: An interhelical residue contact prediction method. *BioMed Research International*, 2020. **2020**: 1–10.

[24] Covell DG and Jernigan RL. Conformations of folded proteins in restricted spaces. *Biochemistry*, 1990. **29**(13): 3287–3294.

[25] Lesk AM. CASP2: Report on ab initio predictions. *Proteins, Structure, Function, and Bioinformatics*, 1997. **29**(S1): 151–166.

[26] Monastyrskyy B, *et al.* Evaluation of residue-residue contact predictions in CASP9. *Proteins*, 2011. **79**(S10): 119–125.

[27] Wang S, *et al.* Folding membrane proteins by deep transfer learning. *Cell Systems*, 2017a. **5**(3): 202–211.e3.

[28] Wang S, *et al.* Accurate de novo prediction of protein contact map by ultra-deep learning model. *PLoS Computational Biology*, 2017b. **13**(1): e1005324.

[29] Xu J. Distance-based protein folding powered by deep learning. *Proceedings of the National Academy of Sciences of the United States of America*, 2019. **116**(34): 16856–16865.

[30] Xu J, *et al.* Improved protein structure prediction by deep learning irrespective of co-evolution information. *Nature Machine Intelligence*, 2021. **3**(7): 601–609.

[31] Xu J and Wang S. Analysis of distance-based protein structure prediction by deep learning in CASP13. *Proteins: Structure, Function and Bioinformatics*, 2019. **87**(12): 1069–1081.

[32] Wang XF, *et al.* Predicting residue-residue contacts and helix-helix interactions in transmembrane proteins using an integrative feature-based random forest approach. *PLoS One*, 2011. **6**(10): e26767.

[33] Adhikari B and Cheng J. Protein residue contacts and prediction methods. *Methods in Molecular Biology*, 2016. **1415**: 463–476.

[34] Duarte JM, *et al.* Optimal contact definition for reconstruction of contact maps. *BMC Bioinformatics*, 2010. **11**(1): 1–10.

[35] Yuan C, *et al.* Effective inter-residue contact definitions for accurate protein fold recognition. *BMC Bioinformatics*, 2012. **13**(1): 1–13.

[36] Monastyrskyy B, *et al.* Evaluation of residue-residue contact prediction in CASP10. *Proteins: Structure, Function and Bioinformatics*, 2014. **82**(S2): 138–153.

[37] Schaarschmidt J, *et al.* Assessment of contact predictions in CASP12: Co-evolution and deep learning coming of age. *Proteins: Structure, Function and Bioinformatics*, 2018. **86**: 51–66.

[38] Shrestha R, *et al.* Assessing the accuracy of contact predictions in CASP13. *Proteins: Structure, Function and Bioinformatics*, 2019. **87**(12): 1058–1068.

[39] Jisna VA and Jayaraj PB. Protein structure prediction: Conventional and deep learning perspectives. *The Protein Journal*, 2021. **40**(4): 522–544.

[40] Latek D and Kolinski A. Contact prediction in protein modeling: scoring, folding and refinement of coarse-grained models. *BMC Structural Biology*, 2008. **8**: 1–15.

[41] Wu S and Zhang Y. A comprehensive assessment of sequence based and template-based methods for protein contact prediction. *Bioinformatics*, 2008. **24**(7): 924–931.

[42] He B, *et al.* NeBcon: Protein contact map prediction using neural network training coupled with naïve Bayes classifiers. *Bioinformatics*, 2017. **33**(15): 2296–2306.

[43] Zhang H, *et al.* Evaluation of residue-residue contact prediction methods: from retrospective to prospective. *PLoS Computational Biology*, 2021. **17**(5): e1009027.

[44] Greener JG, *et al.* A guide to machine learning for biologists. *Nature Reviews. Molecular Cell Biology*, 2021. **23**(1): 40–55.

[45] Xue B, *et al.* Predicting residue-residue contact maps by a two-layer, integrated neural-network method. *Proteins, Structure, Function, and Bioinformatics*, 2009. **76**(1): 176–183.

[46] Sehnal D, *et al.* Mol*Viewer: Modern web app for 3D visualization and analysis of large biomolecular structures. *Nucleic Acids Research*, 2021. **49**(W1): W431–W437.

[47] Adhikari B. DEEPCON: Protein contact prediction using dilated convolutional neural networks with dropout. *Bioinformatics*, 2020. **36**(2): 470–477.

[48] Ding W, *et al.* DeepConPred2: An improved method for the prediction of protein residue contacts. *Computational and Structural Biotechnology Journal*, 2018. **16**: 503–510.

[49] Hanson J, *et al.* Accurate prediction of protein contact maps by coupling residual two-dimensional bidirectional long short-term memory with convolutional neural networks. *Bioinformatics*, 2018. **34**(23): 4039–4045.

[50] Cheng J and Baldi P. Improved residue contact prediction using support vector machines and a large feature set. *BMC Bioinformatics*, 2007. **8**(1): 1–9.

[51] Adhikari B, *et al.* DNCON2: Improved protein contact prediction using two-level deep convolutional neural networks. *Bioinformatics*, 2018. **34**(9): 1466–1472.

[52] Li Y, *et al.* ResPRE: High-accuracy protein contact prediction by coupling precision matrix with deep residual neural networks. *Bioinformatics*, 2019. **35**(22): 4647–4655.

[53] Wu Q, *et al.* Protein contact prediction using metagenome sequence data and residual neural networks. *Bioinformatics*, 2020. **36**(1): 41–48.

[54] Kandathil SM, *et al.* Prediction of interresidue contacts with DeepMetaPSICOV in CASP13. *Proteins: Structure, Function and Bioinformatics*, 2019. **87**(12): 1092–1099.

[55] Wu T, *et al.* DeepDist: Real-value inter-residue distance prediction with deep residual convolutional network. *BMC Bioinformatics*, 2021. **22**(1): 1–17.

[56] Fukuda H and Tomii K. DeepECA: An end-to-end learning framework for protein contact prediction from a multiple sequence alignment. *BMC Bioinformatics*, 2020. **21**(1): 1–15.

[57] Karplus K, *et al.* Predicting protein structure using hidden Markov models. *Proteins: Structure, Function and Genetics*, 1997. **29**(S1): 134–139.

[58] Björkholm P, *et al.* Using multi-data hidden Markov models trained on local neighborhoods of protein structure to predict residue-residue contacts. *Bioinformatics*, 2009. **25**(10): 1264–1270.

[59] Stecking R and Schebesch KB. Informative patterns for credit scoring: support vector machines preselect data subsets for linear discriminant analysis. In *Classification — the Ubiquitous Challenge*, C Weihs & W Gaul (eds.), Berlin Heidelberg, Springer, 2005. 450–457.

[60] Zhao Y and Karypis G. Prediction of contact maps using support vector machines. *Proceedings — 3rd IEEE Symposium on BioInformatics and BioEngineering, BIBE 2003, April 2003*, 2003. 26–33.

[61] Horner DS, *et al.* Correlated substitution analysis and the prediction of amino acid structural contacts. *Briefings in Bioinformatics*, 2008. **9**(1): 46–56.

[62] Li Y, *et al.* Predicting residue-residue contacts using random forest models. *Bioinformatics*, 2011. **27**(24): 3379–3384.

[63] Zheng C, *et al.* An integrative computational framework based on a two-step random forest algorithm improves prediction of zinc-binding sites in proteins. *PLoS One*, 2012. **7**(11): e49716.

[64] Wang Z and Xu J. Predicting protein contact map using evolutionary and physical constraints by integer programming. *Bioinformatics*, 2013. **29**(13): i266–i273.

[65]   Zhang H, *et al.* COMSAT: Residue contact prediction of transmembrane proteins based on support vector machines and mixed integer linear programming. *Proteins: Structure, Function and Bioinformatics*, 2016. **84**(3): 332–348.

[66]   Peng C, *et al.* De novo protein structure prediction by coupling contact with distance profile. *IEEE/ACM Transactions on Computational Biology and Bioinformatics*, 2020. **19**(1): 395–406.

[67]   Hapudeniya M. Artificial neural networks in bioinformatics. *Sri Lanka Journal of Bio-Medical Informatics*, 2010. **1**(2): 104–111.

[68]   Fariselli P and Casadio R. A neural network based predictor of residue contacts in proteins. *Protein Engineering*, 1999. **12**(1): 15–21.

[69]   Shackelford G and Karplus K. Contact prediction using mutual information and neural nets. *Proteins*, 2007. **69**(S8): 159–164.

[70]   Fariselli P, *et al.* Prediction of contact maps with neural networks and correlated mutations. *Protein Engineering*, 2001. **14**(11): 835–843.

[71]   Liu G, *et al.* A study on protein residue contacts prediction by recurrent neural network. *Journal of Bionic Engineering*, 2005. **2**(3): 157–160.

[72]   Schneider R, *et al.* The HSSP database of protein structure-sequence alignments. *Nucleic Acids Research*, 1997. **25**(1): 226–230.

[73]   Torrisi M, *et al.* Deep learning methods in protein structure prediction. *Computational and Structural Biotechnology Journal*, 2020. **18**: 1301–1310.

[74]   Tegge AN, *et al.* NNcon: Improved protein contact map prediction using 2D-recursive neural networks. *Nucleic Acids Research*, 2009. **37**(Suppl 2): W515–W518.

[75]   Di lena P, *et al.* Deep architectures for protein contact map prediction. *Bioinformatics*, 2012. **28**(19): 2449–2457.

[76]   Eickholt J and Cheng J. Predicting protein residue-residue contacts using deep networks and boosting. *Bioinformatics*, 2012. **28**(23): 3066–3072.

[77]   Eickholt J and Cheng J. A study and benchmark of DNcon: A method for protein residue-residue contact prediction using deep networks. *BMC Bioinformatics*, 2013. **14**: S12.

[78]   Ruiz-Serra V, *et al.* Assessing the accuracy of contact and distance predictions in CASP14. *Proteins: Structure, Function, and Bioinformatics*, 2021. **89**(12): 1888–1900.

[79]   Monastyrskyy B, *et al.* New encouraging developments in contact prediction: Assessment of the CASP11 results. *Proteins: Structure, Function and Bioinformatics*, 2016. **84**: 131–144.

[80]   Jones DT, *et al.* MetaPSICOV: Combining coevolution methods for accurate prediction of contacts and long range hydrogen bonding in proteins. *Bioinformatics*, 2015. **31**(7): 999–1006.

[81]   Kuhlman B and Bradley P. Advances in protein structure prediction and design. *Nature Reviews Molecular Cell Biology*, 2019. **20**(11): 681–697.

[82]   Pearce R and Zhang Y. Deep learning techniques have significantly impacted protein structure prediction and protein design. *Current Opinion in Structural Biology*, 2021a. **68**: 194–207.

[83] Zheng W, *et al.* Protein structure prediction using deep learning distance and hydrogen-bonding restraints in CASP14. *Proteins: Structure, Function and Bioinformatics*, 2021. **89**(12): 1734–1751.

[84] Li Y, *et al.* Deducing high-accuracy protein contact-maps from a triplet of coevolutionary matrices through deep residual convolutional networks. *PLoS Computational Biology*, 2021a. **17**(3): e1008865.

[85] Li Y, *et al.* Protein inter-residue contact and distance prediction by coupling complementary coevolution features with deep residual networks in CASP14. *Proteins: Structure, Function and Bioinformatics*, 2021b. **89**(12): 1911–1921.

[86] Graves A, *et al.* Multi-dimensional recurrent neural networks. *Artificial Neural Networks — ICANN 2007*, de Sá JM, *et al.* (eds.), Springer, Berlin, Heidelberg, 2007. 549–558.

[87] Kc DB. Recent advances in sequence-based protein structure prediction. 2017. **18**(6): 1021–1032.

[88] AlQuraishi M. End-to-end differentiable learning of protein structure. *Cell Systems*, 2019. **8**(4): 292–301.e3.

[89] Vaz JM and Balaji S. Convolutional neural networks (CNNs): Concepts and applications in pharmacogenomics. *Molecular Diversity*, 2021. **25**(3): 1569–1584.

[90] Laine E, *et al.* Protein sequence-to-structure learning: Is this the end(-to-end revolution)? *Proteins: Structure, Function and Bioinformatics*, 2021. **89**(12): 1770–1786.

© 2023 World Scientific Publishing Company
https://doi.org/10.1142/9789811258589_0007

Chapter 7

Machine Learning for Protein Inter-Residue Interaction Prediction

Yang Li[‡], Yan Liu[‡] and Dong-Jun Yu[*]

*Department of Computer Science and Engineering,
Nanjing University of Science and Technology, Nanjing, China
[‡]These authors contributed equally to this work.
corresponding author: njyudj@njust.edu.cn

Proteins are essential components of cells and are involved in almost every cellular process. Proteins are large biomolecules composed of one or more amino acid sequences, which fold into a specific topology to perform their biological functions. Determining the structure of proteins is the premise to eventually decipher their functions. Development of novel computational approaches to predict the protein structure from sequence is one of the hottest topics in bioinformatics. The prediction of key topological aspects of structure, e.g., inter-residue contacts and distances, has become one of the most important information sources to help sequence-based protein structure prediction. In this chapter, we provide a systematic review of the representative methods for protein inter-residue geometry interaction prediction, including correlation-based methods, direct coupling analysis methods and their post-processing strategies, classical machine learning methods, and advanced deep learning methods. We also discuss applications of those interactions, especially in protein structure prediction.

1. Introduction

Proteins are the essential components of living cells, acting as one of the most important building blocks of life. Proteins are formed after dehydration and condensation of a sequence of amino acids. The protein sequence can be relatively easy to obtain by the high-throughput sequencing techniques. However, determining its three-dimensional (3D) conformation is a complex task. Experimental methods for structure determination include X-ray analysis [1], nuclear magnetic resonance (NMR) [2] and the emerging cryo-electron microscopy technique [3]. In the post-genomic era, the determination of the 3D structure of proteins by experiment is far from catching up with the high-volume accumulation of protein sequences. Therefore, ways to improve the accuracy of predicting the 3D structure of proteins directly from amino acid sequences becomes a key issue in the field of structure bioinformatics [4].

Due to the limitation of experimental techniques, direct prediction of protein structure from its sequence is gradually gaining attention. Investigations by Anfinsen [5] show that the amino acid sequence of proteins determines its 3D structure, meaning that the structural information of proteins can be fully embedded in its sequences. This hypothesis drives the development of sequence-based protein structure prediction. Conventional approaches for protein structure prediction, e.g., homology modeling [6,7] and threading [8,9], require a reasonably similar structure template to be available in an existing protein database. *Ab initio* prediction methods are becoming increasingly important since the templates are unnecessary for these methods. Usually, the *ab initio* structure prediction methods apply to both physics and knowledge-based potentials to guide the folding simulation.

Predicting coordinates of atoms that make up protein structure directly from the corresponding sequence using machine learning can come naturally to the developers in the field. However, the coordinates are difficult to predict since they will change with rotation and translation transformations while the structure itself will remain unchanged. Alternatively, some invariant representations of the protein structure, e.g., inter-residue contacts, can be predicted using machine learning methods. The patterns of protein contact maps, where each entry indicates whether the corresponding residue pair is in contact, reflect some of the key characteristics of the structures. More importantly, the predicted invariant representations can be used as potentials to build a more reliable force field for the protein structure prediction. Considerable progress has been made to improve the performance of the protein contact map predictions in recent years. Initial approaches focus on analyzing pairwise correlations to address physical contacts based on unsupervised machine learning models [10–15]. Supervised machine learning models were further applied to learn information across Multiple Sequence Alignments (MSAs) of training data [16–22]. Besides the evolution of machine learning models, the invariant geometry terms are also evolving from contacts to distance and orientations [23–26]. This chapter will present a systematic review of representative approaches in those categories.

2. Computational methods for protein inter-residue interaction prediction

2.1 *Definition of geometry terms for inter-residue interactions*

The basic invariant representation for inter-residue interaction is the contacts between their Cβ atoms (Cα for Glycine). Here, two residues are in contact if the Euclidean distance between their Cβ atoms is less than a threshold, i.e., 8 Å, according to the definition considered in the Critical Assessment of protein Structure Prediction (CASP) experiments [27,28]. The inter-residue contacts can further be classified into three categories, i.e., short-, medium- and long-range, corresponding to different sequence separations (distance in the sequence for the contacting residues) of 6 to 11, 12 to 23 and over 24 positions, respectively. Compared to short- or medium-range contacts, long-range contacts have a higher impact on stabilizing a protein topology, thus, are mainly evaluated in CASP.

In addition to the binary contacts, AlphaFold [23] in CASP13 predicts the discretized distance between two atoms, providing more information to assist protein folding. Still, the distance information may not be sufficient to represent a structure, especially its chirality. To address such problem, trRosetta additionally predicts a set of orientation terms, including the ω and θ dihedral angle and φ angle [25]. Recently, DeepPotential [26] considered a set of inter-residue geometry terms by additionally introducing three hydrogen-bond (H-bond) related terms which involve six Cα atoms. Given two residues i and j, two local frames are constructed from their adjacent Cα atoms, respectively. The H-bond terms are the angles between corresponding x-, y-, and z-axes of two constructed frames. The definition of those geometry terms is shown in Figure 1.

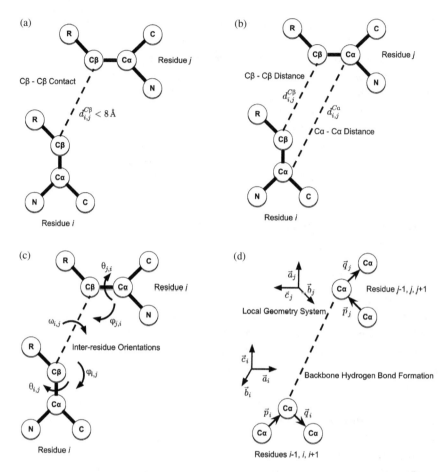

Figure 1.   Definition of inter-residue geometry terms: (a) contacts, (b) distance, (c) orientations, and (d) H-bond terms.

2.2 *Unsupervised methods for contact map prediction*

The unsupervised methods derive information from the coevolution of two residues by analyzing the MSA. Here, MSA is the sequence alignment of multiple homologous sequences that are genetically related to each other to obtain the relevant genetic information of the query sequence. Given an MSA, each row represents a homologous sequence, and each column represents the position of the target protein sequence. The analysis of a single site in the alignment can discern the degree of evolutionary conservation of the site. In fact, amino acids at different positions are not mutated independently and most single-point mutations are deleterious and may disrupt the physical compatibility around the mutation site. Compensatory mutations at neighboring sites can often repair the damage caused by single-point mutations. This phenomenon is generally referred to as coevolution. There are two main approaches to analyze the coevolutionary patterns: the methods based on local correlation and the methods based on Direct Coupling Analysis (DCA). Table 1 summarizes some representative unsupervised methods for inferring the inter-residue contacts. Based on a given MSA, termed as A; we use M to denote the number of sequences in the MSA and L to denote the length of the query sequence. Thus, a_l^m

Table 1. Collection of representative coevolution analysis methods to infer the protein inter-residue contacts. The "Type" column indicates whether the method is local or global. The "Approximation" column indicates the approximation method for Potts model ("NA" for local methods).

Type	Methods	Ref.	Year	Approximation	Availability
Local	Gobel *et al.*	[29]	1994	NA	NA
	Yanofsky *et al.*	[30]	1964	NA	NA
	Korber *et al.*	[31]	1993	NA	NA
	Martin *et al.*	[13]	2005	NA	NA
	OMES	[32]	2002	NA	NA
Global	Lapedes *et al.*	[33]	1999	Monte Carlo	NA
	mpDCA	[34]	2009	Message passing	NA
	mfDCA	[14]	2011	Mean field	NA
	PSICOV	[12]	2011	Graphical lasso	https://github.com/psipred/psicov
	Ricmap	[35]	2019	Graphical ridge	https://zhanggroup.org/ResPRE/
	plmDCA	[11]	2013	Pseudolikelihood	https://github.com/magnusekeberg/plmDCA
	GREMLIN	[36]	2013	Pseudolikelihood	http://gremlin.bakerlab.org/
	CCMpred	[15]	2014	Pseudolikelihood	https://github.com/soedinglab/CCMpred
	clmDCA	[37]	2018	Composite Likelihood	http://protein.ict.ac.cn/clmDCA/

denotes the amino acid type of position l of m-th aligned sequence. There are total of 21 states for each entry in MSA, i.e., 20 natural amino acid types and the additional state representing the gap.

Many local correlation coefficients are used to score the coevolution between a pair of residues. Gobel *et al.* [29] used the Pearson Correlation Coefficient (PCC) to predict residue correlation effects. The mutual information [13,30,31] can further measure a nonlinear relationship between residues in MSA. The Observed Minus Expected Squared (OMES) coefficient [32], which quantifies the difference between the observed and expected frequencies of co-occurrence of residues at any two positions, was also proposed. Nevertheless, the above methods may not lead to satisfactory performance and the primary reason is their inability to model couplings. For example, suppose position i and j, and position j and k are correlated. In that case, the local correlation coefficients will certainly show a high correlation between position i and k, although they may not be physically in contact.

DCA approaches were proposed to model MSA data with the Potts model [38], which considered couplings between positions. Methods based on such models are usually called global models. The generalized Potts Model is usually used:

$$\mathcal{H}(a_1,...,a_L) = -\left(\sum_{i=1}^{L} h_i(a_i) + \sum_{i \neq j}^{L} J_{ij}(a_i, a_j) \right) \tag{1}$$

where $\mathcal{H}(a_1,...,a_L)$ is the Hamiltonian of the system and each position can be considered as one particle; each aligned sequence is one observation. $h_i(a_i)$ and $J_{ij}(a_i,a_j)$ are the local field parameters at position i with residue type a_i and the coupling parameters at positions i and j with residues a_i and a_j, respectively. Accordingly, a global statistical model $P(a_1,...,a_L)$ can be constructed over the whole MSA, in which the probability of a sequence in the MSA can be defined as:

$$P(a_1,...,a_L) = \exp\{-\mathcal{H}(a_1,...,a_L)\} / Z = \frac{1}{Z}\left(\sum_{i=1}^{L} h_i(a_i) + \sum_{i \neq j}^{L} J_{ij}(a_i, a_j) \right) \tag{2}$$

where Z is the normalization constant, i.e., partition function, that guarantees $\sum_a P(a) = 1$. The model needs to be consistent with empirical statistics, i.e., $P(a_i) = f_i(a_i)$ and $P(a_i,a_j) = f_{ij}(a_i,a_j)$. Here $f_i(a_i)$ and $f_{ij}(a_i,a_j)$ are the statistical frequencies of individual and pairwise positions in the MSA, respectively. In addition to the above constraints, the direct coupling analysis requires the parameters *J* as well as *h* to be obtained by minimizing the negative log-likelihood function, as shown in equation (3):

$$S = -\sum_{m=1}^{M} log P\left(a_1^m, ..., a_L^m\right) \qquad (3)$$

Since the summation of 21^L terms in the likelihood function is required to calculate the normalization constant Z, which is computationally intractable, many approximations and alternative strategies have been developed.

Lapedes *et al.* [33] first introduced the Potts model to estimate coupling parameters between positions by approximating the partition function with Monte Carlo (MC). However, for a 21-state Potts model, the convergence can be guaranteed only after exponential times. Ten years later, the concept of DCA was formally introduced by Martin Weigt in 2009 [34], and a message passing-based method was used as an approximate tool, which provides a unique solution.

Morcos *et al.* [14] proposed to use mean fields to approximate the parameters of the Potts model, and the standard mean-field approximation can be used to self-consistently estimate the single point edge values. The coupling parameter *J* can then be obtained by inverting the covariance matrix. Jones *et al.* [12] proposed the sparse Gaussian approximation method PSICOV, to predict protein residue contacts. PSICOV first processed amino acids as binary vectors and then obtained the covariance matrix of the transformed MSA by $C_{ij}(a,b) = f_{ij}(a,b) - f_i(a)f_j(b)$. In this way, the original Potts model can be approximated as a multivariate Gaussian Markov model. The inverse matrix of the covariance matrix can be obtained by minimizing the objective function (4) as follows.

$$\text{tr}\left(C\Theta\right) - \ln \det\left(\Theta\right) + \lambda \sum \left|\Theta_{ij}\right| \qquad (4)$$

where *C* is the covariance matrix and Θ is the positive definite symmetric matrix to be obtained, generally called the precision matrix. $\text{tr}(X)$ is the trace of the matrix X. The coupling parameter $J = -\Theta$ can be estimated. The first two terms can be interpreted as the negative log-likelihood of the multivariate Gaussian distributed data, and the third term is the L1 regularization term of Θ. The mean-field approximation and the Gaussian approximation are the most computational effective schemes to solve the inverse Potts problem. Experimental results on 150 PFAM sequences show that PSICOV outperforms the initial mean-field approach. Li *et al.* [35]

proposed Ricmap, which estimates the precision matrix by ridge regularization. The loss function for precision matrix estimation became:

$$tr(C\Theta) - \ln\det(\Theta) + \lambda\sum\Theta_{ij}^2 \tag{5}$$

In such case, the closed-form solution can be obtained:

$$\Theta = QKQ^T \tag{6}$$

where Q is the eigenvectors of C, and K is a diagonal matrix whose diagonal elements are

$$K_{i,i} = \frac{-\Lambda_{i,i} + \sqrt{\Lambda_{i,i}^2 + 8\lambda}}{4\lambda} \tag{7}$$

and where $\Lambda_{i,i}$ are the eigenvalues of C. Ricmap was benchmarked in a dataset composed of CASP11 and CASP12 targets and was reported to marginally outperform PSICOV [35].

However, the Gaussian approximation ignores the property of labeled data at each location in the MSA and treats the labeled variables as continuous variables by utilizing the one-hot-encoding approach. Gaussian and mean-field approximation cannot converge to the optimal solution even with the infinite number of samples. Magnus Ekeberg *et al.* [11] proposed a pseudolikelihood approximation to infer the Potts model. The pseudolikelihood approximates the probability of a single sequence as the product of the probabilities of the positions in a single sequence:

$$P(a_1,...,a_L) = \prod_{l=1}^{L}P(a_l) \tag{8}$$

where the probability of a single site can be approximated as:

$$P(a_l) = \frac{\exp\left(h_l(a_l) + \sum_{k=1,k\neq l}^{L}J_{lk}(a_l,a_k)\right)}{\sum_{q=1}^{21}\exp\left(h_l(q) + \sum_{k=1,k\neq l}^{L}J_{lk}(q,a_k)\right)} \tag{9}$$

For an MSA with M sequences, the cross-entropy loss function of the pseudolikelihood approximation is:

$$S_{pseudo} = -\sum_{m=1}^{M}\sum_{l=1}^{L}\log\left(\frac{\exp\left(h_l(a_l^m) + \sum_{k=1,k\neq l}^{L}J_{lk}(a_l^m,a_k^m)\right)}{\sum_{q=1}^{21}\exp\left(h_l(q) + \sum_{k=1,k\neq l}^{L}J_{lk}(q,a_k^m)\right)}\right) \tag{10}$$

This loss function is convex; therefore, it can be optimized using L-BFGS [39] or the conjugate gradient method [40]. There are several implementations of DCA methods based on the pseudolikelihood approximation, such as GREMLIN [36], plmDCA [11], and CCMpred [15]. The above implementations use L2 regularization terms to prevent overfitting. GREMLIN attempts to introduce prior information into the DCA. Thus, its performance is slightly better than the other two implementations, but all three methods outperform mfDCA for the mean-field approximation and PSICOV for the Gaussian approximation [15,36].

Zhang *et al.* [37] proposed the clmDCA method, which used Composite Likelihood Maximization (CLM), to better approximate the original likelihood function. Unlike the pseudolikelihood, the composite likelihood function is defined as:

$$\mathcal{L}_{CLM} = -\frac{1}{M}\sum_{m=1}^{M}\sum_{c\in C} \log P(a_c) \tag{11}$$

where C denotes the subset of variables, and the method degenerates to pseudolikelihood when each subset is just one variable. If there is only one subset containing all the variables, then the composite likelihood function is equivalent to the original likelihood function. Therefore, the approximation ability of the composite likelihood function is between the pseudolikelihood and the original likelihood. The clmDCA first calculates the parameters of the pseudolikelihood estimation as the initial parameters of the composite likelihood model and then updates the parameters to obtain the final solution. The experimental results [37] showed that the clmDCA algorithm is superior to the pseudolikelihood method, i.e., plmDCA.

2.3 *Supervised methods for contact map prediction*

The co-evolution-based prediction algorithms have two conspicuous drawbacks. First, the typical direct coupling analysis method only estimates the linear relationship between locations in the MSA. Second, the co-evolution-based prediction algorithm only uses information from its own MSA and does not use information from other sources. These lead to the inability of the coevolutionary algorithms to predict protein residue-residue contact maps more accurately, especially when there are few homologous sequences. Supervised machine learning-based methods can integrate coevolutionary information as features and extra sequence and structure information across samples. The prediction results are often better than those of single coevolutionary algorithms with advanced classification algorithms [17]. A collection of representative supervised machine learning-based methods are shown in Table 2.

Table 2. Overview of the representative supervised machine learning models for inter-residue geometry prediction. The "Raw Coevolution" column indicates whether the model uses raw coevolutionary features as features. The "ML Model" column indicates the machine learning models that the methods use. "SVM", "RF", "ANN", "BNN", "CNN" stand for Support Vector Machine, Random Forest, Artificial Neural Networks, Bayesian Neural Networks, and Convolutional Neural Networks respectively. The "Geometry terms" indicates the predictions of the models.

Name	Ref.	Year	Raw Coevolution	ML Model	Geometry Terms	Availability
SVMcon	[16]	2007	No	SVM	Contact	http://sysbio.rnet.missouri.edu/multicom_toolbox/SVMcon%201.0.html
Li et al.	[41]	2011	No	RF	Contact	NA
MetaPSICOV	[17]	2014	No	ANN	Contact	https://github.com/psipred/metapsicov
plmconv	[42]	2016	Yes	CNN	Contact	NA
Nebcon	[21]	2017	No	BNN	Contact	https://zhanggroup.org/NeBcon/
RaptorX (CASP12)	[43]	2017	No	Residual CNN	Contact	NA
DNCON2	[20]	2017	No	CNN	Contact	https://github.com/multicom-toolbox/DNCON2
DeepContact	[44]	2018	No	CNN	Contact	https://github.com/largelymfs/deepcontact
DeepCOV	[45]	2018	Yes	CNN	Contact	https://github.com/psipred/DeepCov
ResPRE	[35]	2019	Yes	Residual CNN	Contact	https://zhanggroup.org/ResPRE/
RaptorX (CASP13)	[24]	2019	No	Residual CNN	Distance	https://github.com/j3xugit/RaptorX-Contact
AlphaFold	[23]	2020	Yes	Residual CNN	Distance	https://github.com/deepmind/deepmind-research/tree/master/alphafold_casp13
trRosetta	[25]	2020	Yes	Residual CNN	Distance, Orientation	https://github.com/gjoni/trRosetta
TripletRes	[46]	2021	Yes	Residual CNN	Contact	https://zhanggroup.org/TripletRes/
RaptorX (CASP14)	[47]	2021	Yes	Residual CNN	Distance, Orientation	https://github.com/j3xugit/RaptorX-3DModeling
DeepPotential	[26]	2021	Yes	Residual CNN	Distance, Orientation, H-bond	https://zhanggroup.org/DeepPotential/

Classical protein contact prediction algorithms take a single residue pair as a sample and extract the residue pair's contextual features using a sliding window technique. The classifier used by those contact map predictors include random forests [41], support vector machines [16], and fully connected networks [17,21,48]. As one representative method, MetaPSICOV [17] fuses three coevolutionary analysis algorithms as features and then uses artificial neural networks to learn residue contact patterns. In addition to these three coevolutionary features, i.e., PSICOV, CCMpred, and mfDCA [14], MetaPSICOV also includes statistical features extracted from MSA, including single- and two-site statistical features. Moreover, MetaPSICOV employs PSIPRED [49] and SOLVPRED to predict the sequence's secondary structure probability and solvent accessibility as features. The method uses sliding windows of multiple sizes to extract local information for each residue pair. Global features such as amino acid composition distribution probabilities, average protein secondary structure, and average solvent accessibility are also included. The method applies a two-stage training strategy where the first stage contains a hidden layer of 55 nodes and outputs a residue contact probability. The second stage uses the same network model for score correction. The accuracy of the method outperforms other algorithms in CASP11, demonstrating the effectiveness of the supervised machine learning algorithm [50].

In CASP12, Wang *et al.* [43] proposed RaptorX-Contact to predict contact maps using residual Convolutional Neural Networks (CNN), leveraging on the success of residual neural networks (ResNet) in image classification [51]. Since the positive and negative samples in the training data are highly unbalanced, this method uses a weighted cross-entropy loss function to reduce the category bias. The L2 regularization term was also employed to constrain the parameters of the ultra-deep neural network from overfitting. The blind test in CASP12 showed that the RaptorX-Contact significantly improved protein contact map prediction precision [52]. In CASP12, DNCON2 [20] and DeepContact [44] also considered the employment of CNN as the backbone of their deep learning model and thus, also ranked as top methods [52].

The mainstream methods consider post-processing of the coevolutionary analysis as the feature which could, however, lead to possible information loss. For each residue pair, a submatrix with the size of 21 by 21 can be obtained after the coevolutionary analysis. However, the post-processing ignores the weights and the magnetic properties of the entries. A proper way to extract information from the coevolutionary analysis was first adopted by plmconv [42], which combines raw coupling features approximated by pseudolikelihood maximization with CNNs. DeepCOV [45] was proposed later to learn contact patterns directly from the raw covariance matrices. ResPRE [35] employs a ridge estimation of the inverse of the covariance matrix to eliminate the transition noise in the covariance matrix. With a single raw precision matrix, ResPRE outperforms top-performing methods in CASP12 [35]. In CASP13, TripletRes [46] fuses a triplet of raw coevolutionary matrix features. Those features include the coupling matrix of the pseudolikelihood maximized Potts model, the covariance matrix, and the ridge estimation of the precision matrix. A deep residual CNN was employed to fuse the triplet of the features, allowing it to be ranked as one of the top-performing methods for contact map prediction in CASP13 [53]. After CASP13, the strategy of using raw coevolutionary analysis matrix was widely adopted by other representative methods, e.g., trRosetta [25], tFold [54], DeepPotential [26], and other participants in CASP14 [47].

One of the breakthroughs in CASP13 is that the concept of distance prediction was introduced and successfully predicted by AlphaFold [23] and RaptorX [24]. The distance prediction was considered as a multi-class classification problem, and whole neural networks can be trained with cross-entropy loss. The general formulation is that the real-value distance less than a certain threshold was discretized into a distance histogram. One additional class was assigned to those residue pairs whose distances are over the threshold. trRosetta additionally predicts the distributions for the ω and θ dihedral angle, and φ angle. The angles are also discretized into bins with a fixed interval, similar to the distance prediction. The inclusion of such orientation terms brought improved protein structure prediction performance. In CASP14, DeepPotential, which was the extension of TripletRes by including distance, orientation, and H-bond terms [55] with deep multi-task learning, was highly effective for the *ab initio* protein structure prediction of Free-modelling targets [56].

3. Application of protein inter-residue interaction prediction

One of the most impactful direct applications of protein inter-residue interaction prediction is protein structure prediction. The early focus was to use the predicted contact map to assist protein folding and ranking [57–61]. DCAfold [60] and EVfold [61] took the predicted contacts maps as restraints and fed them to the CNS molecular dynamics software suite [62] to find the optimal protein conformation. Such protocol brought significant improvement to the protein folding pipelines when compared to approaches without contact restraints [60,61]. FRAGFOLD designed a new potential function [63] for each residue pair if predicted as contacts by PSICOV, defined as:

$$
E = \begin{cases} -P, \, d \leq d_{con} \\ -Pe^{-(d-d_{con})^2 + P\frac{d-d_{con}}{d}}, d > d_{con} \end{cases} \tag{12}
$$

where P is the predicted probability for the pair as contact. $d_{con} = 8\text{Å}$ is the threshold that defines the contacts. Ovchinnikov *et al.* [64] alternatively used sigmoidal distance restraints to the folding program, Rosetta, in the form of

$$
E = \frac{weight}{1 + \exp(d - d_{cut})} + intercept \tag{13}
$$

here *weight* is proportional to normalized coupling strength and d is the distance between Cβ atoms (Cα in the case of glycine) in the reduced-atom representation of the Rosetta models. d_{cut} and *intercept* are the other two statistical parameters for different modes of Rosetta. Besides, the residue contact information was also used by QUARK [65] and I-TASSER [66] to help protein structure prediction for the CASP competition [67,68]. During the simulation of protein structure using Monte Carlo, the energy term for residue contacts was:

$$
E = \begin{cases} -U_{ij}, & d < d_{cut} \\[2ex] -\dfrac{1}{2}U_{ij}\left[1 - sin\left(\dfrac{d - \left(\dfrac{d_{cut} + D}{2}\right)}{D - d_{cut}}\pi\right)\right], & d_{cut} \leq d < D \\[3ex] \dfrac{1}{2}U_{ij}\left[1 + sin\left(\dfrac{d - \left(\dfrac{D + 80}{2}\right)}{(80 - D)}\pi\right)\right], & D \leq d < 80\text{Å} \\[3ex] U_{ij}, & d \geq 80\text{Å} \end{cases} \tag{14}
$$

where $d_{cut} = 8\text{Å}$ and $D = 8\text{Å} + d_{well}$, where d_{well} is the well width of the first sine function term and 80-D is the well width of the second sine function term. The well width (d_{well}) is a crucial parameter to determine the rate at which residues that are predicted to be in contact are drawn together, and it was tuned based on the length of the training proteins. U_{ij} is the confidence scores for the residue that defines the lower and upper bounds of the energy. With the integration of such contact energy to QUARK and I-TASSER, the resulting methods, C-QUARK and C-I-TASSER, ranked as two of the best servers in CASP12 and CASP13, respectively [67,68].

In CASP13, RaptorX [24] and AlphaFold [23] introduced predicted inter-residue distance and restraints for protein structure prediction. Both models predict discrete distance bins and RaptorX estimates the Mean and Standard Deviation from the distance histogram predictions. Meanwhile, AlphaFold constructs a differentiable potential for a given residue pair by interpolating a cubic spline from the negative log-likelihood of the distance distribution. RaptorX and AlphaFold both achieve promising results in CASP13, and AlphaFold was particularly successful [69] at predicting the most accurate structure for targets rated as the most difficult by the competition organizers.

After CASP13, another deep learning-based protein structure prediction method, trRosetta [25], converted the predicted distance distributions into energy potential following the idea of Dfire [70]. The probability of the last bin was considered as the reference state and the distance energy was:

$$E(i) = -\ln(p_i) + \ln\left(\left(\frac{d_i}{d_N}\right)^\alpha p_N\right), i = 1, 2, \ldots, N \tag{15}$$

where p_i is the probability for the ith distance bin. N is the total number of bins, α is a constant ($= 1.57$) for distance-based normalization, and d_i is the distance for the ith distance bin. For orientation terms, i.e., angles and dihedral angles, the energy is similar but without normalization:

$$E(i) = -\ln(p_i) + \ln(p_N), i = 1, 2, \ldots, N \tag{16}$$

The potentials are smoothened by cubic spline and can be minimized by gradient-descent or its variants, e.g., L-BFGS algorithm [71].

The boundary between protein inter-residue interaction prediction and its main application, protein structure prediction became blurred and undetachable as several end-to-end protein structure prediction methods were proposed, which can directly output the coordinates of the protein. AlphaFold2 [72] was undoubtedly the representative model which ranked first in CASP14 for protein structure prediction [73,74]. In AlphaFold2, the structure of each residue was represented by a global rotation matrix and translation vector in the structure module of AlphaFold2. The side chain of each residue can also be recovered by the positional prediction of Chi angles. AlphaFold2 iteratively updates the residue representations by a geometry-aware attention operation, invariant point attention (IPA), which is an attention-based module that takes the geometry information in the previous conformation in the attention maps. The outputs of the IPA module for each residue are the updates of the rotation matrix (in quaternion) and the translation vector that could be applied to the previous ones, respectively. Once the structure is obtained, a novel loss function, i.e., frame-aligned point error (FAPE), was employed by AlphaFold2 to measure the errors between predicted atom positions and the ground truth coordinates under different alignments. The alignments are defined from the predicted frames, i.e., the rotation matrix and the translation vector, plus the extra frames from side chains. The loss function is invariant to rigid transformations, which makes the optimization of the neural network easier. AlphaFold2 also employed the cutting-edge attention mechanism as the backbone network of the Evoformer module in AlphaFold2. Compared to CNNs that can learn only local information, the attention mechanism has the ability to consider the information from all other variables with proper positional encoding. The Evoformer module also allows the attention-based transformer neural networks to attend arbitrarily over the full MSA, rather than using the (raw) coevolutionary analysis features. Compared to the conventional approaches, the new algorithm focuses on more relevant sequences and extracts richer information from the MSA. Some engineering efforts, e.g., recycling the training process, have also been employed and the resulting model achieves the state-of-the-art performance in CASP14 [72].

4. Discussion

Protein inter-residue interactions, e.g., contacts, can be considered as a simplified representation of protein structure. Predicting protein inter-residue interactions with high accuracy is of great importance for the protein structure prediction. This chapter presents a review of the development of algorithms for the residue inter-residue interaction prediction and the latest research results in this field. The rapid progress in the inter-residue interaction prediction, especially in the deep learning-based methods, has been driving the protein structure prediction problem towards its final solution. We summarize five milestones in this field:

1. Introduction of the global coevolution analysis. The global models, i.e., DCA methods, brought a systematic probabilistic model for MSA data using the Potts model. With a proper approximation, DCA models can efficiently eliminate the transition noise in local coevolutionary analysis methods.
2. Deep convolutional learning of the contact predictions. The prediction of protein residue contacts can be formalized with convolutional neural networks. Such a formula has shown a significant advantage over classical methods. The introduction of residual neural networks further improves the performance.
3. Use of the raw coevolutionary features. Directly feeding the deep learning model with raw coevolutionary correlations can avoid the loss of information. Using raw features without the post-processing corrects the problems of feature extraction procedures in previous models.
4. Consideration of richer geometry terms. Additional geometry terms can provide more information to help better fold the protein structures. In addition, multi-task learning can also contribute to each of the tasks.
5. An effective end-to-end learning model. The AlphaFold2 model is considered to nearly solve the protein structure prediction problem. The end-to-end learning can directly feed the structure error to the model. The powerful attention mechanism also contributes to the success of AlphaFold2.

Nevertheless, there are still a couple of problems that need to be addressed. First, the prediction of inter-residue interactions between domains and chains of multi-domain proteins and protein complexes needs a specific and targeted solution. Second, the research into providing interpretable information during the deep learning-based protein folding prediction require further efforts. We expect to see that the final goal of the protein structure prediction, that is, directly predicting high-accuracy protein structures with amino acid sequences alone, to be solved in the near future.

5. Acknowledgment

This work was supported in part by the National Natural Science Foundation of China (grant: 62072243) and the Natural Science Foundation of Jiangsu (grant: BK20201304)

References

[1]   Kendrew JC, et al. A three-dimensional model of the myoglobin molecule obtained by x-ray analysis. *Nature*, 1958. **181**(4610): 662–666.

[2]   Wüthrich K. The way to NMR structures of proteins. *Nature Structural & Molecular Biology*, 2001. **8**(11): 923.

[3]   Taylor KA and Glaeser RM. Electron diffraction of frozen, hydrated protein crystals. *Science*, 1974. **186**(4168): 1036–1037.

[4]   Baker D and Sali A. Protein structure prediction and structural genomics. *Science*, 2001. **294**(5540): 93–96.

[5]   Anfinsen CB. Principles that govern the folding of protein chains. *Science*, 1973. **181**(4096): 223–230.

[6]   Webb B, et al. Comparative protein structure modeling using MODELLER. *Current Protocols in Bioinformatics*, 2016. **54**(1): 5.6. 1–5.6. 37.

[7]   Schwede T, et al. SWISS-MODEL: An automated protein homology-modeling server. *Nucleic Acids Research*, 2003. **31**(13): 3381–3385.

[8]   Bowie JU, et al. A method to identify protein sequences that fold into a known three-dimensional structure. *Science*, 1991. **253**(5016): 164–170.

[9]   Jones D and Thornton J. Protein fold recognition. *Journal of Computer-Aided Molecular Design*, 1993. **7**(4): 439–456.

[10] Cocco S, et al. Inverse statistical physics of protein sequences: A key issues review. *Reports on Progress in Physics Physical Society*, 2018. **81**(3): 032601.

[11] Ekeberg M, et al. Improved contact prediction in proteins: Using pseudolikelihoods to infer Potts models. *Physical Review E*, 2013. **87**(1): 012707.

[12] Jones DT, et al. PSICOV: Precise structural contact prediction using sparse inverse covariance estimation on large multiple sequence alignments. *Bioinformatics*, 2011. **28**(2): 184–190.

[13] Martin L, et al. Using information theory to search for co-evolving residues in proteins. *Bioinformatics*, 2005. **21**(22): 4116–4124.

[14] Morcos F, et al. Direct-coupling analysis of residue coevolution captures native contacts across many protein families. *Proceedings of the National Academy of Sciences*, 2011. **108**(49): E1293–E1301.

[15] Seemayer S, et al. CCMpred — fast and precise prediction of protein residue–residue contacts from correlated mutations. *Bioinformatics*, 2014. **30**(21): 3128–3130.

[16] Cheng J and Baldi P. Improved residue contact prediction using support vector machines and a large feature set. *BMC Bioinformatics*, 2007. **8**(1): 1–9.

[17] Jones DT, *et al.* MetaPSICOV: Combining coevolution methods for accurate prediction of contacts and long range hydrogen bonding in proteins. *Bioinformatics*, 2014. **31**(7): 999–1006.

[18] Schneider M and Brock O. Combining physicochemical and evolutionary information for protein contact prediction. *Plos One*, 2014. **9**(10): e108438.

[19] Ma J, *et al.* Protein contact prediction by integrating joint evolutionary coupling analysis and supervised learning. *Bioinformatics*, 2015. **31**(21): 3506–3513.

[20] Adhikari B, *et al.* DNCON2: Improved protein contact prediction using two-level deep convolutional neural networks. *Bioinformatics*, 2017. **34**(9): 1466–1472.

[21] He B, *et al.* NeBcon: Protein contact map prediction using neural network training coupled with naïve Bayes classifiers. *Bioinformatics*, 2017. **33**(15): 2296.

[22] Buchan DW and Jones DT. Improved protein contact predictions with the MetaPSICOV2 server in CASP12. *Proteins: Structure, Function, and Bioinformatics*, 2018. **86**: 78–83.

[23] Senior AW, *et al.* Improved protein structure prediction using potentials from deep learning. *Nature*, 2020. **577**(7792): 706–710.

[24] Xu J. Distance-based protein folding powered by deep learning. *Proceedings of the National Academy of Sciences*, 2019. **116**(34): 16856–16865.

[25] Yang J, *et al.* Improved protein structure prediction using predicted interresidue orientations. *Proceedings of the National Academy of Sciences*, 2020. **117**(3): 1496–1503.

[26] Li Y, *et al.* Protein inter-residue contact and distance prediction by coupling complementary coevolution features with deep residual networks in CASP14. *Proteins: Structure, Function, and Bioinformatics*, 2021. **89**(12): 1911–1921.

[27] Moult J. A decade of CASP: Progress, bottlenecks and prognosis in protein structure prediction. *Current Opinion in Structural Biology*, 2005. **15**(3): 285–289.

[28] Kryshtafovych A, *et al.* CASP: A driving force in protein structure modeling. In *Introduction to Protein Structure Prediction: Methods and Algorithms*, Rangwala H & Karypis G (eds.), Wiley, 2010.

[29] Göbel U, *et al.* Correlated mutations and residue contacts in proteins. *Proteins: Structure, Function, and Bioinformatics*, 1994. **18**(4): 309–317.

[30] Yanofsky C, *et al.* Protein structure relationships revealed by mutational analysis. *Science*, 1964. **146**(3651): 1593–1594.

[31] Korber BT, *et al.* Covariation of mutations in the V3 loop of human immunodeficiency virus type 1 envelope protein: An information theoretic analysis. *Proceedings of the National Academy of Sciences*, 1993. **90**(15): 7176–7180.

[32] Kass I and Horovitz A. Mapping pathways of allosteric communication in GroEL by analysis of correlated mutations. *Proteins: Structure, Function, and Bioinformatics*, 2002. **48**(4): 611–617.

[33] Lapedes AS, *et al.* Correlated mutations in models of protein sequences: Phylogenetic and structural effects. *Lecture Notes-Monograph Series*, 1999. **33**: 236–256.

[34] Weigt M, *et al.* Identification of direct residue contacts in protein–protein interaction by message passing. *Proceedings of the National Academy of Sciences*, 2009. **106**(1): 67.

[35] Li Y, *et al.* ResPRE: High-accuracy protein contact prediction by coupling precision matrix with deep residual neural networks. *Bioinformatics*, 2019. **35**(22): 4647–4655.

[36] Kamisetty H, *et al.* Assessing the utility of coevolution-based residue–residue contact predictions in a sequence-and structure-rich era. *Proceedings of the National Academy of Sciences of the United States of America*, 2013. **110**(39): 15674–15679.

[37] Zhang H, *et al.* Predicting protein inter-residue contacts using composite likelihood maximization and deep learning. *BMC Bioinformatics*, 2018. **20**: 537.

[38] Wu F-Y. The potts model. *Reviews of Modern Physics*, 1982. **54**(1): 235.

[39] Liu DC and Nocedal J. On the limited memory BFGS method for large scale optimization. *Mathematical Programming*, 1989. **45**(1–3): 503–528.

[40] Hestenes MR and Stiefel E. Methods of conjugate gradients for solving linear systems. *NBS Washington, DC*, 1952. **49**: 409–435.

[41] Li Y, *et al.* Predicting residue–residue contacts using random forest models. Bioinformatics, 2011. **27**(24): 3379–3384.

[42] Golkov V, *et al.* Protein contact prediction from amino acid co-evolution using convolutional networks for graph-valued images. *Advances in Neural Information Processing Systems*, 2016. **29**: 4222–4230.

[43] Wang S, *et al.* Accurate De Novo prediction of protein contact map by ultra-deep learning model. *Plos Computational Biology*, 2017. **13**(1): e1005324.

[44] Liu Y, *et al.* Enhancing evolutionary couplings with deep convolutional neural networks. *Cell Systems*, 2018. **6**(1): 65.

[45] Jones DT and Kandathil SM. High precision in protein contact prediction using fully convolutional neural networks and minimal sequence features. *Bioinformatics*, 2018. **34**(19): 3308–3315.

[46] Li Y, *et al.* Deducing high-accuracy protein contact-maps from a triplet of coevolutionary matrices through deep residual convolutional networks. *Plos Computational Biology*, 2021. **17**(3): e1008865.

[47] Xu J, *et al.* Improved protein structure prediction by deep learning irrespective of co-evolution information. *Nature Machine Intelligence*, 2021. **3**: 601–609.

[48] Eickholt J and Cheng J. Predicting protein residue–residue contacts using deep networks and boosting. *Bioinformatics*, 2012. **28**(23): 3066–3072.

[49] Mcguffin LJ, *et al.* The PSIPRED protein structure prediction server. *Bioinformatics*, 2000. **16**(4): 404–405.

[50] Monastyrskyy B, *et al.* New encouraging developments in contact prediction: Assessment of the CASP11 results. *Proteins: Structure, Function, and Bioinformatics*, 2016. **84**(S1): 131–144.

[51] He K, *et al.* Deep residual learning for image recognition. *Proceedings of the IEEE Conference on Computer Vision and Pattern Recognition*, Las Vegas, NV, USA, IEEE, 2016.

[52] Schaarschmidt J, *et al.* Assessment of contact predictions in CASP12: Co-evolution and deep learning coming of age. *Proteins: Structure, Function, and Bioinformatics*, 2018. **86**: 51–66.

[53] Li Y, *et al.* Ensembling multiple raw coevolutionary features with deep residual neural networks for contact-map prediction in CASP13. *Proteins: Structure, Function, and Bioinformatics*, 2019. **87**(12): 1082–1091.

[54] Shen T, *et al.* When homologous sequences meet structural decoys: Accurate contact prediction by tFold in CASP14 — (tFold for CASP14 contact prediction). *Proteins: Structure, Function, and Bioinformatics*, 2021. **89**(12): 1901–1910.

[55] Yang J, *et al.* The I-TASSER Suite: Protein structure and function prediction. *Nature Methods*, 2015. **12**(1): 7–8.

[56] Zheng W, *et al.* Protein structure prediction using deep learning distance and hydrogen-bonding restraints in CASP14. *Proteins: Structure, Function, and Bioinformatics*, 2021. **89**(12): 1734–1751.

[57] Sadowski MI, *et al.* Direct correlation analysis improves fold recognition. *Computational Biology and Chemistry*, 2011. **35**(5): 323–332.

[58] Bartlett GJ and Taylor WR. Using scores derived from statistical coupling analysis to distinguish correct and incorrect folds in de-novo protein structure prediction. *Proteins: Structure, Function, and Bioinformatics*, 2008. **71**(2): 950–959.

[59] Taylor WR, *et al.* Protein topology from predicted residue contacts. *Protein Science*, 2012. **21**(2): 299–305.

[60] Sułkowska JI, *et al.* Genomics-aided structure prediction. *Proceedings of the National Academy of Sciences*, 2012. **109**(26): 10340.

[61] Marks DS, *et al.* Protein 3D structure computed from evolutionary sequence variation. *PloS One*, 2011. **6**(12): e28766.

[62] Brunger AT, *et al.* Crystallography & NMR system: A new software suite for macromolecular structure determination. *Acta Crystallographica Section D: Biological Crystallography*, 1998. **54**(5): 905–921.

[63] Kosciolek T and Jones DT. De novo structure prediction of globular proteins aided by sequence variation-derived contacts. *PloS One*, 2014. **9**(3): e92197.

[64] Ovchinnikov S, *et al.* Robust and accurate prediction of residue–residue interactions across protein interfaces using evolutionary information. *eLife*, 2014. **3**: e02030.

[65] Mortuza S, *et al.* Improving fragment-based ab initio protein structure assembly using low-accuracy contact-map predictions. *Nature Communications*, 2021. **12**(1): 1–12.

[66] Zheng W, *et al.* Folding non-homologous proteins by coupling deep-learning contact maps with I-TASSER assembly simulations. *Cell Reports Methods*, 2021. **1**: 100014.

[67] Zhang C, *et al.* Template-based and free modeling of I-TASSER and QUARK pipelines using predicted contact maps in CASP12. *Proteins: Structure, Function, and Bioinformatics*, 2018. **86**: 136–151.

[68] Zheng W, *et al.* Deep-learning contact-map guided protein structure prediction in CASP13. *Proteins: Structure, Function, and Bioinformatics*, 2019. **87**(12): 1149–1164.

[69] Abriata LA, *et al.* A further leap of improvement in tertiary structure prediction in CASP13 prompts new routes for future assessments. *Proteins: Structure, Function, and Bioinformatics*, 2019. **87**(12): 1100–1112.

[70] Zhou H and Zhou Y. Distance-scaled, finite ideal-gas reference state improves structure-derived potentials of mean force for structure selection and stability prediction. *Protein Science*, 2002. **11**(11): 2714–2726.

[71] Chaudhury S, *et al.* PyRosetta: A script-based interface for implementing molecular modeling algorithms using Rosetta. *Bioinformatics*, 2010. **26**(5): 689–691.

[72] Jumper J, *et al.* Highly accurate protein structure prediction with AlphaFold. *Nature*, 2021. **596**(7873): 583–589.

[73] Kinch LN, *et al.* Topology evaluation of models for difficult targets in the 14th round of the critical assessment of protein structure prediction (CASP14). *Proteins: Structure, Function, and Bioinformatics*, 2021. **89**(12): 1673–1686.

[74] Pereira J, *et al.* High-accuracy protein structure prediction in CASP14. *Proteins: Structure, Function, and Bioinformatics*, 2021. **89**(12): 1687–1699.

Chapter 8

Machine Learning for Intrinsic Disorder Prediction

Bi Zhao and Lukasz Kurgan*

*Department of Computer Science,
Virginia Commonwealth University,
Richmond, Virginia, United States*
*corresponding author: lkurgan@vcu.edu

Intrinsic disorder in proteins is manifested by regions that lack stable structure under physiological conditions. Proteins with disordered regions are common across all kingdoms of life. These proteins facilitate many essential cellular functions and contribute to dark proteomes. They are associated with a wide spectrum of human diseases and consequently, are considered as potent drug targets. Disordered regions have unique sequence signatures, making them distinguishable from structured protein sequences. Computational disorder prediction is a vibrant research area with over 40 years of history, which heavily depends on machine learning (ML) algorithms and innovations, such as meta learning and deep learning. We summarize a comprehensive collection of 73 ML-based disorder predictors, detail several most successful methods and survey related resources that predict disorder and disorder functions. We discuss historical trends in the development of disorder predictors, highlighting the shifting focus from traditional ML methods to meta-predictors, and most recently to the deep neural networks. We introduce a wide range of useful resources that support disorder and disorder function predictions including databases, webservers, and methods that provide quality assessment of disorder predictions. The availability of these numerous high-quality methods and resources ensures that the computational disorder predictions will continue to make substantial impact in key areas of research including rational drug design, structural genomics, and medicine.

1. Introduction

Intrinsic disorder in proteins is manifested by the presence of regions that lack stable structure under physiological conditions [1–3]. These intrinsically disordered regions (IDRs) are classified as native coils, native pre-molten globules, and native molten globules, signifying the fact that they can be disordered to different degrees [4,5]. More generally, IDRs can be seen as ensembles of interchanging conformations, with some regions being more expanded (native coils and native pre-molten globules) and others being more compact (native molten globule). Recent bioinformatics studies suggest that intrinsically disordered proteins (IDPs), i.e., proteins with IDRs, are common across all kingdoms of life, with particularly high levels in eukaryotes [6–9], and are distributed across different cellular compartments [10,11]. IDPs are engaged in numerous essential cellular functions that include molecular assembly and recognition, signal transduction, cell cycle regulation, chromosomal packing, transcription, and translation, to name but a few [8,12–22]. They also contribute to the dark proteomes, defined as the non-resolved parts of the protein structure space [23–25].

Dysfunction of IDPs is associated with a wide spectrum of human diseases [26–28]. Examples include the Alzheimer's and Parkinson's diseases, Down's syndrome, prion diseases and dementia [29]. Moreover, IDPs facilitate regulatory and signaling functions that frequently rely on molecular interactions and consequently, their misregulation and misinteractions are linked to cancers, diabetes, and cardiovascular diseases [30–33]. IDPs play also key roles in viral genomes [34–38]. Given the prevalence and importance of IDPs in context of human diseases, IDPs, such as α-synuclein, tau, p53 and BRCA-1 are attractive drug targets [26]. Moreover, novel approaches towards drug rational discovery efforts that target IDPs are being developed [39–43], further highlighting importance of this class of proteins.

Experimentally annotated IDRs can be collected from several databases, such as DisProt [44], PDB [45], IDEAL [46], DIBS [47], and MFIB [48]. The DisProt resource also offers functional annotations for the IDRs. While these resources provide access to very valuable experimental data, their size is relatively small compared to the hundreds of millions of protein sequences that are currently included the UniProt database [49]. For instance, the recent version 8.3 of DisProt covers approximately 2,000 proteins [44]. The substantial and rapidly growing annotation gap has motivated the development of computational predictors of IDRs [50–60]. The ability of these methods to generate accurate predictions of intrinsic disorder stems from the fact that IDRs have distinct sequence signatures compared to structured/ordered regions [61]. For instance, disordered regions typically have high net charge, low hydrophobicity and are depleted in aromatic residues when compared to their structured counterparts [5,62,63]. Availability of these computational methods has made substantial impact on the intrinsic disorder field, driving a rapid increase in the research on IDPs and IDRs [64].

Figure 1(a) shows the disorder annotations for the SIR3 protein from *Saccharomyces cerevisiae* (UniProt ID: P06701) collected from DisProt (DisProt ID: DP00533) [44]. This transcriptional repressor, that facilitates modulation of chromatin structure, includes one IDR (positions 216 to 549) that was shown to interact with proteins and DNA [65]. We use this protein to demonstrate disorder prediction generated flDPnn [66], one of the currently best methods that recently won the Critical Assessment of Intrinsic Protein Disorder (CAID) [67,68]. The disorder predictions were done at the residue level, which mean that they were produced for each amino acid in the input protein sequence. The prediction consists of a numeric propensity score (higher value denotes higher likelihood for disorder) and a binary value (disordered vs. structured). The binary prediction is typically generated from the putative propensities, where amino acids with propensities higher than a threshold are categorized as disordered and the remaining residues are predicted as structured. Figure 1(b) shows the putative propensity of disorder (black line) and binary annotation of disordered residues (green horizontal bar) produced by flDPnn; the corresponding threshold is denoted by the dashed horizontal line. We note that the location of the predicted disordered residues was in close agreement with the experimental data, demonstrating that disorder predictors can produce very accurate results.

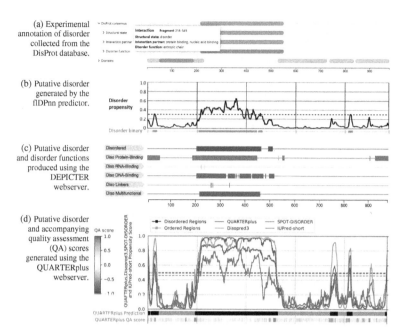

Figure 1.   Experimental and predicted disorder annotations for the SIR3 protein (UniProt ID: P06701, DisProt ID: DP00533). Panel A shows the experimental annotation of the disordered regions (positions 216 to 549) collected from DisProt (https://www.disprot.org/) [44,65]. Panel B shows prediction of disorder generated by the flDPnn method (http://biomine.cs.vcu.edu/servers/flDPnn/) [66] where the black plot gives the putative numerical propensity of disorder and horizonal green bar corresponds to putative disordered regions derived from the propensity values. Panel C is the prediction of disorder and disorder functions produced by the DEPICTER webserver (http://biomine.cs.vcu.edu/servers/DEPICTER/) [69] that include disorder predictions using a consensus of the IUPred2A [70] and SPOT-Disorder-Single methods [71] (gray horizonal bar), putative disordered protein binding (green horizonal bar) by consensus of DisoRDPbind [72,73], ANCHOR2 [70] and fMoRFpred [74], RNA binding (light blue horizonal bar) and DNA binding (dark blue horizonal bar) regions by DisoRDPbind [72,73], disordered linkers (pink horizonal bar) by DFLpred [75], and putative multifunctional/moonlighting disordered regions identified by DMRpred (violet horizonal bar) [76]. Panel D gives the disorder prediction (black and gray horizontal bar) generated by the consensus of SPOT-DISORDER-Single [71], DISOPRED3 [77] and IUPred-short [78] that is accompanied by the quality assessment (QA) scores produced by QUARTERplus (http://biomine.cs.vcu.edu/servers/QUARTERplus/) [79]. The QA scores (color-coded horizontal bar) quantify quality of the consensus disorder prediction, i.e., residues identified with green and yellow colors are more likely to be accurately predicted compared to predictions colored in orange or red.

The flDPnn predictor and a significant majority of other disorder predictors were developed using a machine learning (ML) approach [51,54,60]. This means that the developers of these methods used the available experimental disorder data to train predictive models using ML algorithms. Once trained and properly validated [68,80], the resulting predictive models can be used to produce accurate predictions of disordered residues and regions for the millions of sequences that lack disorder annotations, as illustrated in Figure 1(b). This chapter overviews disorder predictors that rely on the ML models. Here, we produce a comprehensive list of these predictors and discuss the underlying ML algorithms used. Moreover, we provide a detailed description of several most successful methods. Finally, we briefly survey other related resources, including webservers and databases, that facilitate prediction of disorder and disorder functions.

2. Overview of disorder predictors

Over 100 disorder predictors have been developed to date [51,54,60]. They were reviewed in about a dozen surveys [50–60]. The most recent review defines four distinct periods in the development of the disorder predictors [60]:

1. The *first-generation* predictors were developed between 1979 and 2001. The first ML-based method was developed by Romero, Obradovic, and Dunker in 1997 [81]. It relies on a shallow neural network that utilizes physical and chemical characteristics of the protein sequence as its inputs. Relatively few first-generation methods were developed.
2. The *second-generation* predictors date between 2002 and 2006. A significant event during this time period was the inclusion of disorder prediction assessment into the 5th Critical Assessment of Structure Prediction (CASP5) in 2003 [82]. This resulted in rapid popularization of this predictive area [50]. The second-generation predictors are typically based on somewhat simple predictive models, frequently relying on sequence scoring functions and shallow neural networks. The defining innovation was the use of evolutionary profiles produced from the position specific score matrix (PSSM) that is generated from the input protein sequences with the PSI-BLAST program [83,84]. Representative second-generation methods include GlobPlot [85], IUPred [78,86], PONDR predictors [87–90], DISOPRED [91], DisEMBL [92], and RONN [93].

3. The ***third-generation*** predictors were published between 2007 and 2015. One of the defining features of this period was the introduction of meta-predictors, which generated disorder prediction by combining results produced by several disorder predictors. Popular third-generation meta-predictors include MFDp [94], CSpritz [95], PONDR-FIT [96] and DisCoP [97,98]. We also note that the assessment of the disorder predictions continued biannually in the CASP7, CASP8, CASP9 and CASP10 experiments [80,99–101], resulting in a steady stream of new predictors.

4. The ***fourth-generation*** period has started in 2016. This new generation of disorder predictors is defined by the introduction and development of deep learning models. Our analysis reveals that about half of the fourth-generation disorder predictors, 11 out of 23, rely on the deep neural networks [60]. Representative deep learning-based methods include AUCpred [102], SPOT-Disorder [103], SPOT-Disorder-Single [71], SPOT-Disorder2 [104], rawMSA [105] and flDPnn[66]. The focus on designing novel deep network-based predictors culminated in their convincing success in the most recent CAID community assessment [67, 68].

As the above historical overview suggests, disorder predictors utilize a broad spectrum of predictive models. They are typically divided into three categories based on their predictive models [51,52,54,55,58]: (1) sequence scoring function-based methods; (2) machine learning approaches; and (3) meta-predictors. The ***sequence scoring*** function-based predictors use relatively simple additive and/or weighted functions, some of which are grounded in physical principles governing protein folding processes, to process information extracted from the input sequence and sequence-derived evolutionary information. Representative methods in this category include FoldIndex [106], IUPred [78,86], IUPred2A [70] and IUPred3 [107]. The ***machine learning*** methods utilize sophisticated predictive models that are trained from experimental data using a variety of ML algorithms, such as support vector machine, conditional random field, random forest and a variety of neural networks. Well-known ML methods include DisEMBL [92], DISOPRED [91], PONDR [90], PrDOS [108], DISOPRED3 [77], flDPnn [66], SPOT-Disorder2 [104], RawMSA [105] and AUCpred [102]. The ***meta-predictors*** utilize two or more disorder predictions as inputs to re-predict disorder to improve predictive performance when compared to the input predictions. Several empirical studies show that well-designed meta-predictors produce such improvements [97,109–111]. Illustrative meta-predictors include metaPrDOS [112], MFDp [94], Cspritz [95], MFDp2 [113,114], disCoP [97,98] and MobiDB-lite [109]. Moreover, some meta-predictors use ML models to process inputs, which means that they belong to both categories. Corresponding examples include metaPrDOS [112] and MFDp [94].

3. Disorder prediction using machine learning

In this section, we will focus on the machine learning disorder predictors. We searched for these methods using a list of methods that participated in community assessments [68,80,82,99–101,115] and a comprehensive selection of 13 previously published surveys, which also include comparative studies [50–55,58,60,116–119]. This extensive search produced a total of 73 ML-based disorder predictors that are summarized in Table 1. This table identifies when and where these methods were published and reviews the ML models which were utilized.

Table 1. Summary of 73 predictors of intrinsic disorder that use machine learning models. The methods are sorted in the chronological order of their publication.

Disorder Predictor	Year Published	Ref.	Machine Learning Algorithms Used
Predictor by Dunker *et al.*	1997	[81]	Shallow neural network
PONDR CaN-XT	1997	[120]	Shallow neural network
PONDR XL1	1997	[120]	Shallow neural network
PONDR VL-XT	2001	[89]	Shallow neural network
DisEMBL-REM465	2003	[92]	Shallow neural network
DisEMBL-HL	2003	[92]	Shallow neural network
DisEMBL-COIL	2003	[92]	Shallow neural network
DISOPRED	2003	[91]	Shallow neural network
DISOPRED2	2004	[121]	Support vector machine + Shallow neural network
DISpro	2005	[122,123]	Shallow neural network
PONDR VL3	2005	[90]	Shallow neural network
PONDR VL3H	2005	[90]	Shallow neural network

(Continued)

Table 1. (*Continued*)

Disorder Predictor	Year Published	Ref.	Machine Learning Algorithms Used
PONDR VL3E	2005	[90]	Shallow neural network
RONN (JRONN)	2005	[93]	Shallow neural network
PONDR VSL1	2005	[87]	Regression
PROFbval	2006	[124]	Shallow neural network
PONDR VSL2B	2006	[87,88]	Support vector machine
PONDR VSL2P	2006	[87,88]	Support vector machine
Wiggle	2006	[125]	Support vector machine
Distill	2006	[126]	Shallow neural network
Spritz (Spritz3)	2006	[127]	Support vector machine
DisPSSMP	2006	[128]	Radial basis function networks
iPDA (DisPSSMP2)	2007	[129]	Radial basis function networks
POODLE-L	2007	[130]	Support vector machine
POODLE-S	2007	[131]	Support vector machine
POODLE-W	2007	[132]	Nearest neighbor
NORSnet	2007	[133]	Shallow neural network
PrDOS (PrDOS2)	2007	[108]	Support vector machine
Pdisorder	2007	N/A	Shallow neural network
UCON	2007	[133]	Shallow neural network
OnD-CRF (OnD-CRF2)	2008	[134]	Conditional random field
metaPrDOS (metaPrDOS2)	2008	[112]	Support vector machine
PreDisorder	2009	[135]	Shallow neural network
NN-CDF	2009	[136]	Shallow neural network
DRaai	2009	[137]	Random forest
MD	2009	[138]	Shallow neural network
UPforest-L	2009	[139]	Random forest
POODLE-I	2010	[140]	Support vector machine
MFDp	2010	[94]	Support vector machine
PONDR FIT	2010	[96]	Shallow neural network
Cspritz	2011	[95]	Shallow neural network

Table 1. (*Continued*)

Disorder Predictor	Year Published	Ref.	Machine Learning Algorithms Used
Espritz-D	2012	[141]	Shallow neural network
Espritz-N	2012	[141]	Shallow neural network
Espritz-X	2012	[141]	Shallow neural network
SPINE-D	2012	[142]	Shallow neural network
DNdisorder	2013	[143]	Deep neural network (restricted Boltzmann machine)
MFDp2	2013	[113,114]	Support vector machine
disCoP	2014	[97,98]	Regression
DynaMine	2014	[144,145]	Regression
PON-Diso	2014	[146]	Random forest
s2D-1	2015	[147]	Shallow neural network
s2D-2	2015	[147]	Shallow neural network
DISOPRED3	2015	[77]	Support vector machine + Shallow neural network + Nearest neighbor
DisoMCS	2015	[148]	Conditional random field
DeepCNF-D	2015	[149]	Deep neural network (convolutional) + Conditional random field
AUCpred	2016	[102]	Deep neural network (convolutional) + Conditional random field
AUCpred-np	2016	[102]	Deep neural network (convolutional) + Conditional random field
DisPredict (DisPredict2)	2016	[150]	Support vector machine
SPOT-Disorder1	2017	[103]	Deep neural network (recurrent)
SPOT-Disorder-Single	2018	[71]	Deep neural network (hybrid: convolutional + recurrent)
Predictor by Zhao and Xue	2018	[151]	Decision tree + Shallow neural network
IDP-CRF	2018	[152]	Conditional random field
Spark-IDPP	2019	[153]	Support vector machine + Shallow neural network
IDP-FSP	2019	[154]	Conditional random field

(*Continued*)

Table 1. (*Continued*)

Disorder Predictor	Year Published	Ref.	Machine Learning Algorithms Used
rawMSA	2019	[105]	Deep neural network (hybrid: convolutional + recurrent)
SPOT-Disorder2	2019	[104]	Deep neural network (hybrid: convolutional + recurrent)
DisoMine	2020	N/A	Deep neural network (recurrent)
ODiNPred	2020	[155]	Shallow neural network
IDP-Seq2Seq	2020	[156]	Deep neural network (recurrent)
flDPnn	2021	[66]	Deep neural network (feed-forward)
flDPlr	2021	[66]	Regression
RFPR-IDP	2021	[107]	Deep neural network (hybrid: convolutional + recurrent)
Metapredict	2021	[157]	Deep neural network (recurrent)

Table 1 reveals that the first ML predictor was developed in 1997. These methods were developed at a relatively steady pace over the subsequent years. On average, close to three methods were developed annually, with 11 predictors published in the last three years. Figure 2 illustrates these trends. We observed a sharp spike in the development efforts between 2005 and 2007 when 21 methods were released. We speculate that this was fueled by the inclusion of the disorder prediction into CASP5 and CASP6 experiments [82,115], which correspondingly grew from 6 participating methods in CASP5 to 20 in CASP6. Figure 2 also highlights selected popular and/or well-performing methods that were developed over the years (in chronological order): PONDR XL1 [120] (1997), DisEMBL [92] and DISOPRED [91] (2003), PONDR VSL2B [87,88] (2006), PrDOS [108] (2007), MFDp [94] (2010), Espritz [141] (2012), DISOPRED3 [77] (2015), AUCpred [102] (2016), SPOT-Disorder-Single [71] (2018), SPOT-Disorder2 [104] and RawMSA [105] (2019), and flDPnn [66] (2021).

Table 1 shows that the 73 disorder predictors rely on a variety of different ML algorithms. Some of these predictors use an ensemble of models that were produced by multiple different ML algorithms. A case in point is DISOPRED3 that combines support vector machine, shallow neural network and nearest neighbor models [77]. We break down these algorithms

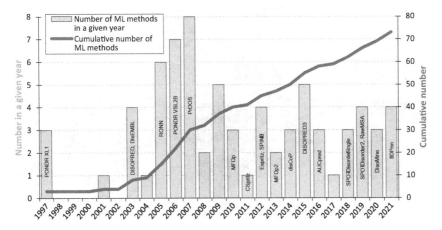

Figure 2. Timeline of the development of the machine learning-based predictors of intrinsic disorder. The green bars show the number of predictors developed in a given year. Selected popular and/or well-performing predictors are named inside the bars. The red line is the cumulative number of the predictors.

in Figure 3. By far, the most popular choices are shallow neural networks and support vector machines, which when combined, account for 60% of all ML algorithms used. The shallow neural networks were particularly popular in early years, with nearly all predictors between 1997 and 2005 utilizing these models (Table 1). The support vector machines were commonly used between 2006 and 2010, when 10 out of 25 predictors applied these models. Interestingly, the most popular ML models in recent years are deep neural networks, which were used by 6 out of 11 disorder predictors that were published since 2019. Deep neural networks differ from the shallow networks by inclusion of multiple hidden layers. The deep networks also often apply to more advanced neuron types, and more sophisticated architectures and connection patterns. The deep learning-based disorder predictors utilize a diverse collection of architecture types that cover feed-forward, recurrent, convolutional, and hybrid topologies (Fig. 3). The latter typically combines convolutional and recurrent topologies (Table 1).

The drive to use the deep networks partly stems from their popularity in related areas of protein bioinformatics [158,159]. Some examples include prediction of protein function [160,161], residue contacts [162–166], residue distances [167], binding residues [168–171], crystallization

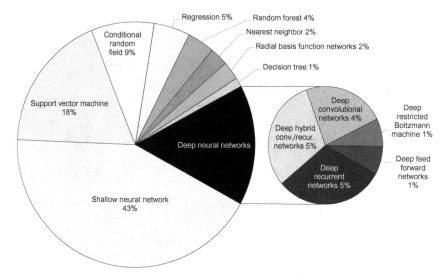

Figure 3. The breakdown of different types of machine learning models used by the disorder predictors. The smaller, gray-scale pie chart shows various types of the deep neural networks.

success [172], solvent accessibility [173,174], secondary structure [174–176], posttranslational modification sites [177], and substrates and cleavage sites [178]. Another key factor contributing to the popularity of the deep learners is their success in recent assessments of disorder predictions [58,68,116]. For example, all of the best-performing methods that participated in the most recent CAID competition [67,68], which include flDPnn [66], SPOT-Disorder2 [104], RawMSA [105] and AUCpred [102], rely on the deep networks. Interestingly, these methods use diverse network architectures including convolutional topology by AUCpred, a hybrid of convolutional and recurrent by SPOT-Disorder2 and RawMSA, and the feed-forward topology by flDPnn.

4. Selected machine learning-based disorder predictors

Several large-scale comparative studies that assessed predictive performance of the intrinsic disorder predictors were published in the past [68,80,82,99–101,115–119]. Among them, we highlight the community assessments where disorder predictors were evaluated based on blind test datasets (i.e., datasets that were not available to the authors of predictors) by an independent group of assessors who do not take part in the competitions. This arguably ensures that the evaluation is fair across the participating predictors. The community assessments include (in chronological order) CASP5 [82], CASP6 [115], CASP7 [99], CASP8 [100], CASP9 [101], CASP10 [80] and CAID [68]. The largest and most recent assessments reported results for 28 disorder predictors in CASP10 [80] and 32 in CAID [68]. The top three methods in these two assessments utilize ML-based models. They include PrDOS [108], DISOPRED3 [77] and MFDp [94] from CASP10 [80]; and flDPnn [66]; SPOT-Disorder2 [104] and AUCpred [102] from CAID [66,68]. Following, we provide detailed description of these six best-performing ML-based disorder predictors.

4.1 *PrDOS*

PrDOS was developed by Dr. Ishida at the University of Tokyo and Dr. Kinoshita at the Tohoku University. This method integrates two predictors: one based on an evolutionary profile of the protein sequence and the other based on the protein structure templates. The result is computed as a weighted average of the outputs from the two predictors. Comparative analysis reveals that this method outperformed all other disorder predictors in the CASP10 experiment [80].

Predictive model: Combination of SVM and the sequence alignments of sequences from PDB structures.

Citation data: 726 citations according to Google Scholar (July 2022); published in 2007 [108].

Availability: Webserver at https://prdos.hgc.jp/cgi-bin/top.cgi.

4.2 *MFDp*

The MFDp predictor was designed by Dr. Kurgan's team, which currently resides at the Virginia Commonwealth University. This is a meta-predictor that combines putative disorder predicted by four complementary methods: IUPred-long [78], IUPred-short [78], DISOPRED2 [121] and DISOclust [179]. The main innovation behind MFDp is the design of the consensus that relies on three SVM models that predict different sizes of IDRs (long, short, and all-size). The subsequent version of this predictor,

MFDp2 [113], calibrates the outputs from the original MFDp using disorder content predictions generated by DisCon [180]. MFDp was ranked third in CASP10 [80], and did not participate in CAID.

Predictive model: Ensemble of three SVMs.

Citation data: 187 citations according to Google Scholar (July 2022); published in 2010 [94].

Availability: MFDp webserver at http://biomine.cs.vcu.edu/servers/ MFDp/; MFDp2 webserver at http://biomine.cs.vcu.edu/servers/MFDp2/.

4.3 *DISOPRED3*

DISOPRED3 was released by Dr. Jones's lab at the University College London. The two prior versions of this popular disorder predictor were released in 2003 (DISOPRED) and 2004 (DISOPRED2) [91,121]. DISOPRED3 integrates its own disorder predictor with outputs generated by DISOPRED2 and two ML-based predictions trained using long IDRs. This method also identifies protein binding sites in the putative IDRs using an additional SVM-based model. DISOPRED3 is a part of the PSIPRED workbench [181]. DISOPRED3 was ranked second in CASP10 and among top 10 predictors in CAID [68,80].

Predictive model: SVM, shallow neural network and nearest neighbor for disorder prediction. SVM for prediction of the disordered protein binding.

Citation data: 642 citations according to Google Scholar (July 2022); published in 2015 [77].

Availability: Webserver at http://bioinf.cs.ucl.ac.uk/psipred/; standalone code at http://bioinfadmin.cs.ucl.ac.uk/downloads/DISOPRED/.

4.4 *AUCpred*

AUCpred was created by Dr. Xu's group at the Toyota Technological Institute in Chicago, in collaboration with researchers from the University of Chicago. One of the innovations was that the underlying predictive model that relies on deep convolutional network was trained to maximize area under the ROC curve (AUC). This method offers two options: prediction with or without the use of the evolutionary profile. The latter prediction is much faster (10 seconds for an average length protein chain) and slightly less accurate than the outputs of the former design. AUCpred was ranked among the top three methods in the CAID experiment [66,68].

Predictive model: Deep convolutional neural network combined with conditional random fields.

Citation data: 81 citations according to Google Scholar (July 2022); published in 2016 [102].

Availability: Webserver at http://raptorx.uchicago.edu/StructureProperty Pred/predict/.

4.5 *SPOT-Disorder2*

The SPOT-Disorder2 method was released by Prof. Zhou's lab, which is now located at the Shenzhen Bay Laboratory. The earlier SPOT-Disorder version of this method was published in 2017 [103]. SPOT-Disorder2 surpasses SPOT-Disorder by integrating several network topologies rather than using one long short-term memory bidirectional recurrent neural network (LSTM-BRNN). The second version utilizes evolutionary profile and predictions from SPOT-1D method [182] as inputs. Besides predicting disorder, SPOT-Disorder2 outputs putative semi-disordered regions which can be used to identify molecular recognition features (MoRFs) [74,183]. CAID results place this method among the top three predictors, however, SPOT-Disorder2 suffers long runtime when compared to its close challengers [66,68].

Predictive model: Hybrid deep neural network that combines convolutional and recurrent topologies.

Citation data: 65 citations according to Google Scholar (July 2022); published in 2019 [104].

Availability: Webserver at https://sparks-lab.org/server/spot-disorder2/; standalone code at https://sparks-lab.org/downloads/.

4.6 *flDPnn*

The flDPnn predictor was developed in Prof. Kurgan's lab, which is currently located at the Virginia Commonwealth University, in collaboration with researchers from the Nankai University. The defining features of this method include innovative predictive inputs that incorporate extended sequences profile and protein-level feature encoding, ability to predict selected functions for the putative IDRs that it predicts, and low runtime. The functions covered by flDPnn include disordered linkers and interaction with proteins, DNA and RNA. FlDPnn produces the disorder and disorder function predictions very quickly, in about five seconds for an average size protein. Comparative analysis in CAID reveals that flDPnn ranks among the top three predictors in that experiment and that it is at least an order of magnitude faster than these competitors [66,68].

Predictive model: Deep feed-forward neural network for the disorder prediction. Ensemble of four random forest models for the prediction of disorder functions.

Citation data: 31 citations according to Google Scholar (July 2022); published in 2021 [66].

Availability: Webserver at http://biomine.cs.vcu.edu/servers/flDPnn/; standalone code at https://gitlab.com/sina.ghadermarzi/fldpnn.

5. Related resources

Nowadays, users are provided with access to webservers and implementations for many disorder predictors. However, making these predictions could be inconvenient, particularly in scenarios where users would like to secure multiple disorder predictions for the same protein or predictions for a large number of proteins. A convenient alternative to making prediction using individual methods is to collect pre-computed predictions from one of the currently available databases: D^2P^2 (Database of Disorder Protein Predictions; https://d2p2.pro/) [184], MobiDB (https://mobidb.bio.unipd.it/) [185,186], and DescribePROT (http://biomine.cs.vcu.edu/servers/DESCRIBEPROT/) [187]. Each of these three databases provides instantaneous access to results generated by several disorder predictors for large datasets of proteins ranging from 1.35 million proteins from 83 genomes in DescribePROT, 10.43 million proteins from 1,765 genomes in D^2P^2, to 219.74 million proteins in MobiDB. One of the key features of the MobiDB resource is the inclusion of the consensus disorder prediction produced by MobiDB-lite method [109] as well as curated experimental annotations of disorder that are collected from several sources including DisProt [44], IDEAL [46], ELM [188], MFIB [48], DIBS [47], FuzDB [189] and PhasePro [190]. While D^2P^2 and MobiDB primarily focus on the disorder predictions, DescribePROT also provides predictions of other structural and functional characteristics of proteins. These include putative solvent accessibility predicted by ASAquick [191], putative disordered linked by DFLpred [75], putative protein-binding residues by DisoRDPbind [72,73,192], MoRFChibi [193] and SCRIBER [194,195], putative DNA-binding and RNA-binding residues by DisoRDPbind and DRNApred [196], secondary structure by PSIPRED [197], signal peptides by SignalP [198,199], disorder by PONDR VSL2B [87,88], and alignment profiles produced by MMseqs2 [200,201]. In total, the most recent release 1.4 of DescribePROT provides access to over 7.8 billion residue-level predictions.

However, we note that users must still rely on disorder predictors when they want to predict sequences that are not included in a given database. A very useful resource in that case is DEPICTER (DisorderEd PredictIon CenTER; http://biomine.cs.vcu.edu/servers/DEPICTER/) [69]. This unique webserver generates a comprehensive collection of disorder and disorder function predictions. It provides consensus for disorder predictions using results produced by the fast IUPred-short [78], IUPred-long [78] and SPOT-Disorder-Single [71] methods. These predictions are accompanied by the disordered linker predictions made by DFLpred [75], putative disordered regions that interact with proteins and nucleic acids that are predicted by combining results of fMoRFpred [74], DisoRDPbind [72] and ANCHOR2 [70], and putative disordered multifunctional (moonlighting) regions generated by DMRpred [76]. Figure 1(c) shows results computed by DEPICTER for the SIR3 protein (UniProt ID: P06701; DisProt ID: DP00533). DEPICTER suggests that the putative IDRs (grey horizontal bar) is multifunctional (violet horizontal bar), and that it binds DNA (dark blue horizontal bar) and proteins (green horizontal bar). These predictions are in good agreement with the experimental disorder annotations shown in Figure 1(a).

Another recent advancement is the development of methods that provide interpretable residue-level quality assessment scores [202]: QUARTER (http://biomine.cs.vcu.edu/servers/QUARTER/) [203,204] and QUARTERplus (http://biomine.cs.vcu.edu/servers/QUARTERplus/) [79]. The scores produced by these tools can be used to identify regions where the quality of the disorder predictions generated by several popular methods, such as DISOPRED3, IUPred, PONDR VSL2B and disEMBL, is high. QUARTERplus relies on a deep convolutional neural network to make accurate, consensus-based disorder predictions accompanied by the quality assessment scores, which allow the users to easily pinpoint which disorder predictions are more trustworthy. We illustrate this in Figure 1(d) where the disorder predictions, shown using the black and gray horizontal bar, are annotated with the quality assessment scores, i.e., color-coded horizontal bar, where residues identified with green and yellow colors are more likely to be accurately predicted compared to predictions colored in orange or red. We note that the short putative IDRs that were identified at both sequence termini are marked in red/orange, which suggests that these predictions are likely incorrect. In contrast, the long putative IDRs in the middle of the sequence is marked in yellow/green, suggesting that this disorder prediction is likely accurate. These color-coded annotations concur with the experimental disorder annotations from Figure 1(a), signifying the usefulness of the quality assessment scores that QUARTERplus produces.

We also briefly overview computational predictors of disorder functions. There are well over a dozen predictors of disordered protein-binding regions [205], including recent methods, such as OPAL+ [206], MoRFPred_en [207], FLIPPER [208] and SPOT-MoRF [209]. Moreover, users can utilize DisoRDPbind [72,73,192] and DeepDISOBind [169] to identify putative IDRs that interact with DNA, RNA and proteins, as well as DisoLipPred [170] that predicts for the IDRs that bind lipids. There are also two methods that predict disordered linker regions, DFLpred [75] and IPOD [210]. We stress that the prediction of binding IDRs (i.e, IDRs that bind proteins, DNA, RNA and small ligands) was recently assessed in the CAID experiment [68]. CAID found that ANCHOR2 [70], DisoRDPbind [72,73,192] and MoRFCHiBi [193] are the most accurate predictors of binding IDRs. However, this assessment concluded that "*disordered binding regions remain hard to predict*", suggesting that disorder function predictors should be further improved [68]. Interested readers can find more details in several recent surveys on this topic [51,205,211].

6. Summary

Disorder prediction is a vibrant and very active research area that relies heavily on ML models and innovations, including meta learning and deep learning. We identify 73 ML-based disorder predictors that were developed in the last four decades. We find that the original focus on traditional ML methods, such as shallow neural networks and support vector machines, which dominated this field until mid-2000s has shifted towards the meta-predictors in late 2010s. This was subsequently followed by a transition to the deep neural networks in around 2015. Given the success of deep learners in the recent CAID experiment [66,68] and their popularity in the broader protein bioinformatics area [158,159], we anticipate that the development of deep neural network-based disorder predictors will continue in the near future. We also highlight the availability of many useful resources that support disorder and disorder function predictions including databases such as D^2P^2 [184], MobiDB [185,186], and DescribePROT [187], comprehensive webservers such as DEPICTER [69], and methods that provide quality assessment of disorder predictions such as QUARTERplus [79]. The easy access to these numerous methods and resources ensures that the computational disorder predictions will continue to make substantial impact in other key areas of research, such as rational drug design [42,43,212–214], structural genomics [24,92,215], study of human diseases [216], and systems medicine [59].

7. Funding

This work was funded in part by the National Science Foundation (grants 2146027 and 2125218) and the Robert J. Mattauch Endowment funds to L.K.

References

[1] Dunker AK, *et al.* What's in a name? Why these proteins are intrinsically disordered. *Intrinsically Disordered Proteins*, 2013. **1**(1): e24157.

[2] Lieutaud P, *et al.* How disordered is my protein and what is its disorder for? A guide through the "dark side" of the protein universe. *Intrinsically Disordered Proteins*, 2016. **4**(1): e1259708.

[3] Oldfield CJ, *et al.* Introduction to intrinsically disordered proteins and regions. In *Intrinsically Disordered Proteins*, Salvi N (ed.), Academic Press, 2019. 1–34.

[4] Uversky VN. Unusual biophysics of intrinsically disordered proteins. *Biochimica et Biophysica Acta*, 2013. **1834**(5): 932–951.

[5] Oldfield CJ, *et al.* Introduction to intrinsically disordered proteins and regions. In *Intrinsically Disordered Proteins*, Salvi N (ed.), Academic Press, 2019. 1–34.

[6] Xue B, *et al.* Orderly order in protein intrinsic disorder distribution: Disorder in 3500 proteomes from viruses and the three domains of life. *Journal of Biomolecular Structure and Dynamics*, 2012. **30**(2): 137–149.

[7]   Peng Z, *et al.* Exceptionally abundant exceptions: Comprehensive characterization of intrinsic disorder in all domains of life. *Cellular and Molecular Life Sciences*, 2015. **72**(1): 137–151.

[8]   Wang C, *et al.* Disordered nucleiome: Abundance of intrinsic disorder in the DNA- and RNA-binding proteins in 1121 species from Eukaryota, Bacteria and Archaea. *Proteomics*, 2016. **16**(10): 1486–1498.

[9]   Peng Z, *et al.* Genome-scale prediction of proteins with long intrinsically disordered regions. *Proteins*, 2014. **82**(1): 145–158.

[10] Zhao B, *et al.* IDPology of the living cell: Intrinsic disorder in the subcellular compartments of the human cell. *Cellular and Molecular Life Sciences*, 2021. **78**(5): 2371–2385.

[11] Meng F, *et al.* Compartmentalization and functionality of nuclear disorder: Intrinsic disorder and protein-protein interactions in intra-nuclear compartments. *International Journal of Molecular Sciences*, 2015. **17**(1): 24.

[12] Uversky VN, *et al.* Showing your ID: Intrinsic disorder as an ID for recognition, regulation and cell signaling. *Journal of Molecular Recognition*, 2005. **18**(5): 343–384.

[13] Liu J, *et al.* Intrinsic disorder in transcription factors. *Biochemistry*, 2006. **45**(22): 6873–6888.

[14] Peng Z, *et al.* A creature with a hundred waggly tails: Intrinsically disordered proteins in the ribosome. *Cellular and Molecular Life Sciences*, 2014. **71**(8): 1477–1504.

[15] Babu MM. The contribution of intrinsically disordered regions to protein function, cellular complexity, and human disease. *Biochemical Society Transactions*, 2016. **44**(5): 1185–1200.

[16] Peng ZL, *et al.* More than just tails: Intrinsic disorder in histone proteins. *Molecular Biosystems*, 2012. **8**(7): 1886–1901.

[17] Xie HB, *et al.* Functional anthology of intrinsic disorder. 3. Ligands, post-translational modifications, and diseases associated with intrinsically disordered proteins. *Journal of Proteome Research*, 2007. **6**(5): 1917–1932.

[18] Xie HB, *et al.* Functional anthology of intrinsic disorder. 1. Biological processes and functions of proteins with long disordered regions. *Journal of Proteome Research*, 2007. **6**(5): 1882–1898.

[19] Staby L, *et al.* Eukaryotic transcription factors: Paradigms of protein intrinsic disorder. *Biochemical Journal*, 2017. **474**(15): 2509–2532.

[20] Hu G, *et al.* Functional analysis of human hub proteins and their interactors involved in the intrinsic disorder-enriched interactions. *International Journal of Molecular Sciences*, 2017. **18**(12): 2761.

[21] Na I, *et al.* Autophagy-related intrinsically disordered proteins in intra-nuclear compartments. *Molecular Biosystems*, 2016. **12**(9): 2798–2817.

[22] Peng Z, *et al.* Resilience of death: Intrinsic disorder in proteins involved in the programmed cell death. *Cell Death & Differentiation*, 2013. **20**(9): 1257–1267.

[23] Bhowmick A, *et al.* Finding our way in the dark proteome. *Journal of the American Chemical Society*, 2016. **138**(31): 9730–9742.

[24] Hu G, *et al.* Taxonomic landscape of the dark proteomes: Whole-proteome scale interplay between structural darkness, intrinsic disorder, and crystallization propensity. *Proteomics*, 2018. e1800243.

[25] Kulkarni P and Uversky VN. Intrinsically disordered proteins: The dark horse of the dark proteome. *Proteomics*, 2018. **18**: 21–22.

[26] Uversky VN, *et al.* Intrinsically disordered proteins in human diseases: Introducing the D2 concept. *Annual Review of Biophysics*, 2008. **37**: 215–246.

[27] Midic U, *et al.* Protein disorder in the human diseasome: Unfoldomics of human genetic diseases. *BMC Genomics*, 2009. **10**(Suppl 1): S12.

[28] Xie H, *et al.* Functional anthology of intrinsic disorder. 3. Ligands, posttranslational modifications, and diseases associated with intrinsically disordered proteins. *Journal of Proteome Research*, 2007. **6**(5): 1917–1932.

[29] Uversky VN. The triple power of D(3): Protein intrinsic disorder in degenerative diseases. *Frontiers in Bioscience (Landmark Edition)*, 2014. **19**: 181–258.

[30] Uversky VN, *et al.* Unfoldomics of human diseases: Linking protein intrinsic disorder with diseases. *BMC Genomics*, 2009. **10**(Suppl 1): S7.

[31] Cheng Y, *et al.* Abundance of intrinsic disorder in protein associated with cardiovascular disease. *Biochemistry*, 2006. **45**(35): 10448–10460.

[32] Iakoucheva LM, *et al.* Intrinsic disorder in cell-signaling and cancer-associated proteins. *Journal of Molecular Biology*, 2002. **323**(3): 573–584.

[33] Al-Jiffri OH, *et al.* Intrinsic disorder in biomarkers of insulin resistance, hypoadiponectinemia, and endothelial dysfunction among the type 2 diabetic patients. *Intrinsically Disordered Proteins*, 2016. **4**(1): e1171278.

[34] Fan X, *et al.* The intrinsic disorder status of the human hepatitis C virus proteome. *Molecular Biosystems*, 2014. **10**(6): 1345–1363.

[35] Xue B, *et al.* Protein intrinsic disorder as a flexible armor and a weapon of HIV-1. *Cellular and Molecular Life Sciences*, 2012. **69**(8): 1211–1259.

[36] Xue B and Uversky VN. Intrinsic disorder in proteins involved in the innate antiviral immunity: Another flexible side of a molecular arms race. *Journal of Molecular Biology*, 2014. **426**(6): 1322–1350.

[37] Alshehri MA, *et al.* On the prevalence and potential functionality of an intrinsic disorder in the MERS-CoV proteome. *Viruses*, 2021. **13**(2): 339.

[38] Xue B, *et al.* Structural disorder in viral proteins. *Chemical Reviews*, 2014. **114**(13): 6880–6911.

[39] Cheng Y, *et al.* Rational drug design via intrinsically disordered protein. *Trends in Biotechnology*, 2006. **24**(10): 435–442.

[40] Dunker AK and Uversky VN. Drugs for 'protein clouds': Targeting intrinsically disordered transcription factors. *Current Opinion in Pharmacology*, 2010. **10**(6): 782–788.

[41] Ghadermarzi S, *et al.* Sequence-derived markers of drug targets and potentially druggable human proteins. *Frontiers in Genetics*, 2019. **10**: 1075.

[42] Hu G, *et al.* Untapped potential of disordered proteins in current druggable human proteome. *Current Drug Targets*, 2016. **17**(10): 1198–1205.

[43] Uversky VN. Intrinsically disordered proteins and novel strategies for drug discovery. *Expert Opinion on Drug Discovery*, 2012. **7**(6): 475–488.

[44] Hatos A, *et al.* DisProt: Intrinsic protein disorder annotation in 2020. *Nucleic Acids Research*, 2020. **48**(D1): D269–D276.

[45] Le Gall T, *et al.* Intrinsic disorder in the protein data bank. *Journal of Biomolecular Structure and Dynamics*, 2007. **24**(4): 325–342.

[46] Fukuchi S, *et al.* IDEAL in 2014 illustrates interaction networks composed of intrinsically disordered proteins and their binding partners. *Nucleic Acids Research*, 2014. **42**(D1): D320–D325.

[47] Schad E, *et al.* DIBS: A repository of disordered binding sites mediating interactions with ordered proteins. *Bioinformatics*, 2018. **34**(3): 535–537.

[48] Ficho E, *et al.* MFIB: A repository of protein complexes with mutual folding induced by binding. *Bioinformatics*, 2017. **33**(22): 3682–3684.

[49] The UniProt Consortium. UniProt: The universal protein knowledgebase in 2021. *Nucleic Acids Research*, 2021. **49**(D1): D480–D489.

[50] He B, *et al.* Predicting intrinsic disorder in proteins: An overview. *Cell Research*, 2009. **19**(8): 929–949.

[51] Meng F, *et al.* Comprehensive review of methods for prediction of intrinsic disorder and its molecular functions. *Cellular and Molecular Life Sciences*, 2017. **74**(17): 3069–3090.

[52] Meng F, *et al.* Computational prediction of intrinsic disorder in proteins. *Current Protocols in Protein Science*, 2017. **88**: 2.16.1–2.16.14

[53] Deng X, *et al.* A comprehensive overview of computational protein disorder prediction methods. *Molecular Biosystems*, 2012. **8**(1): 114–121.

[54] Liu Y, *et al.* A comprehensive review and comparison of existing computational methods for intrinsically disordered protein and region prediction. *Briefings in Bioinformatics*, 2019. **20**(1): 330–346.

[55] Li J, *et al.* An overview of predictors for intrinsically disordered proteins over 2010–2014. *International Journal of Molecular Sciences*, 2015. **16**(10): 23446–23462.

[56] Pryor EE Jr and Wiener MC. A critical evaluation of in silico methods for detection of membrane protein intrinsic disorder. *Biophysical Journal*, 2014. **106**(8): 1638–1649.

[57] Dosztanyi Z, *et al.* Bioinformatical approaches to characterize intrinsically disordered/unstructured proteins. *Briefings in Bioinformatics*, 2010. **11**(2): 225–243.

[58] Katuwawala A, *et al.* Accuracy of protein-level disorder predictions. *Briefings in Bioinformatics*, 2020. **21**(5): 1509–1522.

[59] Kurgan L, *et al.* The methods and tools for intrinsic disorder prediction and their application to systems medicine. In *Systems Medicine*, Wolkenhauer O (Ed.), 2021. Oxford: Academic Press, 159–169.

[60] Zhao B and Kurgan L. Surveying over 100 predictors of intrinsic disorder in proteins. *Expert Review of Proteomics*, 2021. **18**(12): 1019–1029.

[61] Campen A, *et al.* TOP-IDP-scale: A new amino acid scale measuring propensity for intrinsic disorder. *Protein & Peptide Letters*, 2008. **15**(9): 956–963.

[62] Uversky VN. A decade and a half of protein intrinsic disorder: Biology still waits for physics. *Protein Science*, 2013. **22**(6): 693–724.

[63] Yan J, *et al.* Structural and functional analysis of "non-smelly" proteins. *Cellular and Molecular Life Sciences*, 2020. **77**(12): 2423–2440.

[64] Kurgan L, *et al.* On the importance of computational biology and bioinformatics to the origins and rapid progression of the intrinsically disordered proteins field. *Biocomputing*, 2020. **2020**: 149–158.

[65] McBryant SJ, *et al.* Domain organization and quaternary structure of the Saccharomyces cerevisiae silent information regulator 3 protein, Sir3p. *Biochemistry*, 2006. **45**(51): 15941–15948.

[66] Hu G, *et al.* flDPnn: Accurate intrinsic disorder prediction with putative propensities of disorder functions. *Nature Communications*, 2021. **12**(1): 4438.

[67] Lang B and Babu MM. A community effort to bring structure to disorder. *Nature Methods*, 2021. **18**(5): 454–455.

[68] Necci M, *et al.* Critical assessment of protein intrinsic disorder prediction. *Nature Methods*, 2021. **18**(5): 472–481.

[69] Barik A, *et al.* DEPICTER: Intrinsic disorder and disorder function prediction server. *Journal of Molecular Biology*, 2020. **432**(11): 3379–3387.

[70] Meszaros B, *et al.* IUPred2A: Context-dependent prediction of protein disorder as a function of redox state and protein binding. *Nucleic Acids Research*, 2018. **46**(W1): W329–W337.

[71] Hanson J, *et al.* Accurate single-sequence prediction of protein intrinsic disorder by an ensemble of deep recurrent and convolutional architectures. *Journal of Chemical Information and Modeling*, 2018. **58**(11): 2369–2376.

[72] Peng Z and Kurgan L. High-throughput prediction of RNA, DNA and protein binding regions mediated by intrinsic disorder. *Nucleic Acids Research*, 2015. **43**(18): e121.

[73] Peng Z, *et al.* Prediction of disordered RNA, DNA, and protein binding regions using DisoRDPbind. *Methods in Molecular Biology*, 2017. **1484**: 187–203.

[74] Yan J, *et al.* Molecular recognition features (MoRFs) in three domains of life. *Molecular Biosystems*, 2016. **12**(3): 697–710.

[75] Meng F and Kurgan L. DFLpred: High-throughput prediction of disordered flexible linker regions in protein sequences. *Bioinformatics*, 2016. **32**(12): i341–i350.

[76] Meng F and Kurgan L. High-throughput prediction of disordered moonlighting regions in protein sequences. *Proteins*, 2018. **86**(10): 1097–1110.

[77] Jones DT and Cozzetto D. DISOPRED3: Precise disordered region predictions with annotated protein-binding activity. *Bioinformatics*, 2015. **31**(6): 857–863.

[78] Dosztanyi Z, *et al.* IUPred: Web server for the prediction of intrinsically unstructured regions of proteins based on estimated energy content. *Bioinformatics*, 2005. **21**(16): 3433–3434.

[79] Katuwawala A, *et al.* QUARTERplus: Accurate disorder predictions integrated with interpretable residue-level quality assessment scores. *Computational and Structural Biotechnology Journal*, 2021. **19**: 2597–2606.

[80] Monastyrskyy B, *et al.* Assessment of protein disorder region predictions in CASP10. *Proteins*, 2014. **82**(Suppl 2): 127–137.

[81] Romero P, *et al.* Identifying disordered regions in proteins from amino acid sequence. *Proceedings of International Conference on Neural Networks (ICNN'97)*, Houston, USA, 1997. **1**: 90–95.

[82] Melamud E and Moult J. Evaluation of disorder predictions in CASP5. *Proteins*, 2003. **53**(Suppl 6): 561 565.

[83] Altschul SF, *et al.* Gapped BLAST and PSI-BLAST: A new generation of protein database search programs. *Nucleic Acids Research*, 1997. **25**(17): 3389–3402.

[84] Hu G and Kurgan L. Sequence similarity searching. *Current Protocols in Protein Science*, 2019. **95**(1): e71.

[85] Linding R, *et al.* GlobPlot: Exploring protein sequences for globularity and disorder. *Nucleic Acids Research*, 2003. **31**(13): 3701–3708.

[86] Dosztanyi Z, *et al.* The pairwise energy content estimated from amino acid composition discriminates between folded and intrinsically unstructured proteins. *Journal of Molecular Biology*, 2005. **347**(4): 827–839.

[87] Obradovic Z, *et al.* Exploiting heterogeneous sequence properties improves prediction of protein disorder. *Proteins*, 2005. **61**(Suppl 7): 176–182.

[88] Peng K, *et al.* Length-dependent prediction of protein intrinsic disorder. *BMC Bioinformatics*, 2006. **7**: 208.

[89] Romero P, *et al.* Sequence complexity of disordered protein. *Proteins*, 2001. **42**(1): 38–48.

[90] Peng K, *et al.* Optimizing long intrinsic disorder predictors with protein evolutionary information. *Journal of Bioinformatics and Computational Biology*, 2005. **3**(1): 35–60.

[91] Jones DT and Ward JJ. Prediction of disordered regions in proteins from position specific score matrices. *Proteins*, 2003. **53**(Suppl 6): 573–578.

[92] Linding R, *et al.* Protein disorder prediction: Implications for structural proteomics. *Structure*, 2003. **11**(11): 1453–1459.

[93] Yang ZR, *et al.* RONN: The bio-basis function neural network technique applied to the detection of natively disordered regions in proteins. *Bioinformatics*, 2005. **21**(16): 3369–3376.

[94] Mizianty MJ, *et al.* Improved sequence-based prediction of disordered regions with multilayer fusion of multiple information sources. *Bioinformatics*, 2010. **26**(18): i489–i496.

[95] Walsh I, *et al.* CSpritz: Accurate prediction of protein disorder segments with annotation for homology, secondary structure and linear motifs. *Nucleic Acids Research*, 2011. **39**(Web Server issue): W190–W196.

[96] Xue B, *et al.* PONDR-FIT: A meta-predictor of intrinsically disordered amino acids. *Biochimica et Biophysica Acta*, 2010. **1804**(4): 996–1010.

[97] Fan X and Kurgan L. Accurate prediction of disorder in protein chains with a comprehensive and empirically designed consensus. *Journal of Biomolecular Structure and Dynamics*, 2014. **32**(3): 448–464.

[98] Oldfield CJ, *et al.* Computational prediction of intrinsic disorder in protein sequences with the disCoP meta-predictor. *Methods in Molecular Biology*, 2020. **2141**: 21–35.

[99] Bordoli L, *et al.* Assessment of disorder predictions in CASP7. *Proteins*, 2007. **69**(Suppl 8): 129–136.

[100] Noivirt-Brik O, *et al.* Assessment of disorder predictions in CASP8. *Proteins*, 2009. **77**(Suppl 9): 210–216.

[101] Monastyrskyy B, *et al.* Evaluation of disorder predictions in CASP9. *Proteins*, 2011. **79**(Suppl 10): 107–118.

[102] Wang S, *et al.* AUCpreD: Proteome-level protein disorder prediction by AUC-maximized deep convolutional neural fields. *Bioinformatics*, 2016. **32**(17): i672–i679.

[103] Hanson J, *et al.* Improving protein disorder prediction by deep bidirectional long short-term memory recurrent neural networks. *Bioinformatics*, 2017. **33**(5): 685–692.

[104] Hanson J, *et al.* SPOT-Disorder2: Improved protein intrinsic disorder prediction by ensembled deep learning. *Genomics Proteomics Bioinformatics*, 2019. **17**(6): 645–656.

[105] Mirabello C and Wallner B. rawMSA: End-to-end deep learning using raw multiple sequence alignments. *PLoS One*, 2019. **14**(8): e0220182.

[106] Prilusky J, *et al.* FoldIndex: A simple tool to predict whether a given protein sequence is intrinsically unfolded. *Bioinformatics*, 2005. **21**(16): 3435–3438.

[107] Liu Y, *et al.* RFPR-IDP: Reduce the false positive rates for intrinsically disordered protein and region prediction by incorporating both fully ordered proteins and disordered proteins. *Briefings in Bioinformatics*, 2021. **22**(2): 2000–2011.

[108] Ishida T and Kinoshita K. PrDOS: Prediction of disordered protein regions from amino acid sequence. *Nucleic Acids Research*, 2007. **35**(Web Server issue): W460–W464.

[109] Necci M, *et al.* MobiDB-lite: Fast and highly specific consensus prediction of intrinsic disorder in proteins. *Bioinformatics*, 2017. **33**(9): 1402–1404.

[110] Peng Z and Kurgan L. On the complementarity of the consensus-based disorder prediction. *Pacific Symposium on Biocomputing*, 2012. 176–187.

[111] Katuwawala A, *et al.* DISOselect: Disorder predictor selection at the protein level. *Protein Science*, 2020. **29**(1): 184–200.

[112] Ishida T and Kinoshita K. Prediction of disordered regions in proteins based on the meta approach. *Bioinformatics*, 2008. **24**(11): 1344–1348.

[113] Mizianty MJ, *et al.* MFDp2: Accurate predictor of disorder in proteins by fusion of disorder probabilities, content and profiles. *Intrinsically Disordered Proteins*, 2013. **1**(1): e24428.

[114] Mizianty MJ, *et al.* Prediction of intrinsic disorder in proteins using MFDp2. *Methods in Molecular Biology*, 2014. **1137**: 147–162.

[115] Jin Y and Dunbrack RL Jr. Assessment of disorder predictions in CASP6. *Proteins*, 2005. **61**(Suppl 7): 167–175.

[116] Katuwawala A and Kurgan L. Comparative assessment of intrinsic disorder predictions with a focus on protein and nucleic acid-binding proteins. *Biomolecules*, 2020. **10**(12): 1636.

[117] Peng ZL and Kurgan L. Comprehensive comparative assessment of in-silico predictors of disordered regions. *Current Protein & Peptide Science*, 2012. **13**(1): 6–18.

[118] Walsh I, *et al.* Comprehensive large-scale assessment of intrinsic protein disorder. *Bioinformatics*, 2015. **31**(2): 201–208.

[119] Necci M, *et al.* A comprehensive assessment of long intrinsic protein disorder from the DisProt database. *Bioinformatics*, 2018. **34**(3): 445–452.

[120] Romero O and Dunker K. Sequence data analysis for long disordered regions prediction in the calcineurin family. *Genome Informatics. Workshop on Genome Informatics*, 1997. **8**: 110–124.

[121] Ward JJ, *et al.* The DISOPRED server for the prediction of protein disorder. *Bioinformatics*, 2004. **20**(13): 2138–2139.

[122] Hecker J, *et al.* Protein disorder prediction at multiple levels of sensitivity and specificity. *BMC Genomics*, 2008. **9**(Suppl 1): S9.

[123] Cheng JL, *et al.* Accurate prediction of protein disordered regions by mining protein structure data. *Data Mining and Knowledge Discovery*, 2005. **11**(3): 213–222.

[124] Schlessinger A, *et al.* PROFbval: Predict flexible and rigid residues in proteins. *Bioinformatics*, 2006. **22**(7): 891–893.

[125] Gu J, *et al.* Wiggle — Predicting functionally flexible regions from primary sequence. *PLoS Computational Biology*, 2006. **2**(7): 769–785.

[126]   Bau D, *et al.* Distill: A suite of web servers for the prediction of one-, two- and three-dimensional structural features of proteins. *BMC Bioinformatics*, 2006. **7**: 402.

[127]   Vullo A, *et al.* Spritz: A server for the prediction of intrinsically disordered regions in protein sequences using kernel machines. *Nucleic Acids Research*, 2006. **34**(Web Server issue): W164–W168.

[128]   Su CT, *et al.* Protein disorder prediction by condensed PSSM considering propensity for order or disorder. *BMC Bioinformatics*, 2006. **7**: 319.

[129]   Su CT, *et al.* iPDA: Integrated protein disorder analyzer. *Nucleic Acids Research*, 2007. **35**: W465–W472.

[130]   Hirose S, *et al.* POODLE-L: A two-level SVM prediction system for reliably predicting long disordered regions. *Bioinformatics*, 2007. **23**(16): 2046–2053.

[131]   Shimizu K, *et al.* POODLE-S: Web application for predicting protein disorder by using physicochemical features and reduced amino acid set of a position-specific scoring matrix. *Bioinformatics*, 2007. **23**(17): 2337–2338.

[132]   Shimizu K, *et al.* Predicting mostly disordered proteins by using structure-unknown protein data. *BMC Bioinformatics*, 2007. **8**(1): 78.

[133]   Schlessinger A, *et al.* Natively unstructured regions in proteins identified from contact predictions. *Bioinformatics*, 2007. **23**(18): 2376–2384.

[134]   Wang L and Sauer UH. OnD-CRF: Predicting order and disorder in proteins using [corrected] conditional random fields. *Bioinformatics*, 2008. **24**(11): 1401–1402.

[135]   Deng X, *et al.* PreDisorder: Ab initio sequence-based prediction of protein disordered regions. *BMC Bioinformatics*, 2009. **10**: 436.

[136]   Xue B, *et al.* CDF it all: Consensus prediction of intrinsically disordered proteins based on various cumulative distribution functions. *FEBS Letters*, 2009. **583**(9): 1469–1474.

[137]   Han PF, *et al.* Predicting disordered regions in proteins using the profiles of amino acid indices. *BMC Bioinformatics*, 2009. **10**: S42.

[138]   Schlessinger A, *et al.* Improved disorder prediction by combination of orthogonal approaches. *PLoS One*, 2009. **4**(2): e4433.

[139]   Han, PF *et al.* Large-scale prediction of long disordered regions in proteins using random forests. *BMC Bioinformatics*, 2009. **10**: 8.

[140]   Hirose S, *et al.* POODLE-I: Disordered region prediction by integrating POODLE series and structural information predictors based on a workflow approach. *In Silico Biology*, 2010. **10**(3): 185–191.

[141]   Walsh I, *et al.* ESpritz: Accurate and fast prediction of protein disorder. *Bioinformatics*, 2012. **28**(4): 503–509.

[142]   Zhang T, *et al.* SPINE-D: Accurate prediction of short and long disordered regions by a single neural-network based method. *Journal of Biomolecular Structure and Dynamics*, 2012. **29**(4): 799–813.

[143] Eickholt J and Cheng J. DNdisorder: Predicting protein disorder using boosting and deep networks. *BMC Bioinformatics*, 2013. **14**: 88.

[144] Cilia E, *et al.* From protein sequence to dynamics and disorder with DynaMine. *Nature Communications*, 2013. **4**: 2741.

[145] Cilia E, *et al.* The DynaMine webserver: Predicting protein dynamics from sequence. *Nucleic Acids Research*, 2014. **42**(Web Server issue): W264–W270.

[146] Ali H, *et al.* Performance of protein disorder prediction programs on amino acid substitutions. *Human Mutation*, 2014. **35**(7): 794–804.

[147] Sormanni P, *et al.* The s2D method: Simultaneous sequence-based prediction of the statistical populations of ordered and disordered regions in proteins. *Journal of Molecular Biology*, 2015. **427**(4): 982–996.

[148] Wang Z, *et al.* DisoMCS: Accurately predicting protein intrinsically disordered regions using a multi-class conservative score approach. *PLoS One*, 2015. **10**(6): e0128334.

[149] Wang S, *et al.* DeepCNF-D: Predicting protein order/disorder regions by weighted deep convolutional neural fields. *International Journal of Molecular Sciences*, 2015. **16**(8): 17315–17330.

[150] Iqbal S and Hoque MT. DisPredict: A predictor of disordered protein using optimized RBF kernel. *PLoS One*, 2015. **10**(10): e0141551.

[151] Zhao B and Xue B. Decision-tree based meta-strategy improved accuracy of disorder prediction and identified novel disordered residues inside binding motifs. *International Journal of Molecular Sciences*, 2018. **19**(10): 3052.

[152] Liu YM, *et al.* IDP-CRF: Intrinsically disordered protein/region identification based on conditional random fields. *International Journal of Molecular Sciences*, 2018. **19**(9): 2483.

[153] Malysiak-Mrozek B, *et al.* Spark-IDPP: High-throughput and scalable prediction of intrinsically disordered protein regions with spark clusters on the cloud. *Cluster Computing-The Journal of Networks Software Tools and Applications*, 2019. **22**(2): 487–508.

[154] Liu Y, *et al.* Identification of intrinsically disordered proteins and regions by length-dependent predictors based on conditional random fields. *Molecular Therapy Nucleic Acids*, 2019. **17**: 396–404.

[155] Dass R, *et al.* ODiNPred: Comprehensive prediction of protein order and disorder. *Scientific Reports*, 2020. **10**(1): 14780.

[156] Tang YJ, *et al.* IDP-Seq2Seq: Identification of intrinsically disordered regions based on sequence to sequence learning. *Bioinformatics*, 2021. **36**(21): 5177–5186.

[157] Emenecker RJ, *et al.* Metapredict: A fast, accurate, and easy-to-use predictor of consensus disorder and structure. *Biophysical Journal*, 2021. **120**(20): 4312–4319.

[158] Torrisi M, *et al.* Deep learning methods in protein structure prediction. *Computational and Structural Biotechnology Journal,* 2020. **18**: 1301–1310.

[159] Pakhrin SC, *et al.* Deep learning-based advances in protein structure prediction. *International Journal of Molecular Sciences,* 2021. **22**(11): 5553.

[160] Zhang F, *et al.* DeepFunc: A deep learning framework for accurate prediction of protein functions from protein sequences and interactions. *Proteomics,* 2019. **19**(12): e1900019.

[161] Gligorijevic V, *et al.* deepNF: Deep network fusion for protein function prediction. *Bioinformatics,* 2018. **34**(22): 3873–3881.

[162] Di Lena P, *et al.* Deep architectures for protein contact map prediction. *Bioinformatics,* 2012. **28**(19): 2449–2457.

[163] Wang S, *et al.* Accurate de novo prediction of protein contact map by ultra-deep learning model. *PLoS Computational Biology,* 2017. **13**(1): e1005324.

[164] Wu T, *et al.* Analysis of several key factors influencing deep learning-based inter-residue contact prediction. *Bioinformatics,* 2020. **36**(4): 1091–1098.

[165] Wang S, *et al.* Analysis of deep learning methods for blind protein contact prediction in CASP12. *Proteins,* 2018. **86**(Suppl 1): 67–77.

[166] Schaarschmidt J, *et al.* Assessment of contact predictions in CASP12: Co-evolution and deep learning coming of age. *Proteins,* 2018. **86**(Suppl 1): 51–66.

[167] Wu TQ, *et al.* DeepDist: Real-value inter-residue distance prediction with deep residual convolutional network. *BMC Bioinformatics,* 2021. **22**(1): 30.

[168] Sun Z, *et al.* To improve the predictions of binding residues with DNA, RNA, carbohydrate, and peptide via multi-task deep neural networks. *IEEE/ACM Transactions on Computational Biology and Bioinformatics,* 2021. doi: 10.1109/TCBB.2021.3118916.

[169] Zhang F, *et al.* DeepDISOBind: Accurate prediction of RNA-, DNA- and protein-binding intrinsically disordered residues with deep multi-task learning. *Briefings in Bioinformatics,* 2022. **23**(1): bbab521.

[170] Katuwawala A, *et al.* DisoLipPred: Accurate prediction of disordered lipid binding residues in protein sequences with deep recurrent networks and transfer learning. *Bioinformatics,* 2022. **38**(1): 115–124.

[171] Pan XY and Shen HB. Predicting RNA-protein binding sites and motifs through combining local and global deep convolutional neural networks. *Bioinformatics,* 2018. **34**(20): 3427–3436.

[172] Elbasir A, *et al.* DeepCrystal: A deep learning framework for sequence-based protein crystallization prediction. *Bioinformatics,* 2019. **35**(13): 2216–2225.

[173] Zhang B, *et al.* Protein solvent-accessibility prediction by a stacked deep bidirectional recurrent neural network. *Biomolecules,* 2018. **8**(2): 33.

[174] Heffernan R, *et al.* Improving prediction of secondary structure, local backbone angles, and solvent accessible surface area of proteins by iterative deep learning. *Scientific Reports,* 2015. **5**: 11476.

[175] Zhang B, *et al.* Prediction of 8-state protein secondary structures by a novel deep learning architecture. *BMC Bioinformatics*, 2018. **19**(1): 293.

[176] Spencer M, *et al.* A deep learning network approach to ab initio protein secondary structure prediction. *IEEE/ACM Transactions on Computational Biology and Bioinformatics*, 2015. **12**(1): 103–112.

[177] Wang D, *et al.* MusiteDeep: A deep-learning framework for general and kinase-specific phosphorylation site prediction. *Bioinformatics*, 2017. **33**(24): 3909–3916.

[178] Li F, *et al.* DeepCleave: A deep learning predictor for caspase and matrix metalloprotease substrates and cleavage sites. *Bioinformatics*, 2020. **36**(4): 1057–1065.

[179] McGuffin LJ. Intrinsic disorder prediction from the analysis of multiple protein fold recognition models. *Bioinformatics*, 2008. **24**(16): 1798–1804.

[180] Mizianty MJ, *et al.* In-silico prediction of disorder content using hybrid sequence representation. *BMC Bioinformatics*, 2011. **12**: 245.

[181] Buchan DWA and Jones DT. The PSIPRED protein analysis workbench: 20 years on. *Nucleic Acids Research*, 2019. **47**(W1): W402–W407.

[182] Hanson J, *et al.* Improving prediction of protein secondary structure, backbone angles, solvent accessibility and contact numbers by using predicted contact maps and an ensemble of recurrent and residual convolutional neural networks. *Bioinformatics*, 2019. **35**(14): 2403–2410.

[183] Mohan A, *et al.* Analysis of molecular recognition features (MoRFs). *Journal of Molecular Biology*, 2006. **362**(5): 1043–1059.

[184] Oates ME, *et al.* D(2)P(2): Database of disordered protein predictions. *Nucleic Acids Research*, 2013. **41**(Database issue): D508–D516.

[185] Piovesan D, *et al.* MobiDB 3.0: More annotations for intrinsic disorder, conformational diversity and interactions in proteins. *Nucleic Acids Research*, 2018. **46**(D1): D471–D476.

[186] Piovesan D, *et al.* MobiDB: Intrinsically disordered proteins in 2021. *Nucleic Acids Research*, 2021. **49**(D1): D361–D367.

[187] Zhao B, *et al.* DescribePROT: Database of amino acid-level protein structure and function predictions. *Nucleic Acids Research*, 2021. **49**(D1): D298–D308.

[188] Dinkel H, *et al.* ELM 2016–data update and new functionality of the eukaryotic linear motif resource. *Nucleic Acids Research*, 2016. **44**(D1): D294–D300.

[189] Miskei M, *et al.* FuzDB: Database of fuzzy complexes, a tool to develop stochastic structure-function relationships for protein complexes and higher-order assemblies. *Nucleic Acids Research*, 2017. **45**(D1): D228–D235.

[190] Meszaros B, *et al.* PhaSePro: The database of proteins driving liquid-liquid phase separation. *Nucleic Acids Research*, 2020. **48**(D1): D360–D367.

[191] Faraggi E, *et al.* Fast and accurate accessible surface area prediction without a sequence profile. *Prediction of Protein Secondary Structure*, 2017. **1484**: 127–136.

[192] Oldfield CJ, *et al.* Disordered RNA-binding region prediction with DisoRDPbind. *Methods in Molecular Biology*, 2020. **2106**: 225–239.

[193] Malhis N, *et al.* MoRFchibi system: Software tools for the identification of MoRFs in protein sequences. *Nucleic Acids Research*, 2016. **44**(W1): W488–W493.

[194] Zhang J, *et al.* Prediction of protein-binding residues: Dichotomy of sequence-based methods developed using structured complexes versus disordered proteins. *Bioinformatics*, 2020. **36**(18): 4729–4738.

[195] Zhang J and Kurgan L. SCRIBER: Accurate and partner type-specific prediction of protein-binding residues from proteins sequences. *Bioinformatics*, 2019. **35**(14): i343–i353.

[196] Yan J and Kurgan L. DRNApred, fast sequence-based method that accurately predicts and discriminates DNA- and RNA-binding residues. *Nucleic Acids Research*, 2017. **45**(10): e84.

[197] McGuffin LJ, *et al.* The PSIPRED protein structure prediction server. *Bioinformatics*, 2000. **16**(4): 404–405.

[198] Teufel F, *et al.* SignalP 6.0 predicts all five types of signal peptides using protein language models. *Nature Biotechnology*, 2022. https://doi.org/10.1038/s41587-021-01156-3.

[199] Almagro Armenteros JJ, *et al.* SignalP 5.0 improves signal peptide predictions using deep neural networks. *Nature Biotechnology*, 2019. **37**(4): 420–423.

[200] Mirdita M, *et al.* MMseqs2 desktop and local web server app for fast, interactive sequence searches. *Bioinformatics*, 2019. **35**(16): 2856–2858.

[201] Steinegger M and Soding J. MMseqs2 enables sensitive protein sequence searching for the analysis of massive data sets. *Nature Biotechnology*, 2017. **35**(11): 1026–1028.

[202] Wu Z, *et al.* Exploratory Analysis of Quality Assessment of Putative Intrinsic Disorder in Proteins. *6th international conference on artificial intelligence and soft computing*, Rutkowski L, *et al.* (eds.), Zakopane, Poland, Springer International Publishing, 2017. 722–732.

[203] Wu Z, *et al.* Prediction of intrinsic disorder with quality assessment using quarter. *Methods in Molecular Biology*, 2020. **2165**: 83–101.

[204] Hu G, *et al.* Quality assessment for the putative intrinsic disorder in proteins. *Bioinformatics*, 2019. **35**(10): 1692–1700.

[205] Katuwawala A, *et al.* Computational prediction of MoRFs, short disorder-to-order transitioning protein binding regions. *Computational and Structural Biotechnology Journal*, 2019. **17**: 454–462.

[206] Sharma R, *et al.* OPAL+: Length-specific MoRF prediction in intrinsically disordered protein sequences. *Proteomics*, 2019. **19**(6): e1800058.

[207] Fang C, *et al.* MoRFPred_en: Sequence-based prediction of MoRFs using an ensemble learning strategy. *Journal of Bioinformatics and Computational Biology*, 2019. **17**(6): 1940015.

[208] Monzon AM, *et al.* FLIPPER: Predicting and characterizing linear interacting peptides in the protein data bank. *Journal of Molecular Biology*, 2021. **433**(9): 166900.

[209] Hanson J, *et al.* Identifying molecular recognition features in intrinsically disordered regions of proteins by transfer learning. *Bioinformatics*, 2020. **36**(4): 1107–1113.

[210] Peng Z, *et al.* APOD: Accurate sequence-based predictor of disordered flexible linkers. *Bioinformatics*, 2020. **36**(Suppl 2): i754–i761.

[211] Katuwawala A, *et al.* Computational prediction of functions of intrinsically disordered regions. *Progress in Molecular Biology and Translational Science*, 2019. **166**: 341–369.

[212] Hosoya Y and Ohkanda J. Intrinsically disordered proteins as regulators of transient biological processes and as untapped drug targets. *Molecules*, 2021. **26**(8): 2118.

[213] Biesaga M, *et al.* Intrinsically disordered proteins and biomolecular condensates as drug targets. *Current Opinion in Chemical Biology*, 2021. **62**: 90–100.

[214] Ambadipudi S and Zweckstetter M. Targeting intrinsically disordered proteins in rational drug discovery. *Expert Opinion on Drug Discovery*, 2016. **11**(1): 65–77.

[215] Oldfield CJ, *et al.* Utilization of protein intrinsic disorder knowledge in structural proteomics. *Biochimica et Biophysica Acta*, 2013. **1834**(2): 487–498.

[216] Deng X, *et al.* An overview of practical applications of protein disorder prediction and drive for faster, more accurate predictions. *International Journal of Molecular Sciences*, 2015. **16**(7): 15384–15404.

Chapter 9

Sequence-Based Predictions of Residues that Bind Proteins and Peptides

Qianmu Yuan and Yuedong Yang*

*School of Computer Science and Engineering,
Sun Yat-sen University, Guangzhou 510000, China*
*corresponding author: yangyd25@mail.sysu.edu.cn

Protein-protein interactions (PPI) and protein-peptide interactions (PPepI) are essential for many cellular processes. Identifications of PPI and PPepI sites are important steps to investigate molecular-level mechanisms underlying diseases and for rational drug design. Since experimental approaches for PPI and PPepI site identifications are difficult, expensive and time-consuming, and a growing amount of proteomics data have been collected due to the development of public databases, many computational predictors of these sites have been developed to complement the experimental efforts. However, most of these predictors are not publicly available and many are limited to proteins with known experimental structures. Here, we provide a comprehensive survey of studies concerning sequence-based predictions of residues that interact with proteins and peptides. We introduce commonly used biological databases and sequence-derived features that can be employed to learn computational models for predicting PPI and PPepI sites. To understand their relative merits, we also discuss different validation schemes and performance evaluation metrics. Finally, we investigate the development history and algorithm design of the recently released sequence-based PPI site and PPepI site prediction methods.

1. Introduction

Protein-protein interactions (PPI) play crucial roles in many cellular processes including transport, metabolism, and signal transduction [1]. Identification of the residues involved in physical contacts in protein-protein complexes, namely PPI sites, contributes to the construction of protein-protein interaction networks [2], prediction of protein functions [3], investigations of molecular mechanisms in diseases [4], and rational design of novel drugs [5,6]. Up to 40% of the protein-protein interactions are mediated by protein-peptide interactions [7]. Peptides also play important roles by binding with other proteins to participate in key cellular processes including gene expression regulation and programmed cell death [8,9]. Owing to the safety and favorable tolerability in human bodies, peptides have become desirable starting points for the design of novel therapeutics [10]. Thus, accurately identifying protein-peptide interactions (PPepI) and PPepI sites is crucial for the therapeutics invention process. However, conventional experimental methods for the PPI and PPepI site detection, such as two-hybrid assay and affinity purification, are costly and time-consuming [11,12]. Moreover, experimental identification of PPepI is even more challenging due to the small peptide sizes [13] and weak binding affinity [14]. Despite these challenges, there is still a steady progress in the collection of protein-protein and protein-peptide complex structures determined by experimental techniques, which are publicly available in Protein Data Bank (PDB) [15]. Therefore, the functional importance, the challenges in the experimental detection, and the availability of the experimental data motivate the development of complementary computational methods capable of making high-throughput and accurate PPI and PPepI site predictions.

Majority of the existing PPI and PPepI site prediction methods are machine learning-based techniques since sequence conservation [16] and interaction energy scoring [17] can only capture part of the determinants for protein-protein and protein-peptide interactions, and since performance of the template-based methods is severely restricted when the query proteins have no high-quality templates [18]. Machine learning methods for the protein binding residue prediction can be categorized into two classes according to the information used for the prediction. The first class is the structure-based approaches that infer protein functional sites from known structures. For example, GraphPPIS [19] considers proteins as undirected graphs and the PPI site prediction as graph node classification problem, where the pairwise amino acid distances are used to construct the adjacency matrix. Afterwards, a deep graph convolutional network is employed to refine the protein structural context topology. Alternatively, BiteNet [20] maps protein atoms into 3D voxels and employs 3D convolutional neural networks (3DCNN) to extract features from the neighborhood of the target residue for the PPepI site prediction. Albeit powerful, these structure-based methods cannot be applied to majority of proteins that do not have known tertiary structures due to the cost and the difficulties to determine protein structures experimentally [21]. As of October 2021, PDB holds 183,584 biological macromolecular structures that cover only 70,099 proteins (at identity 90%). By comparison, the second class of the machine learning predictors is the sequence-based approaches that make predictions solely from the widely available protein sequences. Based on the UniProt resource [22], the number of proteins with known sequences is at about 109 million (at identity 90%) as of October 2021. Although predictive accuracies of the sequence-based methods are usually lower than the structure-based methods due to the lack of the tertiary structure information [23], the sequence-based methods have a wider range of applications including large-scale annotation of the human proteome.

The sequence-based methods employ sequence-derived features to represent the query proteins and characterize the protein binding patterns. These features generally include evolutionary information, such as position-specific scoring matrix (PSSM) and hidden Markov models (HMM) profile, predicted secondary structure, and predicted relative solvent accessibility (RSA). Additional features were also explored for this problem including one-hot amino acid vector, amino acid embedding such as ProtVec [24], high-scoring segment pair (HSP) [25], relative amino acid binding propensity, physical and physicochemical characteristics, and protein intrinsic disorder. With these well-designed features, various machine learning algorithms have been adopted, including Naïve Bayes classifier in PSIVER [26], random forest in IntPred [27] and EL-SMURF [28], SVM in SVMpep [12] and SPRINT [29], XGBoost in IHT-XGB [30], logistic regression in SCRIBER [31] and neural network in ProNA2020 [32]. Some of the recent methods utilize deep contextual learning. For example, convolutional neural network (CNN) is used to explore the context of consecutive residues [33,34]. DLPred [35] develop a simplified long short-term memory (LSTM) network to perform multi-task learning for simultaneous prediction of PPI sites and RSA. DELPHI [36] implements an ensemble framework with a CNN and a recurrent neural network (RNN) component. MTDsite [37] employs bidirectional LSTM (BiLSTM) and multi-task learning to concurrently predict binding residues with DNA, RNA, carbohydrate and peptide.

In this survey, we put emphasis on the investigation of the up-to-date machine learning methods for the sequence-based PPI site and PPepI site prediction. The survey is organized as follows. First, we delineate the background knowledge including commonly used biological databases and sequence-derived features that can be employed to learn computational models for predicting PPI and PPepI sites. Then, we discuss different validation schemes and metrics to evaluate the predictive performance. Finally, we comprehensively investigate the development history and algorithm design of the recently published sequence-based PPI site and PPepI site prediction tools.

2. Commonly used biological databases

Supervised machine learning algorithms require a large amount of labeled data for model training. Owing to the extensive efforts to establish biological databases, it has become easier to obtain protein-protein and protein-peptide complex structures. This section introduces two commonly used databases in this field: PDB and BioLiP [38]. Most of the datasets used in the studies of PPI and PPepI site prediction were collected and filtered from these two databases.

2.1 *Protein Data Bank (PDB)*

PDB, an important resource in structural biology, was created by Brookhaven National Laboratory in the United States in 1971 and maintained by Research Collaboratory for Structural Bioinformatics (RCSB). The PDB database is updated weekly and can be freely accessed. As of October 2021, PDB has stored about 184,000 experimental structures.

The three-dimensional structures of the biological macromolecules (proteins, nucleic acid polysaccharides, lipids and viruses) in PDB are mainly determined by experimental methods such as the X-ray crystal diffraction, nuclear magnetic resonance (NMR), and electron diffraction. The contents of the PDB files generally include the atomic coordinates of the macromolecules, references, and primary and secondary structure information. The PDB database allows users to search in various ways. The searchable fields include PDB ID, UniProt ID, authors, references, functional category, structural characteristics, experimental methods, resolution, source of species, storage time, molecular formula, etc. PDB also supports search for similar sequences by sequence similarity. The PDB files can be read and processed by software packages, such as Biopython [39], and visualized through software applications, such as PyMOL[40] or Mercury [41].

2.2 *BioLiP*

BioLiP is a semi-manually generated database of biologically relevant ligand-protein interactions. Not all ligands present in PDB are biologically relevant and should be used to develop and test machine learning predictors, as small molecules are often used as additives to resolve protein structures. In order to facilitate template-based ligand-protein docking, virtual ligand screening and prediction of protein-protein and protein-peptide interactions, BioLiP has developed a hierarchical procedure to evaluate the biological relevance of ligands present in the PDB structure, involving a four-step biological feature filtering and carefully manual verification. More specifically, the requirement for the ligand to be considered biologically relevant is that the ligand is not on the artificial additives list and repeats fewer than 15 times in the same PDB file, the number of the receptor binding sites that interacts with the ligand is not less than two, and the binding sites are not continuous. If the ligand is on the artificial additive list but meets the rest of the above requirements and has PubMed evidence, it is also considered as biologically relevant.

BioLiP was officially released on October 18, 2012, and the database is generally updated once a week. The current version (October 15, 2021) contains 529,047 entries. Each entry in BioLiP contains the following annotations: ligand-binding residues, ligand-binding affinity, catalytic sites, Enzyme Commission numbers, Gene Ontology terms, and is cross-linked to other databases. As of October 2021, BioLiP contains 109,998 proteins from PDB, 57,095 DNA/RNA ligands, 25,960 peptide (less than 30 residues) ligands, 146,969 metal ligands, and 299,051 regular ligands.

3. Biological characteristics and sequence-based representation of proteins

Proteins can be characterized from biological and biophysical perspectives and these representations are commonly used by machine learning models as input features to perform prediction tasks. At present, the biological sequence-derived features of proteins generally include protein sequence, evolutionary information, predicted structural features, and physicochemical characteristics of the amino acids that compose the sequence.

3.1 *Protein sequence-derived information*

3.1.1 *One-hot encoding*

Since the protein sequence generally contains 20 amino acid types, which are discrete features, one-hot encoding can be performed. Each amino acid is represented by a 20-bit status register. Only the bit of the register that corresponds to the given amino acid is recorded as 1, and the remaining registers are recorded as 0. Thus, the protein sequence can be represented by a $L \times 20$ matrix, where L is the sequence length.

3.1.2 *Pre-trained embedding*

In neural networks, the one-hot protein sequence matrix can be input to an embedding layer to get continuous sequence embedding, whose parameters are learned together with the network. Conversely, self-supervised pre-trained protein sequence embedding from other studies can also be directly employed to represent sequences. For example, the 3-mer amino acid embedding ProtVec [24] uses word2vec [42] to construct a one hundred dimensional embedding for each amino acid 3-mer. ProtVec was applied to several diverse problems including protein-protein interaction prediction, protein family classification, structure prediction, disordered protein identification, and protein visualization [24]. TAPE (Tasks Assessing Protein Embeddings) [43] is another widely-used protein embedding based on a transformer model. These methods can compute the sequence embedding features in a short time. For instance, on a Titan Xp, TAPE can process around 200 sequences per second. However, in many cases, features learned by pretraining still lag behind features extracted by state-of-the-art non-neural techniques, such as evolutionary information that is produced from a multiple-sequence alignment.

3.2 *Evolutionary information*

Evolutionarily conserved residues may contain functionally relevant motifs, such as those involved in the protein and peptide binding. The PSSM and HMM profile extracted from multi-sequence alignment are widely-used and powerful evolutionary features. Concretely, PSSM can be generated by running PSI-BLAST [44] to search the query sequence against the UniRef90 database [45], while HMM profile can be produced by running HHblits [46] to align the query sequence against the UniClust30 database [47]. Using these approaches, amino acids are encoded into a 20-dimensional vector in PSSM or HMM, and the values are often normalized using Min-Max normalization or z-score standardization.

3.3 *Predicted structural features of proteins*

Putative protein structural features can be generated using a variety of predictive emethods, such as SPIDER3 [48], whose inputs include the protein sequence and the PSSM and HMM profiles. The outputs of SPIDER3 include: 1. Solvent accessible surface area (ASA); 2. The sine and cosine values of the 4 protein backbone torsion angles: θ, ϕ, ψ, and τ; 3. Contact number (CN), which is the number of neighboring residues in three-dimensional space within a specific distance cutoff; 4. Half sphere exposure (HSE), which adds directionality to CN by differentiating between the counts in the top and bottom half of the sphere. HSE uses the $C\alpha$-$C\alpha$ direction vector and $C\alpha$-$C\beta$ direction vector to determine the boundary of the two hemispheres; 5. The predicted probabilities of the three secondary structures (i.e., α-helix, β-sheet and random coil). Besides, ASAquick [49] is a popular program that is widely adopted to predict relative solvent accessibility (RSA). It does not use sequence alignment-based features, which significantly shortens its prediction runtime.

3.4 *Amino acid physicochemical characteristics*

Several physicochemical properties of amino acid [50] are widely used in protein prediction-related tasks. They typically include steric parameters, hydrophobicity, volume, polarizability, isoelectric point, helix probability, and sheet probability. In addition, the number of atoms, electrostatic charges and potential hydrogen bonds for amino acid were also employed to predict the protein-protein interactions [35].

3.5 Other features relevant to the prediction of PPI and PPepI sites

Other features that are useful for the PPI and PPepI site prediction include protein disorder-based features, HSPs, and relative amino acid propensity (RAAP). Recent study [12] applied IUPred [51] to predict intrinsic disorder from protein sequence, including the short and the long intrinsic disordered regions. Then, nine sliding-window-based features were derived for each residue, indicating the structural flexibility of the query residue and its neighbors, which were used to predict PPepI sites. SPOT-disorder [52] is another protein disorder predicting method which can be used for this purpose. Additionally, ANCHOR [53] is a program predicting protein binding regions in disordered proteins and its predictions were also be applied to PPI site prediction task [36]. An HSP is a pair of similar sub-sequences between two proteins, which can be computed by SPRINT [54]. The similarity between two amino acids in two sub-sequences in the training and test sets are measured by scoring matrices, such as PAM, which was used to generate features for predicting PPI sites [36]. RAAP is a widely used feature, defined as the relative difference in abundance of a given amino acid type between binding residues and the corresponding non-binding residues on protein surface. RAAP for each amino acid type was computed in a recent study [55] using Composition Profiler [56].

4. Performance evaluation

As an important step to empirically compare different computational methods, performance evaluation involves several aspects including experimental data preparation, validation scheme and evaluation metrics.

4.1 Experimental data preparation

Since BioLiP is a well-annotated database with binding partners and pre-computed binding sites according to experimentally determined complex structures from PDB, the preparation of protein-peptide interaction data is straightforward and explicit. However, BioLiP does not include any protein-protein interaction data, so in general, the datasets for PPI sites were directly collected from PDB and refined by several filtering steps to reserve high-quality and nonredundant transient PPI. As performed in [26,57,58], the filtering processes include the exclusion of structures with more than 30% missing residues; removal of chains with identical UniprotKB/Swiss-Prot accessions; removal of transmembrane proteins; removal of oligomeric structures (higher than dimeric); removal of proteins with buried surface accessibility <500 Å or ≥2500 Å and interface polarity ≤25%, and finally removal of redundant proteins. In some cases, interaction data for the same protein that spans across multiple complexes in PDB was combined together [1,31].

4.2 Validation scheme

Once the experimental data is available, the next step is to select an appropriate validation scheme for performance evaluation, where the data is normally divided into training and validation data. The purpose of the training data is to train the computational models while the validation data is to tune the hyperparameters of the models. Popular validation schemes contain hold-out validation, K-fold cross-validation (CV) and leave-one-out CV.

4.2.1 Hold-out validation

Hold-out validation is the simplest implementation of validation in which data are randomly split into training and validation sets (8:2 in common). The evaluation obtained from hold-out validation may have a large variance, since the model performance is heavily determined by the training-validation split, especially when the dataset is small.

4.2.2 K-fold CV

K-fold CV divides the data randomly into K groups, and then repeats the hold-out validation K times. For each time, one group is used as the validation data, and the remaining (K − 1) groups are used as training data. The average performance of the K validation groups is used as the overall validation performance. K-fold CV depends less on the division of experimental data compared with hold-out validation. However, the computational cost of K-fold CV is K times as much as that of hold-out validation.

4.2.3 Leave-one-out CV

The extreme case of K-fold CV is leave-one-out CV (LOOCV) where K is equal to the number of proteins in the dataset. The advantage of LOOCV is that the evaluation results are more stable and reliable. However, the computational cost of LOOCV increases dramatically when the dataset is large-scale.

4.3 Evaluation metrics

Accuracy (ACC), precision, recall, specificity, F1-score (F1), Matthews correlation coefficient (MCC), area under the receiver operating characteristic curve (AUC), and area under the precision-recall curve (AUPR) are generally used to quantitatively measure the performance of PPI and PPepI site predicting methods:

$$ACC = \frac{TP+TN}{TP+TN+FP+FN} \tag{1}$$

$$Precision = \frac{TP}{TP+FP} \tag{2}$$

$$Recall = \frac{TP}{TP+FN} \tag{3}$$

$$Specificity = \frac{TN}{TN+FP} \tag{4}$$

$$F1 = 2 \times \frac{Precision \times Recall}{Precision+Recall} \tag{5}$$

$$MCC = \frac{TP \times TN - FN \times FP}{\sqrt{(TP+FP) \times (TP+FN) \times (TN+FP) \times (TN+FN)}} \tag{6}$$

where true positives (TP) and true negatives (TN) denote the number of binding and non-binding residues identified correctly, and false positives (FP) and false negatives (FN) denote the number of incorrectly predicted binding and non-binding residues, respectively. MCC is a conventional and widely-used metrics which is within the [−1,1] range. It takes into account TP, TN, FP and FN and is generally regarded as a balanced measure which can be used even if the classes are of very different sizes [59]. F1-score is the harmonic mean of precision and recall, which ranges from 0 to 1. F1-score depends on the definition of positive samples, and the switch of the "positive" and "negative" will change the value, while MCC is invariant to such switch [60]. The receiver operating characteristic (ROC) curve is a curve of recall versus 1 — specificity given a predetermined threshold, and AUC is the area under the ROC curve. However, AUC may present an over-optimistic impression about the performance of computational models for imbalanced dataset [61]. To overcome this, AUPR could be considered as an alternative to AUC, as it is more sensitive and informative and it emphasizes more on the minority class in the imbalanced two-class classification tasks [62]. AUC and AUPR are independent of thresholds, thus revealing the overall performance of a model, while the other metrics are calculated using a threshold to convert predicted binding probabilities to binary predictions.

5. Computational methods for protein-protein interaction (PPI) site prediction

In this section, we describe five representative and publicly available sequence-based methods for the PPI site prediction. They are briefly summarized in Table 1, in the order of their publication time. In general, the datasets used by these methods have been getting larger over time, due to the increased experimentally resolved protein-protein complex structures in PDB. Moreover, the adopted features and algorithms have also been getting more and more sophisticated and powerful. Note that Dset_186 and Dset_72 introduced in [26] and Dset_164 in [57] were widely used as benchmark datasets in this problem, and PSSM and RSA were almost always employed by these methods as features because the protein-binding residues tend to locate on the protein surface and are typically evolutionarily conserved. As for availability, we suggest that a web server, which can help users submit their sequences and get prediction results without any programming background, is necessary for the PPI site prediction task. The rest of this section provides detailed descriptions of the designs for these methods.

5.1 *PSIVER*

PSIVER [26] is the first Naïve Bayes classifier (NBC)-based method that predicts protein-binding residues using two features: PSSM and predicted

Table 1. Summary of the sequence-based methods for PPI site prediction in recent years.

Method	Year	Dataset Size	Feature[1]	Algorithm	Availability[2]
PSIVER	2010	Train:186 Test:72	PSSM, RSA	Naïve Bayes classifier	S
LORIS	2014	Train:186 Test:72,164	PSSM, hydropathy, RSA	logistic regression	C
DLPred	2019	Train:5860 Test:186,72,164	sequence coding, PSSM, hydropathy, PHY, PKx, ECO, 3D-1D scores	simplified LSTM	S, C
SCRIBER	2019	Train:843 Test:448	RSA, ECO, RAAP, PHY, ID, SS, residue position	logistic regression	S
DELPHI	2020	Train:9982 Test:448,186, 72,164	HSP, ProtVec1D, PSSM, ECO, RSA, RAAP, hydropathy, ID, PHY, PKx, position information	CNN+GRU	S, C

[1]PHY denotes physicochemical properties; ECO denotes evolutionary conservation; RAAP denotes relative amino acid propensity; ID denotes intrinsic disorder; SS denotes putative secondary structure.
[2]S and C correspond to the availability of the web server and source code, respectively.

RSA (obtained by SABLE [63]). The PSSM and RSA features of the target residue and its sequence neighbors in a sliding window are concatenated and then input to two NBCs with kernel density estimation (KDE) separately, from which two predicted scores are generated. These two scores are normalized using a sigmoid function and then averaged to get the final prediction. Lastly, a post-processing procedure is performed to filter out isolated residues based on the observation that about 98% of the binding residues in the training set have at least one additional binding site and about 76% have at least four binding sites in a window of nine consecutive residues (four on each side), indicating that binding residues tend to cluster in sequence. As a result, PSIVER converts all isolated positive predictions with ≤ 2 other positive residues in a window of 11 residues, and the MCC improved from 0.140 to 0.151 in LOOCV.

5.2 *LORIS*

In addition to Dset_186 and Dset_72 presented by PSIVER, LORIS [57] collected another independent test consisting of 164 newly resolved proteins in PDB (Dset_164), which had been filtered under the same criteria as in Dset_186 and Dset_72 (described in Section 4.1). Similar to PSIVER, LORIS adopts a sliding window strategy to extract sequence features including PSSM, RSA and averaged cumulative hydropathy (ACH). The RSA is predicted by Sann [64], and ACH is a 5-dimensional feature representing the average hydropathy over window sizes of 1, 3, 5, 7 and 9. Finally, L1-regularized logistic regression is applied to identify PPI sites.

5.3 *DLPred*

The contributions of DLPred [35] mainly lies in three aspects. Firstly, a simplified long short-term memory (SLSTM) network is implemented to reduce the model parameters and the equations are described as follows:

$$i_t = \sigma(W_i x_t + b_i) \tag{7}$$

$$o_t = \sigma(W_o x_t + b_o) \tag{8}$$

$$X_t = \sigma(W_x x_t + b_x) \tag{9}$$

$$z_t = \tanh(W_z x_t + U_z h_{t-1} + b_z) \tag{10}$$

$$c_t = (1 - i_t) \odot c_{t-1} + i_t \odot z_t \tag{11}$$

$$h_t = (1 - o_t) \odot X_t + o_t \odot c_t \tag{12}$$

where x_t and h_t are input and output vectors at time t respectively; σ is the sigmoid function; i, o, and c are the input gate, output gate and cell activation vectors, respectively. The parameters of this lightweight variant of LSTM are only 61.4% of LSTM and 81.7% of gated recurrent units (GRU), but the performance of SLSTM is better than LSTM and comparable to GRU. Secondly, in order to cope with the imbalance issue, a new penalization factor is added in the loss function to enhance the penalization on the misclassified non-binding sites. Thirdly, multi-task learning for concurrent binding site and ASA prediction is performed through a shared network structure. Since only the surface residues have the potential to become binding sites, binding sites have larger solvent accessibility. By providing multi-level supervised information of two related tasks, the performance of the main task was boosted. In addition, DLPred also introduces some new data and features. Concretely, the training set of DLPred contains a total of 5860 proteins collect from PISCES [65] and cons-PPISP [66], and the independent tests consisted of Dset_186, Dset_72 and Dset_164. 1-dimensional PKx, 18-dimensional 3D-1D scores, and 1-dimensional evolutionary conservation score (ECO) are used in this study for the first time. PKx is the negative of the logarithm of the dissociation constant for any other group in the molecule, and 3D-1D scores represent the side-chain environment proposed by [67]. ECO is a compact evolutionary feature derived from the amino acid frequency distribution in the corresponding column of multiple-sequence alignment (MSA) developed by [68].

5.4 *SCRIBER*

A comparative study [1] discovered that recent PPI site predictors are unable to separate residues that bind to other molecules, such as DNA, RNA and small ligands, from PPI sites. This suggests that these methods essentially predict all binding residues, instead of making specific predictions according to the binding partners. This is because they utilized biased training datasets including only protein-binding proteins. Moreover, these methods are quite slow since they employ MSA-based features computed by PSI-BLAST. SCRIBER [31] is a fast and relatively accurate predictor that minimizes cross-predictions, by introducing three innovations: 1. It uses a comprehensive dataset that covers multiple types of binding residues; 2. It uses novel types of feature groups that ensure fast prediction; 3. It implements a two-layer logistic regression-based architecture tailored to reduce the cross-predictions.

Specifically, the training and test sets of SCRIBER are mainly sourced from BioLiP and contain 843 and 448 proteins, respectively. These two datasets include proteins that bind to proteins, RNA, DNA and small ligands, and provide high coverage of native binding residues by combining annotations across multiple complexes that share the same protein (same UniProt ID). As for sequence profile, SCRIBER is the first to use protein disorder and secondary structure for the PPI site prediction. More specifically, the RSA is predicted by the fast ASAquick method and the ECO is computed from the output generated by the fast and sensitive HHblits.

The RAAP and ANCHOR-derived putative protein-binding disorder are described in Section 3.5, which are also computationally efficient. Secondary structure is predicted by the fast version of PSIPRED [69] which does not need MSA. The physicochemical properties are derived from AAindex [70]. Conservative estimates show that SCRIBER is at least three times faster than the other methods. In algorithm design, SCRIBER utilizes a two-layer architecture where the first layer generates predictions of protein-binding, RNA-binding, DNA-binding and small ligand-binding propensities through sliding window strategy and logistic regression. The second layer re-predicts protein-binding propensity according to the results from the first layer, to reduce overlap between protein-binding residues and the other types of binding residues.

5.5 *DELPHI*

DELPHI [36] is the latest sequence-based PPI site predictor. Its idea is quite straightforward, namely that using more training data, more informative features and complex deep learning architecture may improve the predictive performance. The training data of DELPHI are obtained from [55] and contain 9,982 proteins. The independent test sets consist of Dset_186, Dset_72, Dset_164, and Dset_448 from SCRIBER. A subset of Dset_448 (Dset_355) was further constructed since Dset_448 contains 93 proteins sharing more than 40% similarity with the training set of one competing method (DLPred). DELPHI uses 11 feature groups including HSP, ProtVec1D, position information, PSSM, ECO, RSA, RAAP, hydropathy, protein-binding disorder, physicochemical properties and PKx, in which HSP, ProtVec1D and position information are firstly introduced in the PPI site prediction. HSP and ProtVec are described in Section 3.5. ProtVec1D is a one-dimensional value summed from the 100-dimensional vector by ProtVec in order to enhance the training speed. Inspired by [71], DELPHI uses position information to provide global information, which is computed by the residue position divided by the protein's length. DELPHI applies an ensemble architecture that combines a CNN and an RNN component with fine tuning technique. Concretely, a window size of 31 is employed to extract a subsequence around the target residue, and then a 2D feature vector is constructed. The 2D protein profile is considered as an image with one channel and is input to a CNN model to refine local information. The same feature matrix is also input to a bidirectional

GRU to extract long-range information. Finally, the outputs of these two modules are concatenated and passed to two fully connected layers. During training, the CNN and GRU networks are firstly tuned separately, and then their weights are frozen and the fully connected layers are trained in the ensemble architecture. DELPHI obtained an AUC and AUPR of 0.746 and 0.326, respectively, on Dset_355, surpassing all other methods as indicated in [36].

6. Computational methods for protein-peptide interaction (PPepI) site prediction

In this section, we introduce four representative and publicly available sequence-based methods for the PPepI site prediction. We summarize these methods in Table 2, in the order of their publication time. Although there is no acknowledged and widely used benchmark dataset as in PPI site prediction problem, all PPepI site datasets used by these methods were derived from BioLiP. Similarly, PSSM is also widely adopted in this problem, indicating that evolutionary information is important for the peptide binding site detection. The algorithm designs of these methods have also

Table 2. Summary of the sequence-based methods for PPepI site prediction in recent years.

Method	Year	Dataset size	Feature[1]	Algorithm	Availability[2]
SPRINT	2016	Train:1199 Test:80	one-hot sequence coding, PSSM, RSA, SS, PHY	SVM	S
PepBind	2018	Train:640 Test:639	PSSM, HMM, SS, ID	SVM+ template-based method	S
Visual	2020	Train:1116 Test:125	PSSM, HSE, SS, ASA, torsion angles, PHY	CNN	C
MTDsite	2021	Train:1115 Test:125	PSSM, HMM, CN, HSE, RSA, torsion ·angles	LSTM+ multi-task	S, C

[1]PHY denotes physicochemical properties; ID denotes intrinsic disorder; SS denotes putative secondary structure; HSE denotes half sphere exposure; CN denotes contact number.
[2]S and C correspond to the availability of the web server and source code, respectively.

been getting more complex, from machine learning to deep learning. As for availability, we also recommend that web server is necessary for the PPepI site prediction task. The remainder of this section describes the designs of these four methods.

6.1 SPRINT

SPRINT [29] is the first machine-learning method for the sequence-based prediction of PPepI sites. The training and test sets of SPRINT are collected from BioLiP and contain 1,199 and 80 proteins, respectively. However, to avoid bias caused by the imbalanced problem, SPRINT did not employ whole protein chains, but constructs a balanced dataset using under-sampling technique by randomly selecting a number of non-binding residues which is equal to the number of binding residues. The final training set contains 15,688 binding and non-binding residues while the independent test set contains 1,056 binding and non-binding residues. In the feature preparation, the one-hot sequence encoding and physicochemical properties are computed and employed, as described in Section 3. The RSA and secondary structure are predicted by SPIDER2 [72] first, and then feature groups, such as local average RSA that represent the local structural properties, are derived based on these two features with the sliding window strategy. The PSSM-based features not only contain PSSM, but also include the information entropy S_E and Close Neighbor Correlation Coefficient (CNCC) [73] based on the probability matrix P_{ij} from PSI-BLAST where $j = 1, 2, ..., 20$ and $i = 1, ..., L$ (L is the protein length):

$$S_E = \sum_{j=1}^{20} P_{ij} \times \ln\left(P_{ij}\right) \tag{13}$$

$$\text{CNCC}_{ik} = \frac{\sum_{j=1}^{20} P_{ij} P_{kj}}{\sqrt{\sum_{j=1}^{20} \left(P_{ij}\right)^2 \sum_{j=1}^{20} \left(P_{kj}\right)^2}} \tag{14}$$

where CNCC_{ik} is the Pearson correlation coefficient between the sequence profile of the given residue i and the adjacent residue k in a sliding window. The features are optimized using greedy sequential forward selection, which are finally concatenated and input to an SVM for the PPepI site prediction (Fig. 1).

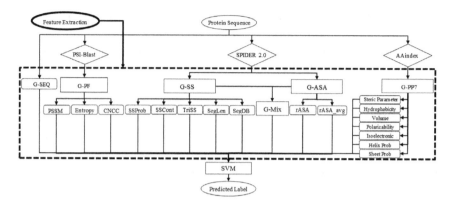

Figure 1.   The architecture of SPRINT.

6.2 *PepBind*

PepBind [12] is a consensus method consisting of an *ab initio* method SVMpep and two template-based methods, S-SITE and TM-SITE. This is the first predictor that introduces intrinsic disorder into this task. The 1,270 peptide-binding proteins used in PepBind are the same as in SPRINT, while PepBind re-divides them into a training and test set consisting of 640 and 639 proteins, respectively. Unlike SPRINT, PepBind retains complete protein chains without using any down-sampling technique. Besides intrinsic disorder computed by IUPred as described in Section 3.5, PepBind also adopts sliding window to extract local secondary-structure-based features by SPIDER2, HMM-based features including HMM and CNCC, and PSSM-based features including PSSM, CNCC and relative entropy (RE) similar to SPRINT:

$$RE_i = \sum_{j=1}^{20} P_{ij} \log_2 \frac{P_{ij}}{b_j} \qquad (15)$$

where *i* is the *i*th residue, *j* represents one of the 20 standard residues, and b_j is the Robinson background frequency [74] of the residue *j*. All these features are fed into SVM to train SVMpep. In addition, two template-based methods TM-SITE and S-SITE for ligand-binding site detection [75] are adopted, to transfer the binding annotations from homologous templates from BioLiP to the query sequence based on structure alignment (TM-SITE) and sequence profile-profile alignment (S-SITE). The protein structures required by TM-SITE are predicted by I-TASSER Suite [76]. The predictions of SVMpep, S-SITE and TM-SITE are selectively combined according to the confidence scores of these two complementary template-based methods, which derived the final prediction of PepBind.

6.3 *Visual*

Visual [34] is a CNN-based method which transforms protein sequences into images. The training and test sets of Visual contains 1,116 and 125 proteins, respectively, which are all extracted from BioLiP. The input features consist of commonly used PSSM, physicochemical properties and the features predicted by SPIDER2, including HSE, ASA, backbone torsion angles and secondary structure. These features are concatenated into a 38-dimensional vector for each amino acid, and then a sliding window with the size of seven was employed to construct an image-like representation with 7×38 pixels, which was input to a two-layer CNN for the PPepI site prediction.

6.4 *MTDsite*

Although many successes have been achieved for the protein binding residue predictions, the existing methods have relatively low accuracies due to the limited available experimental structures for training. As different types of molecules partially share common chemical mechanisms and interaction patterns, the prediction for one molecule type should benefit from the binding information of other molecule types. MTDsite [37] is the first method to use multi-task learning for simultaneous predictions of multiple ligand binding sites, which learns common information through shared networks while retaining task-related output layers. MTDsite applies a training set of 309 DNA-binding proteins and 157 RNA-binding proteins from [77], 1,155 peptide-binding proteins from BioLiP, and 157 carbohydrate-binding proteins from [78], and a test set of 125 proteins was used to evaluate the PPepI site prediction performance. In feature preparation, PSSM, HMM and SPIDER3-based features CN, HSE, RSA, and torsion angles (described in Section 3.3) are concatenated into a 54-dimensional feature vector for each amino acid. As shown in Figure 2, MTDsite consists of two parts. The first part is a shared bidirectional LSTM (BiLSTM) network that is used to collect the common information of long-range residues

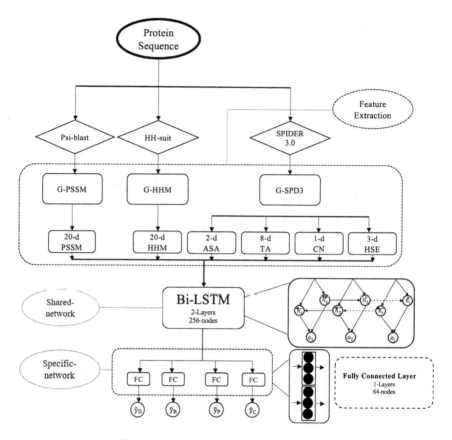

Figure 2.   The architecture of MTDsite.

in the protein chain. The second part is four specific fully-connected networks that are trained respectively for four individual ligand types to extract ligand-specific properties. Each protein is used to train the shared networks and the individual network for its binding type t without affecting other three individual networks by using a loss function:

$$L_t(y_c, P_c) = -\delta_{c,t} \sum_{i=1}^{N_c} \left(y_i \log(P_i) + (1 - y_i) \log(1 - P_i) \right) \qquad (16)$$

where y_i and P_i are the binding label and predicted binding probability for residue i in the input protein chain c with length N_c. $\delta_{c,t}$ is set to 1 if chain c is known binding type t, otherwise 0. In such architecture, the shared networks can be better trained with a larger dataset.

7. Summary

Looking at the history of the PPI and PPepI site predictions, generally, the datasets used for training and testing are getting larger over time. This is due to the increase in the experimentally resolved protein-protein and protein-peptide complex structures in PDB. Besides, the adopted features are also getting more complex and powerful, and the algorithm designs are getting more sophisticated, moving from machine learning to deep learning. Consensus methods, ensemble learning and multi-task learning are also explored. However, there is still room for further improvements on these tasks, since the performance of these sequence-based methods still greatly lags behind that of the structure-based methods. As reported in [58], the structure-based PPI site predictor GraphPPIS surpasses the state-of-the-art sequence-based method DELPHI by 34.5% in AUPR, although GraphPPIS uses less training data and features. As proteins are composed of amino acids located in the three-dimensional space, GraphPPIS represents proteins as two-dimensional connected graphs defined by the distance maps of amino acids, and adopts graph convolution networks to efficiently learn the structural environments around the target residues. This suggests that not only the residue-wise structural features such as RSA and secondary structure, but also the entire protein structural context topology is very important for identifying protein binding residues.

Due to the significance of protein structures, many sequence-based protein-related predictors first use protein structure predicting tools including RaptorX [79] and SPOT-Contact [80] to predict the protein structure models or protein distance maps, and then convert the problems to structure-based predictions, such as the protein function predictor GAT-GO [81] and protein solubility predictor GraphSol [82]. Recently, protein structure prediction is experiencing a breakthrough with the emergence of AlphaFold2 [83], which has incorporated physical and biological knowledge about protein structure, information of multi-sequence alignment, and the sophisticated design of the deep learning algorithm. This method demonstrated accuracy competitive with experiment in a majority of cases in the challenging 14th Critical Assessment of protein Structure Prediction (CASP14) [84]. At roughly the same time, RoseTTAFold [85] was published, which performs slightly worse than AlphaFold2 but is more computationally efficient. Such breakthroughs will undoubtedly benefit downstream protein function studies, including binding site predictions. In fact, recently, AlphaFold2 has been successfully applied to protein-DNA binding site prediction in GraphSite [86]. Input with protein sequence, GraphSite employs AlphaFold2 to produce the single representation and the predicted protein structure, from which the distance map and DSSP are extracted. Then, the single representation, DSSP and sequence-derived features PSSM and HMM are concatenated to form the node feature vector, which is then input to a graph transformer model with k-nearest mask by the distance map to learn DNA-binding site patterns. GraphSite significantly outperforms other sequence-based methods, and even shows better or comparable performance to other structure-based methods [86]. We expect that the mainstream sequence-based PPI and PPepI site prediction methods in the future will first predict the protein structures, and then extract geometric context of the predicted structures using graph-based deep learning algorithms. Meanwhile, the computational costs of AlphaFold2 and RoseTTAFold are still relatively high, and thus model compression and optimization techniques as well as knowledge distillation [87] should be explored to compact these complex models for potential application for the genome-scale protein binding site prediction.

8. Acknowledgment

This study has been supported by the National Key R&D Program of China [2020YFB0204803], National Natural Science Foundation of China [61772566, 62041209], Guangdong Key Field R&D Plan [2019B020228001, 2018B010109006], Introducing Innovative and Entrepreneurial Teams [2016ZT06D211], and Guangzhou S&T Research Plan [202007030010].

References

[1] Zhang J and Kurgan L. Review and comparative assessment of sequence-based predictors of protein-binding residues. *Briefings in Bioinformatics*, 2018. **19**(5): 821–837.

[2] De Las Rivas J and Fontanillo C. Protein–protein interaction networks: Unraveling the wiring of molecular machines within the cell. *Briefings in Functional Genomics*, 2012. **11**(6): 489–496.

[3] Orii N and Ganapathiraju MK. Wiki-pi: A web-server of annotated human protein-protein interactions to aid in discovery of protein function. *PloS One*, 2012. **7**(11): e49029.

[4] Kuzmanov U and Emili A. Protein-protein interaction networks: Probing disease mechanisms using model systems. *Genome Medicine*, 2013. **5**(4): 1–12.

[5] Wells JA and McClendon CL. Reaching for high-hanging fruit in drug discovery at protein–protein interfaces. *Nature*, 2007. **450**(7172): 1001–1009.

[6] Xu M *et al.* De novo molecule design through the molecular generative model conditioned by 3D information of protein binding sites. *Journal of Chemical Information and Modeling*, 2021. **61**(7): 3240–3254.

[7] Petsalaki E and Russell RB. Peptide-mediated interactions in biological systems: New discoveries and applications. *Current Opinion in Biotechnology*, 2008. **19**(4): 344–350.

[8] Lee AC-L, *et al.* A comprehensive review on current advances in peptide drug development and design. *International Journal of Molecular Sciences*, 2019. **20**(10): 2383.

[9] Fosgerau K and Hoffmann T. Peptide therapeutics: Current status and future directions. *Drug Discovery Today*, 2015. **20**(1): 122–128.

[10] Lei Y, *et al.* A deep-learning framework for multi-level peptide–protein interaction prediction. *Nature Communications*, 2021. **12**(1): 1–10.

[11] Shoemaker BA and Panchenko AR. Deciphering protein–protein interactions. Part I. Experimental techniques and databases. *PLoS Computational Biology*, 2007. **3**(3): e42.

[12] Zhao Z, *et al.* Improving sequence-based prediction of protein–peptide binding residues by introducing intrinsic disorder and a consensus method. *Journal of Chemical Information and Modeling*, 2018. **58**(7): 1459–1468.

[13] Vlieghe P, *et al.* Synthetic therapeutic peptides: Science and market. *Drug Discovery Today*, 2010. **15**(1–2): 40–56.

[14] Dyson HJ and Wright PE. Intrinsically unstructured proteins and their functions. *Nature Reviews Molecular Cell Biology*, 2005. **6**(3): 197–208.

[15] Berman HM, *et al.* The protein data bank. *Nucleic Acids Research*, 2000. **28**(1): 235–242.

[16] Capra JA and Singh M. Predicting functionally important residues from sequence conservation. *Bioinformatics*, 2007. **23**(15): 1875–1882.

[17] Liang S, *et al.* Protein binding site prediction using an empirical scoring function. *Nucleic Acids Research*, 2006. **34**(13): 3698–3707.

[18] Shoemaker BA, *et al.* IBIS (Inferred Biomolecular Interaction Server) reports, predicts and integrates multiple types of conserved interactions for proteins. *Nucleic Acids Research*, 2012. **40**(D1): D834–D840.

[19] Yuan Q, *et al.* Structure-aware protein–protein interaction site prediction using deep graph convolutional network. *Bioinformatics*, 2021. **38**(1): 125–132.

[20] Kozlovskii I and Popov P. Protein–peptide binding site detection using 3D convolutional neural networks. *Journal of Chemical Information and Modeling*, 2021. **61**(8): 3814–3823.

[21] Nagarajan R, *et al.* Novel approach for selecting the best predictor for identifying the binding sites in DNA binding proteins. *Nucleic Acids Research*, 2013. **41**(16): 7606–7614.

[22]   Apweiler R, *et al.* UniProt: The universal protein knowledgebase. *Nucleic Acids Research*, 2004. **32**(Suppl 1): D115–D119.

[23]   Esmaielbeiki R, *et al.* Progress and challenges in predicting protein interfaces. *Briefings in Bioinformatics*, 2015. **17**(1): 117–131.

[24]   Asgari E and Mofrad MR. Continuous distributed representation of biological sequences for deep proteomics and genomics. *PloS One*, 2015. **10**(11): e0141287.

[25]   Li Y and Ilie L. SPRINT: Ultrafast protein-protein interaction prediction of the entire human interactome. *BMC Bioinformatics*, 2017. **18**(1): 485.

[26]   Murakami Y and Mizuguchi K. Applying the Naïve Bayes classifier with kernel density estimation to the prediction of protein–protein interaction sites. *Bioinformatics*, 2010. **26**(15): 1841–1848.

[27]   Northey TC, *et al.* IntPred: A structure-based predictor of protein–protein interaction sites. *Bioinformatics*, 2017. **34**(2): 223–229.

[28]   Wang X, *et al,* Protein–protein interaction sites prediction by ensemble random dom forests with synthetic minority oversampling technique. *Bioinformatics*, 2019. **35**(14): 2395–2402.

[29]   Taherzadeh G, *et al.* Sequence-based prediction of protein–peptide binding sites using support vector machine. *Journal of Computational Chemistry*, 2016. **37**(13): 1223–1229.

[30]   Deng A, *et al.* Developing computational model to predict protein-protein interaction sites based on the XGBoost algorithm. *International Journal of Molecular Sciences*, 2020. **21**(7): 2274.

[31]   Zhang J and Kurgan L. SCRIBER: Accurate and partner type-specific prediction of protein-binding residues from proteins sequences. *Bioinformatics*, 2019. **35**(14): i343–i353.

[32]   Qiu J, *et al.* ProNA2020 predicts protein–DNA, protein–RNA, and protein–protein binding proteins and residues from sequence. *Journal of Molecular Biology*, 2020. **432**(7): 2428–2443.

[33]   Zhu H, *et al.* ConvsPPIS: Identifying protein-protein interaction sites by an ensemble convolutional neural network with feature graph. *Current Bioinformatics*, 2020. **15**(4): 368–378.

[34]   Wardah W, *et al.* Predicting protein-peptide binding sites with a deep convolutional neural network. *Journal of Theoretical Biology*, 2020. **496**: 110278.

[35]   Zhang B, *et al.* Sequence-based prediction of protein-protein interaction sites by simplified long short-term memory network. *Neurocomputing*, 2019. **357**: 86–100.

[36]   Li Y, *et al.* DELPHI: Accurate deep ensemble model for protein interaction sites prediction. *Bioinformatics*, 2020. **37**(7): 896–904.

[37]   Sun Z, *et al.* To improve the predictions of binding residues with DNA, RNA, carbohydrate, and peptide via multi-task deep neural networks. *IEEE/ACM Transactions on Computational Biology and Bioinformatics*, 2021.

[38]   Yang J, *et al.* BioLiP: A semi-manually curated database for biologically relevant ligand–protein interactions. *Nucleic Acids Research*, 2012. **41**(D1): D1096–D1103.

[39] Cock PJ, *et al.* Biopython: Freely available Python tools for computational molecular biology and bioinformatics. *Bioinformatics*, 2009. **25**(11): 1422–1423.

[40] DeLano WL. The PyMOL Molecular Graphics System, Version 2.0 Schrödinger, LLC.

[41] Macrae CF, *et al.* Mercury 4.0: From visualization to analysis, design and prediction. *Journal of Applied Crystallography*, 2020. **53**(1): 226–235.

[42] Mikolov T, *et al.* Distributed representations of words and phrases and their compositionality. *Advances in Neural Information Processing Systems*, Burges CJ, *et al.* (eds.), Curran Associates, Inc., Red Hook, NY, US, 2013. **2**: 3111–3119.

[43] Rao R, *et al.* Evaluating protein transfer learning with tape. *Advances in Neural Information Processing Systems*, 2019. **32**: 9689.

[44] Altschul SF, *et al.* Gapped BLAST and PSI-BLAST: A new generation of protein database search programs. *Nucleic Acids Research*, 1997. **25**(17): 3389–3402.

[45] Suzek BE, *et al.* UniRef: Comprehensive and non-redundant UniProt reference clusters. *Bioinformatics*, 2007. **23**(10): 1282–1288.

[46] Remmert M, *et al.* HHblits: Lightning-fast iterative protein sequence searching by HMM-HMM alignment. *Nature Methods*, 2012. **9**(2): 173–175.

[47] Mirdita M, *et al.* Uniclust databases of clustered and deeply annotated protein sequences and alignments. *Nucleic Acids Research*, 2017. **45**(D1): D170–D176.

[48] Heffernan R, *et al.* Capturing non-local interactions by long short-term memory bidirectional recurrent neural networks for improving prediction of protein secondary structure, backbone angles, contact numbers and solvent accessibility. *Bioinformatics*, 2017. **33**(18): 2842–2849.

[49] Faraggi E, *et al.* Accurate single-sequence prediction of solvent accessible surface area using local and global features. *Proteins: Structure, Function, and Bioinformatics*, 2014. **82**(11): 3170–3176.

[50] Meiler J, *et al.* Generation and evaluation of dimension-reduced amino acid parameter representations by artificial neural networks. *Molecular Modeling Annual*, 2001. **7**(9): 360–369.

[51] Dosztányi Z, *et al.* IUPred: Web server for the prediction of intrinsically unstructured regions of proteins based on estimated energy content. *Bioinformatics*, 2005. **21**(16): 3433–3434.

[52] Hanson J, *et al.* Improving protein disorder prediction by deep bidirectional long short-term memory recurrent neural networks. *Bioinformatics*, 2017. **33**(5): 685–692.

[53] Dosztányi Z, *et al.* ANCHOR: Web server for predicting protein binding regions in disordered proteins. *Bioinformatics*, 2009. **25**(20): 2745–2746.

[54] Li Y and Ilie L. SPRINT: Ultrafast protein-protein interaction prediction of the entire human interactome. *BMC Bioinformatics*, 2017. **18**(1): 1–11.

[55] Zhang J, *et al.* Comprehensive review and empirical analysis of hallmarks of DNA-, RNA-and protein-binding residues in protein chains. *Briefings in Bioinformatics*, 2019. **20**(4): 1250–1268.

[56] Vacic V, *et al*. Composition profiler: A tool for discovery and visualization of amino acid composition differences. *BMC Bioinformatics*, 2007. **8**(1): 1–7.

[57] Dhole K, *et al*. Sequence-based prediction of protein–protein interaction sites with L1-logreg classifier. *Journal of Theoretical Biology*, 2014. **348**: 47–54.

[58] Yuan Q, *et al*. Structure-aware protein–protein interaction site prediction using deep graph convolutional network. *Bioinformatics*, 2021. **38**(1): 125–132.

[59] Boughorbel S, *et al*. Optimal classifier for imbalanced data using Matthews correlation coefficient metric. *PloS One*, 2017. **12**(6): e0177678.

[60] Chicco D and Jurman G. The advantages of the Matthews correlation coefficient (MCC) over F1 score and accuracy in binary classification evaluation. *BMC Genomics*, 2020. **21**(1): 1–13.

[61] Davis J and Goadrich M. The relationship between precision-recall and ROC curves. *Proceedings of the 23rd International Conference on Machine Learning*, Cohen W & Moore A (eds.), Association for Computing Machinery, New York, NY, USA, 2006. 233–240.

[62] Saito T and Rehmsmeier M. The precision-recall plot is more informative than the ROC plot when evaluating binary classifiers on imbalanced datasets. *PloS One*, 2015. **10**(3). e0118432.

[63] Wagner M, *et al*. Linear regression models for solvent accessibility prediction in proteins. *Journal of Computational Biology*, 2005. **12**(3): 355–369.

[64] Joo K, *et al*. Sann: Solvent accessibility prediction of proteins by nearest neighbor method. *Proteins: Structure, Function, and Bioinformatics*, 2012. **80**(7): 1791–1797.

[65] Wang G and Dunbrack Jr RL. PISCES: A protein sequence culling server. *Bioinformatics*, 2003. **19**(12): 1589–1591.

[66] Chen H and Zhou HX. Prediction of interface residues in protein–protein complexes by a consensus neural network method: test against NMR data. *Proteins: Structure, Function, and Bioinformatics*, 2005. **61**(1): 21–35.

[67] Bowie JU, *et al*. A method to identify protein sequences that fold into a known three-dimensional structure. *Science*, 1991. **253**(5016): 164–170.

[68] Quan L, *et al*. STRUM: Structure-based prediction of protein stability changes upon single-point mutation. *Bioinformatics*, 2016. **32**(19): 2936–2946.

[69] Buchan DW, *et al*. Scalable web services for the PSIPRED protein analysis workbench. *Nucleic Acids Research*, 2013. **41**(W1): W349–W357.

[70] Kawashima S, *et al*. AAindex: Amino acid index database, progress report 2008. *Nucleic Acids Research*, 2007. **36**(Suppl 1): D202–D205.

[71] Zeng M, *et al*. Protein–protein interaction site prediction through combining local and global features with deep neural networks. *Bioinformatics*, 2020. **36**(4): 1114–1120.

[72] Yang Y, *et al*. Spider2: A package to predict secondary structure, accessible surface area, and main-chain torsional angles by deep neural networks. In

Prediction of Protein Secondary Structure, Zhou Y, *et al.* (eds.), Springer, 2017. 55–63.

[73] Cheng J and Baldi P. Improved residue contact prediction using support vector machines and a large feature set. *BMC Bioinformatics*, 2007. **8**(1): 1–9.

[74] Robinson AB and Robinson LR. Distribution of glutamine and asparagine residues and their near neighbors in peptides and proteins. *Proceedings of the National Academy of Sciences*, 1991. **88**(20): 8880–8884.

[75] Yang J, *et al.* Protein–ligand binding site recognition using complementary binding-specific substructure comparison and sequence profile alignment. *Bioinformatics*, 2013. **29**(20): 2588–2595.

[76] Yang J, *et al.* The I-TASSER Suite: Protein structure and function prediction. *Nature Methods*, 2015. **12**(1): 7–8.

[77] Yan J and Kurgan L. DRNApred, fast sequence-based method that accurately predicts and discriminates DNA-and RNA-binding residues. *Nucleic Acids Research*, 2017. **45**(10): e84–e84.

[78] Zhao H, *et al.* Carbohydrate-binding protein identification by coupling structural similarity searching with binding affinity prediction. *Journal of Computational Chemistry*, 2014. **35**(30): 2177–2183.

[79] Källberg M, *et al.* Template-based protein structure modeling using the RaptorX web server. *Nature Protocols*, 2012. **7**(8): 1511–1522.

[80] Hanson J, *et al.* Accurate prediction of protein contact maps by coupling residual two-dimensional bidirectional long short-term memory with convolutional neural networks. *Bioinformatics*, 2018. **34**(23): 4039–4045.

[81] Lai B and Xu J. Accurate protein function prediction via graph attention networks with predicted structure information. *Briefings in Bioinformatics*, 2021. **23**(1): bbab502.

[82] Chen J, *et al.* Structure-aware protein solubility prediction from sequence through graph convolutional network and predicted contact map. *Journal of Cheminformatics*, 2021. **13**(1): 1–10.

[83] Jumper J, *et al.* Highly accurate protein structure prediction with AlphaFold. *Nature*, 2021. **596**(7873): 583–589.

[84] Pereira J, *et al.* High-accuracy protein structure prediction in CASP14. *Proteins: Structure, Function, and Bioinformatics*, 2021. **89**(12): 1687–1699.

[85] Baek M, *et al.* Accurate prediction of protein structures and interactions using a three-track neural network. *Science*, 2021. **373**(6557): 871–876.

[86] Yuan Q, *et al.* AlphaFold2-aware protein–DNA binding site prediction using graph transformer. *Briefings in Bioinformatics*, 2022. **23**(2): bbab564.

[87] Hinton G, *et al.* Distilling the knowledge in a neural network. arXiv, 2015.

© 2023 World Scientific Publishing Company
https://doi.org/10.1142/9789811258589_0010

Chapter 10

Machine Learning Methods for Predicting Protein-Nucleic Acids Interactions

Min Li*,‡, Fuhao Zhang* and Lukasz Kurgan†,‡

**Hunan Provincial Key Lab on Bioinformatics,*
School of Computer Science and Engineering,
Central South University,
Changsha, China, 410083.
†*Department of Computer Science,*
Virginia Commonwealth University,
Richmond, Virginia, United States
‡*corresponding authors: Min Li at limin@mail.csu.edu.cn;*
Lukasz Kurgan at lkurgan@vcu.edu

Protein-nucleic acids interactions drive many key cellular functions, such as regulation of gene expression, transcription, and translation. Experimental characterization of the molecular-level details of these interactions is relatively expensive and time-consuming since it requires application of complex and labor-intensive methods, such as X-ray crystallography and/or NMR. Given the relatively low coverage of the experimental molecular-level data on the protein-nucleic acids interactions, many computational methods that predict these interactions from the readily available protein sequences were developed. We introduce and describe a comprehensive collection of 51 methods that predict nucleic acid interacting amino acids in protein sequences. These methods include 20 DNA-binding predictors, 20 RNA-binding predictors and 11 methods that predict both DNA- and RNA-binding residues. We briefly summarize their inputs, predictive architectures, outputs and availability. We find that most of these methods were trained using protein-nucleic acids structures, compared to a more limited number of methods that predict these interactions in the intrinsically disordered regions. We observe that these methods rely almost exclusively on classical/shallow machine learning and deep learning algorithms. Finally, we endorse five recent, readily available and arguably more useful predictors.

1. Introduction

Proteins carry out their cellular functions by interacting with nucleic acids [1–12], proteins [13–17] and a variety of other ligands including small molecules and lipids [18–20]. The protein-nucleic acid interactions are instrumental for a wide spectrum of cellular functions, such as regulation of gene expression, transcription, and translation, to name but a few. Molecular-level knowledge of these interactions is largely derived from structural studies of protein-nucleic acid complexes, which are often sourced from the Protein Data Bank database [21]. The structural data are used to categorize the protein-nucleic acids interactions, characterize the underlying physics, and decipher patterns that define molecular recognition and specificity of interactions [22–25]. Moreover, recent studies reveal that the protein-DNA and protein-RNA interactions are also a common function of intrinsically disordered regions (IDRs) [11,26–30], which are defined as sequence segments that lack a stable equilibrium structure under physiological conditions [31–33].

The experimental methods to study these interactions are relatively time-consuming and labor-intensive, and consequently, they cannot keep up with the rapid accumulation of protein sequences. One solution is to use the available experimental data on protein-nucleic acids interactions to devise computational models that accurately predict these interactions from protein sequences [6,34–42]. These computational methods can be classified into two categories: protein-level versus residue-level. The protein-level methods identify whether a given protein sequence binds DNA and/or RNA, while the residue-level methods predict whether and which residues in a given protein sequence interact with DNA and/or RNA. We focus on the residue-level methods that provide a higher level of details. The sequence-based predictors of nucleic-acid binding residues require only a protein sequence as the input and thus, are able to provide predictions for the over 200 million of currently available protein sequences [43].

We survey and describe over 50 sequence-based predictors of the DNA and/or RNA binding residues. We identify these methods by scanning past surveys [6,35,37–40,42,44,45] supplemented with a manual literature search. These methods output binary predictions and numeric propensities for each residue in an input protein sequence. The binary predictions denote whether a given residue interacts with DNA and/or RNA (0 for non-binding residue versus 1 means for a DNA and/or RNA binding residue), while the propensities express likelihood of these interactions. Figure 1 shows an example result generated by DeepDISOBind, one of the most recent methods that predict DNA and RNA interactions in IDRs [46] for the silent information regulator Sir3p from budding yeast (DisProt: DP00533; UniProt: P06701). This protein is instrumental for modulating chromatin [47,48]. It includes disordered DNA-binding region (positions 216 to 549) that is flanked by structured segments that extend to both sequence termini, as shown based on the experimental annotations

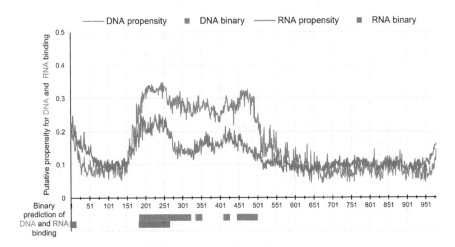

Figure 1. Prediction of the DNA binding and RNA-binding residues generated by DeepDISObind for the silent information regulator Sir3p from yeast (DisProt: DP00533; UniProt: P06701). The prediction was generated using the DeepDISObind's webserver located at https://www.csuligroup.com/DeepDISOBind/.

available in the DisProt database [49]. The blue and red plots in Figure 1 represent the predicted propensities for interactions with DNA and RNA partners, respectively. The binary predictions are shown underneath using horizontal color-coded bars. This example illustrates the format and value of the sequence-based predictions. In this particular case, DeepDISOBind identifies DNA-binding residues in the segments between positions 190 and 500, which is in good agreement with the location of the experimentally identified DNA-binding IDR (positions 216 to 549). At the same time, the RNA-binding prediction generated by DeepDISOBind suggests a much smaller likelihood of interactions with RNA for this protein, i.e., the propensities shown in red are relatively low.

2. Prediction of the protein-nucleic acid binding residues from sequence

Majority of the sequence based methods for the predictions of protein-nucleic binding residues rely on predictive models that are generated from training data using classical/shallow machine learning algorithms and deep learning algorithms. The training process employs the experimentally annotated data, which is typically collected from publicly available databases, such as PDB [21], BioLip [50] and DisProt [49,51], to optimize the architectures and parameters of the machine learning-generated models. This is done by minimizing differences between the outputs of these models and the native annotations. After the training process is completed, the models can be used to predict the DNA-binding and/or RNA-binding residues for proteins sequences outside of the training set.

Figure 2 shows a timeline of the sequence-based predictors, which are categorized into methods that rely on training data collected from structured protein-RNA and protein-DNA complexes, typically collected from PDB [21,52] or BioLip [50] (in orange) versus those that use training data concerning interactions in IDRs, which are usually obtained from DisProt [49,51]. The first predictors were developed around 2004 [53,54] and over 50 were published since. This results in the average rate of three new methods per year. Figure 2 reveals that these development efforts are relatively even across the years. However, we note that only two predictors focus on the interactions in the disordered regions, DisoRDPbind that was released in 2015 [55–57] and DeepDISObind that was published in 2021 [46]. Interestingly, recent research that investigates structure-trained versus

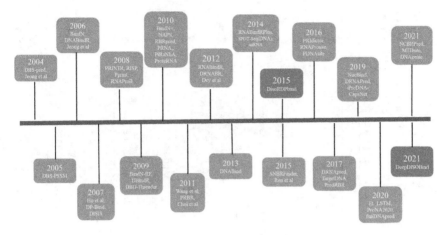

Figure 2.   Timeline of the sequence-based predictors of the protein-nucleic acid binding residues. The structure-trained predictors are in orange while the disorder-trained predictors are in blue.

disorder-trained methods for the prediction of the protein-binding residues shows that these two classes of methods produce complementary results [58]. Similar observations are expected in the case of the protein-nucleic acids interactions, highlighting the importance of developing both classes of predictors.

2.1 Overview of the sequence-based predictors

Table 1 summarizes the key aspects of the 51 sequence-based predictors of the DNA and/or RNA binding residues, including their predictive targets (DNA binding, RNA binding, and DNA and RNA binding) and availability. This comprehensive table reveals several interesting insights. First, significant majority of these computational methods target prediction of either DNA-binding or RNA-binding residues. To be more exact, we found 20 methods that predict the DNA-binding residues and another 20 predictors of the RNA-binding residues. However, we also identified 11 methods that concomitantly predict RNA-binding and DNA-binding residues, with the first one being BindN that was published in 2006 [59]. These methods are arguably more convenient to use since they provide both predictions, typically with a single code execution, and since their use does not require (sometimes painful) conversions between different formats of input and

Table 1. Summary of the 51 sequence-based predictors of the protein-nucleic acid binding residue. The table considers the prediction target DNA binding, RNA binding, and DNA and RNA binding) and availability. The availability indicates whether the published predictors are provided as web servers, source code or in both modes. The URL provides the location of the implementation/web server that was published in the original article. The "accessible" column indicates whether the URL was available as 15 December 2021. N/A means that a given predictor did not provide web server and source codes information when it was published.

Method	Ref	Predictive Target	Availability Type	URL	Accessible
DBS-pred	[54]	DNA	web server	http://www.abren.net/dbs-pred/	NO
Jeong et al. 2004	[53]	RNA	N/A	N/A	N/A
DBS-PSSM	[65]	DNA	web server	http://dxpssm.netasa.org	NO
BindN	[59]	DNA and RNA	web server	http://bioinfo.ggc.org/bindn	NO
DNABindR	[66]	DNA	web server	http://turing.cs.iastate.edu/PredDNA/index.html	NO
Jeong et al. 2006	[67]	RNA	N/A	N/A	N/A
Ho et al.	[68]	DNA	N/A	N/A	N/A
DP-Bind	[69]	DNA	web server	http://1g.rit.albany.edu/dp-bind	YES
DISIS	[70]	DNA	web server	http://cubic.bioc.columbia.edu/services/disis	NO
PRINTR	[71]	RNA	web server	http://210.42.106.80/printr/	NO
RISP	[72]	RNA	web server	http://grc.seu.edu.cn/RISP	NO
Pprint	[73]	RNA	web server	http://www.imtech.res.in/raghava/pprint/	YES
RNAProB	[74]	RNA	N/A	N/A	N/A
BindN-RF	[75]	DNA	web server	http://bioinfo.ggc.org/bindn-rf/	NO
DBindR	[76]	DNA	web server	http://www.cbi.seu.edu.cn/DBindR/DBindR.htm	NO
DBD-Threader	[77]	DNA	web server	http://cssb.biology.gatech.edu/skolnick/webservice/DBD-Threader/index.html	NO
BindN+	[78]	DNA and RNA	web server	http://bioinfo.ggc.org/bindn+/	NO
NAPS	[79]	DNA and RNA	web server	http://proteomics.bioengr.uic.edu/NAPS/	NO
PiRaNhA	[80]	RNA	web server	http://www.bioinformatics.sussex.ac.uk/PIRANHA	NO
ProteRNA	[81]	RNA	N/A	N/A	N/A
RBRpred	[10]	RNA	N/A	N/A	N/A
PRNA	[82]	RNA	source code	http://www.aporc.org/doc/wiki/PRNA http://www.sysbio.ac.cn/datatools.asp	NO

Method	Ref.	Nucleic acid	Access	URL	Available
Wang et al	[83]	RNA	N/A	N/A	N/A
PRBR	[84]	RNA	web server	http://www.cbi.seu.edu.cn/PRBR/	NO
Choi et al	[85]	RNA	N/A	N/A	N/A
RNABindR	[86]	RNA	web server	http://einstein.cs.iastate.edu/RNABindR/	NO
DNABR	[87]	DNA	web server	http://www.cbi.seu.edu.cn/DNABR/	NO
Dey et al.	[88]	DNA	N/A	N/A	N/A
DNABind	[89]	DNA	web server	http://mleg.cse.sc.edu/DNABind/	YES
RNABindRPlus	[90]	RNA	web server	http://einstein.cs.iastate.edu/RNABindRPlus/	NO
SPOT-Seq (DNA)	[91]	DNA	web server	http://sparks-lab.org	YES
aaRNA	[92]	RNA	web server	http://sysimm.ifrec.osaka-u.ac.jp/aarna/	YES
SNBRFinder	[93]	DNA and RNA	web server	http://ibi.hzau.edu.cn/SNBRFinder	NO
DisoRDPbind	[57]	DNA and RNA	web server	http://biomine.ece.ualberta.ca/DisoRDPbind/	YES
Ren et al.	[94]	RNA	N/A	N/A	N/A
PRIdictor	[95]	RNA	web server	http://bclab.inha.ac.kr/pridictor	YES
RNAProSite	[96]	RNA	web server	http://lilab.ecust.edu.cn/NABind	YES
PDNAsite	[97]	DNA	web server	http://hlt.hitsz.edu.cn:8080/PDNAsite/	NO
DRNApred	[98]	DNA and RNA	web server	http://biomine.cs.vcu.edu/servers/DRNApred/	YES
TargetDNA	[99]	DNA	web server	http://csbio.njust.edu.cn/bioinf/TargetDNA/	YES
PredRBR	[100]	RNA	N/A	N/A	N/A
NucBind	[60]	DNA and RNA	web server	http://yanglab.nankai.edu.cn/NucBind	YES
DNAPred	[101]	DNA	web server	http://csbio.njust.edu.cn/bioinf/dnapred/	YES
iProDNA-CapsNet	[64]	DNA	web server and source code	http://45.117.83.253/problem-iProDNA-CapsNet https://github.com/ngphubinh/iProDNA-CapsNet	YES
EL_LSTM	[102]	DNA	source code	http://hlt.hitsz.edu.cn/EL_LSTM/	NO
ProNA2020	[61]	DNA and RNA	web server and source code	https://github.com/Rostlab/ProNA2020.git http://www.predictprotein.org	YES
funDNApred	[103]	DNA	web server	http://biomine.cs.vcu.edu/servers/funDNApred/	YES
NCBRPred	[62]	DNA and RNA	web server and source code	http://bliulab.net/NCBRPred	YES
MTDsite	[63]	DNA and RNA	web server	http://biomed.nscc-gz.cn/server/MTDsite/	YES
DNAgenie	[104]	DNA	web server	http://biomine.cs.vcu.edu/servers/DNAgenie/	YES
DeepDISOBind	[46]	DNA and RNA	web server	https://csuligroup.com/DeepDISOBind/	YES

outputs (which is often required when using different individual methods). Interestingly, 5 of these 11 predictors were released in the last three years including NucBind[60], ProNA2020 [61], NCBRPred [62], MTDsite [63] and DeepDISOBind [46]. Moreover, four methods simultaneously predict protein/peptide-binding, DNA-binding and RNA-binding residues: DisoRDPbind [55–57], ProNA2020 [61], MTDsite [63], and DeepDISOBind [46], providing further advantages when compared to the other currently available options.

Table 1 also lists the URLs where the 51 methods are available, as described in the original articles. Most of these predictors are provided as web servers or source code. Only 11 predictors did not provide web servers nor source code at the time of the publication. The web servers are arguably more convenient and primarily target users who are not programmers or computer experts. The predictions are performed on the web server side, which means that users do not need to install software or run the predictions on their local hardware. The users simply utilize a web browser to arrive at a given URL, input their proteins sequence(s), typically in the FASTA format, click start, and collect the resulting predictions after the web server completes the work. The results are returned via the web page and/or are sent back by email. However, one of the common limitations of web servers is the number of input protein sequences, which is often restricted to one or a few for a single request. On the other hand, the source code is better suited for programmers or bioinformaticians. It has to be downloaded, installed and run on user's hardware. This option is particularly attractive if the predictions must be done on a larger scale (large protein families or genomes) and when embedding these predictors into larger bioinformatics platforms. Table 1 reveals that source code is available for only five methods and three of them, iProDNA-CapsNet [64], ProNA2020 [61] and NCBRPred [62], are also available as web servers. These three tools were released in three years. We found that while 38 of the 51 predictors provided web servers when they were originally published, only half the URLs of these web servers were working as of December 2021 when we tested them. We note that in some cases, these methods were possibly moved to another URL while we rely on the addresses provided in the original article.

2.2 *Architectures of the sequence-based predictors*

Table 2 provides further insights concerning the 51 sequence-based predictors of the DNA and/or RNA binding residues by summarizing and

Table 2. Summary of predictive models, inputs and outputs of the 51 sequence-based predictors of the protein-nucleic acid binding residues. The inputs include amino acid sequence (AAS), secondary structure (SS) predicted from sequence, solvent accessibility (RSA) predicted from sequence and evolutionary information (EVO) computed from multiple-sequence alignment. The predictive models are divided into two categories: classical/shallow machine learning (ML) and deep learning (DL). Specific ML algorithms include logistic regression (LR), neural network (NN), support vector machine (SVM), and random forest (RF). N/A in the outputs means that the information could not be checked since the implementation is not available.

Method	Ref	Inputs				Predictive Model	Outputs	
		AAS	EVO	SS	RSA		Binary Prediction	Predictive Propensity
DBS-pred	[54]	√	×	×	×	ML(NN)	N/A	N/A
Jeong et al. 2004	[53]	√	×	√	×	ML(NN)	N/A	N/A
DBS-PSSM	[65]	√	√	×	×	ML(NN)	√	√
BindN	[59]	√	×	×	×	ML(SVM)	√	√
DNABindR	[66]	√	×	×	×	ML(Naïve Bayes)	√	×
Jeong et al. 2006	[67]	√	×	×	×	ML(NN)	N/A	N/A
Ho et al.	[68]	√	×	×	×	ML(SVM)	N/A	N/A
DP-Bind	[69]	√	√	×	×	ML(ensemble learning)	√	√
DISIS	[70]	√	√	√	√	ML(SVM, NN)	N/A	N/A
PRINTR	[71]	√	×	√	×	ML(SVM)	N/A	N/A
RISP	[72]	√	√	×	×	ML(NN)	N/A	N/A
Pprint	[73]	√	√	×	×	ML(SVM)	√	√
RNAProB	[74]	√	√	×	×	ML(SVM)	N/A	N/A

(Continued)

Table 2. (*Continued*)

Method	Ref	Inputs				Predictive Model	Outputs	
		AAS	EVO	SS	RSA		Binary Prediction	Predictive Propensity
BindN-RF	[75]	✓	✓	×	×	ML(RF)	N/A	N/A
DBindR	[76]	✓	✓	✓	×	ML(RF)	N/A	N/A
DBD-Threader	[77]	✓	×	×	×	Template-based	✓	×
BindN+	[78]	✓	✓	×	×	ML(SVM)	✓	✓
NAPS	[79]	✓	✓	×	×	ML(C4.5)	N/A	N/A
PiRaNhA	[80]	✓	✓	×	✓	ML(SVM)	N/A	N/A
ProteRNA	[81]	✓	✓	✓	×	ML(SVM)	N/A	N/A
RBRpred	[10]	✓	✓	✓	✓	ML(SVM)	N/A	N/A
PRNA	[82]	✓	×	✓	×	ML(RF)	✓	✓
Wang et al.	[83]	✓	✓	×	✓	ML(SVM)	N/A	N/A
PRBR	[84]	✓	✓	✓	×	ML(RF)	✓	✓
Choi et al.	[85]	✓	✓	×	×	ML(SVM), Homology-Based	✓	✓
RNABindR	[86]	✓	×	✓	×	ML(SVM)	✓	×
DNABR	[87]	✓	✓	×	×	ML(SVM)	✓	✓
Dey et al.	[88]	✓	✓	×	×	ML(RF)	N/A	N/A
DNABind	[89]	✓	✓	✓	×	ML(SVM)	N/A	N/A
RNABindRPlus	[90]	✓	✓	✓	✓	ML, Template-based	✓	✓

Name	Ref					Method		
SPOT-Seq (DNA)	[91]	✓	×	×	×	Template-based	✓	✓
aaRNA	[92]	✓	✓	✓	✓	ML(NN)	✓	✓
SNBRFinder	[93]	✓	✓	✓	✓	ML(HMM,SVM), Template-based	N/A	N/A
DisoRDPbind	[57]	✓	×	×	✓	ML(LR)	✓	✓
Ren et al.	[94]	×	×	×	×	ML(ensemble learning)	N/A	N/A
PRIdictor	[95]	✓	×	✓	×	ML(SVM)	✓	✓
RNAProSite	[96]	✓	✓	✓	✓	ML(RF)	✓	✓
PDNAsite	[97]	✓	✓	✓	✓	ML(ensemble learning)	N/A	N/A
DRNApred	[98]	✓	✓	✓	✓	ML(ensemble learning)	✓	✓
TargetDNA	[99]	✓	✓	✓	✓	ML(SVM)	✓	✓
PredRBR	[100]	✓	✓	✓	✓	ML(Gradient Boosting)	✓	✓
NucBind	[60]	✓	✓	✓	×	ML(SVM), homologous templates	✓	✓
DNAPred	[101]	✓	✓	✓	✓	ML(SVM)	✓	✓
iProDNA-CapsNet	[64]	✓	×	×	×	ML(NN)	✓	✓
EL_LSTM	[102]	✓	✓	✓	✓	ML(NN, Bagging)	✓	✓
ProNA2020	[61]	✓	✓	✓	✓	ML(SVM, NN)	✓	✓
funDNApred	[103]	✓	✓	✓	✓	ML(FCM)	×	×
NCBRPred	[62]	✓	✓	✓	✓	DL	✓	✓
MTDsite	[63]	✓	×	×	×	DL	✓	✓
DNAgenie	[104]	✓	✓	✓	✓	ML(SVM)	✓	✓
DeepDISOBind	[46]	✓	✓	✓	✓	DL	✓	✓

comparing their predictive architectures. This includes details about the predictive inputs that are produced from the protein sequence, predictive models, and the outputs that they produce.

Based on a recent study [6], we identified several common types of inputs that these predictors use including the amino acid sequence (AAS) itself; the evolutionary information (EVO) that is derived from the sequence using multiple sequence alignment algorithms [105], such as PSI-BLAST [106] or HHblits [107]; and the secondary structure (SS) and relative solvent accessibility (RSA) that are predicted from the input sequence using third-party predictors, such as PSIPRED [108], ASAquick [109], SPINE X[110], SPOT-1D [111] and SPIDER [112]. Recent surveys of the sequence-based SS and RSA predictors provide a broader overview of these methods [113–120]. The AAS input is typically expressed using one-hot encoding, where each amino acid is represented by 20-dimensional binary vector. We find that each of the 51 predictors utilizes at least one of these four input types, and virtually all of them use the AAS input. Furthermore, Table 2 reveals only 1 out of the 16 methods that were published until 2009 uses all four inputs [53,54,59,65,67–77]. In contrast, 10 out of the 13 methods that were published in the last five years (since 2017) utilize the four inputs [46,60–64,98–104]. This transition clearly demonstrates that these four inputs are very useful for the prediction of the protein-nucleic acid interactions.

Interestingly, we note that nearly all the 51 methods rely on machine learning algorithms to produce their predictive models. The only two exceptions are DBD-Threader [77] and SPOT-Seq [91] that predict protein-DNA binding residues based on template-based/homology prediction. The most commonly used classical/shallow machine learning methods are support vector machines (24 predictors), neural networks (11 predictors), random forest (7 predictors) and logistic regression (3 predictors). We also find that 10 methods utilize more than one machine learning algorithm. Moreover, several methods, including DNABind [89], SNBRFinder [93] and NucBind [60] combine a machine learning-generated model with the template-based approach.

Three recent methods utilize deep learning [46,62,63]. The deep learning algorithms produce neural networks with multiple/many hidden layers which often rely on sophisticated network topologies that include recurrent, convolutional and transformer modules. Nowadays, deep learners are used to develop predictors of numerous aspects of protein structure and function. Examples include the state-of-the-art protein structure predictor, AlphaFold [121,122], methods that predict residue-residue contacts [123], secondary structure [111,124–126], protein function [127–129], protein-drug interactions [130,131], and functional sites [58,132]. The introduction of the deep learning into the prediction of the nucleic-acid binding residues stems from the recent popularity of these models, as shown above, but also from the fact that recent works find them to produce more accurate models when compared to the shallow/classical machine learning algorithms [46,62].

Table 2 also summarizes the outputs of the 51 predictors, which may include the binary score and/or numeric propensity, as explained in the introduction. We cannot identify the outputs for 20 methods since we have no access to their implementations or web servers. Among the remaining predictors, over 80% (27 out of 31) produce both types of outputs. DNABindR [66], DBD-Threader [77] and the method by Choi *et al.* [85] output just the binary predictions. While funDNApred [103] provides only the propensities, user can utilize these scores to produce binary predictions, i.e., residues with propensity greater than a given threshold can be predicted in binary as binding. Altogether, we suggest that predictors should generate both types of outputs since the predictive propensities provide a useful context for the arguably easier to comprehend binary predictions.

3. Summary

This chapter summarizes a comprehensive collection of 51 sequence-based predictors of protein-nucleic acid binding residues. We find that while the early methods would typically target prediction of the DNA-binding or RNA-binding residues, five methods that were published in the last three years simultaneously predict DNA and RNA binding residues [46,60–63]. Moreover, three of these methods, including ProNA2020, MTDsite [63] and DeepDISOBind [46], extend this scope by predicting protein/peptide-binding residues. This observation suggests a recent trend to develop tools that offer a wider scope of predictions.

We observe that many methods do not have source code or web servers. Some are unavailable because the authors did not maintain the originally published URLs. This problem is especially acute for the methods published before 2016. To better serve the community, we encourage the authors to keep the web servers running and to provide source code for the end users.

We also identify a shortage of methods that predict protein-nucleic acids interactions in intrinsically disordered regions, with only two currently available methods: DisoRDPbind [57] and DeepDISObind [46]. This is an important issue since recent work demonstrates that structure-trained and disorder-trained predictors of the protein-binding residues produce complementary results [58]. While a comparable study for the prediction of the nucleic acid binding residues is missing, we believe that similar conclusions would be drawn.

Given the above observations, we endorse a few predictors, focusing on the methods that cover multiple types of ligands (DNA, RNA and proteins/peptides), are currently available, and which consider both structure- and disorder-annotated interactions. Our recommendations include NCBRPred [62], DisoRDPbind [55–57], MTDsite [63], ProNA2020 [61] and DeepDISOBind [46].

We conclude this chapter with a brief discussion of future directions for this active research areas. We find that a few recent methods apply deep learners [46,62,63]. Given the recent empirical results which demonstrate that deep learners provide more accurate predictions compared to the shallow machine learning algorithms [46,62], we anticipate that future methods will continue to utilize deep neural networks. We note that three of our recommended predictors, namely NCBRPred [62], MTDsite [63] and DeepDISOBind [46], use deep networks. The use of more sophisticated deep learners could help to combat the cross-prediction problem [60,62,98], which means that residues that are predicted to bind RNA, in fact, often interact with DNA, and vice versa. In other words, current methods relatively often mis-predict the type of the interacting ligand. The cross-predictions also happens between protein-binding and nucleic acids-binding residues [58,133–135]. Lastly, the current predictors are agnostic to the DNA and RNA types, with one exception, DNAgenie [104]. DNAgenie is capable of accurately identifying the type of interacting DNA, covering A-DNA, B-DNA and single-stranded DNA. This means DNAgenie predicts which residues bind A-DNA, B-DNA versus single stranded DNA, providing more details when compared to the other current tools. Tools that would provide RNA-type specific predictions would be a welcomed addition. This was motivated by a recent study that found that current RNA type-agnostic methods are deficient for the predictions of some RNA types, such as tRNA [42].

References

[1]   Siggers T and Gordan R. Protein-DNA binding: Complexities and multi-protein codes. *Nucleic Acids Research*, 2014. **42**(4): 2099–2111.

[2]   Cook KB, *et al.* High-throughput characterization of protein-RNA interactions. *Briefings in Functional Genomics*, 2015. **14**(1): 74–89.

[3]   Sathyapriya R, *et al.* Insights into protein-DNA interactions through structure network analysis. *Plos Computational Biology*, 2008. **4**(9): e1000170.

[4]   Steffen NR, *et al.* DNA sequence and structure: Direct and indirect recognition in protein-DNA binding. *Bioinformatics*, 2002. **18**(Suppl 1): S22–30.

[5]   Pugh BF and Gilmour DS. Genome-wide analysis of protein-DNA interactions in living cells. *Genome Biology*, 2001. **2**(4): REVIEWS1013.

[6]   Zhang J, *et al.* Comprehensive review and empirical analysis of hallmarks of DNA-, RNA- and protein-binding residues in protein chains. *Briefings in Bioinformatics*, 2019. **20**(4): 1250–1268.

[7]   Marchese D, *et al.* Advances in the characterization of RNA-binding proteins. *Wiley Interdisciplinary Reviews RNA*, 2016. **7**(6): 793–810.

[8]   Nahalka, J., Protein-RNA recognition: Cracking the code. *Journal of Theoretical Biology*, 2014. **343**: 9–15.

[9]   Konig J, *et al.* Protein-RNA interactions: New genomic technologies and perspectives. *Nature Reviews Genetics*, 2012. **13**(2): 77–83.

[10] Zhang T, *et al.* Analysis and prediction of RNA-binding residues using sequence, evolutionary conservation, and predicted secondary structure and solvent accessibility. *Current Protein & Peptide Science*, 2010. **11**(7): 609–628.

[11] Chowdhury S, *et al.* In silico prediction and validation of novel RNA binding proteins and residues in the human proteome. *Proteomics*, 2018. e1800064.

[12] Cozzolino F, *et al.* Protein-DNA/RNA interactions: An overview of investigation methods in the -omics era. *Journal of Proteome Research*, 2021. **20**(6): 3018–3030.

[13] Sudha G, *et al.* An overview of recent advances in structural bioinformatics of protein-protein interactions and a guide to their principles. *Progress in Biophysics and Molecular Biology*, 2014. **116**(2–3): 141–150.

[14] Vakser IA. Protein-protein docking: From interaction to interactome. *Biophysical Journal*, 2014. **107**(8): 1785–1793.

[15] Hsu WL, *et al.* Intrinsic protein disorder and protein-protein interactions. *Pacific Symposium on Biocomputing*, 2012: 116–127.

[16] De Las Rivas J and Fontanillo C. Protein-protein interaction networks: Unraveling the wiring of molecular machines within the cell. *Briefings in Functional Genomics*, 2012. **11**(6): 489–496.

[17]   Meng F, *et al.* Compartmentalization and functionality of nuclear disorder: Intrinsic disorder and protein-protein interactions in intra-nuclear compartments. *International Journal of Molecular Sciences,* 2015. **17**(1): 24.

[18]   Chen K and Kurgan L. Investigation of atomic level patterns in protein-small ligand interactions. *PLoS One,* 2009. **4**(2): e4473.

[19]   Dudev T and Lim C. Competition among metal ions for protein binding sites: Determinants of metal ion selectivity in proteins. *Chemical Reviews,* 2014. **114**(1): 538–556.

[20]   Peng T, *et al.* Turning the spotlight on protein-lipid interactions in cells. *Current Opinion in Chemical Biology,* 2014. **21**: 144–153.

[21]   wwPDB consortium, Protein data bank: The single global archive for 3D macromolecular structure data. *Nucleic Acids Research,* 2019. **47**(D1): D520–D528.

[22]   Nagarajan R, *et al.* Structure based approach for understanding organism specific recognition of protein-RNA complexes. *Biology Direct,* 2015. **10**: 8.

[23]   Ellis JJ, *et al.* Protein-RNA interactions: Structural analysis and functional classes. *Proteins,* 2007. **66**(4): 903–911.

[24]   Prabakaran P, *et al.* Classification of protein-DNA complexes based on structural descriptors. *Structure,* 2006. **14**(9): 1355–1367.

[25]   Lejeune D, *et al.* Protein-nucleic acid recognition: statistical analysis of atomic interactions and influence of DNA structure. *Proteins,* 2005. **61**(2): 258–271.

[26]   Wang C, *et al.* Disordered nucleiome: Abundance of intrinsic disorder in the DNA- and RNA-binding proteins in 1121 species from Eukaryota, Bacteria and Archaea. *Proteomics,* 2016. **16**(10): 1486–1498.

[27]   Wu Z, *et al.* In various protein complexes, disordered protomers have large per-residue surface areas and area of protein-, DNA- and RNA-binding interfaces. *FEBS Letters,* 2015. **589**(19 Pt A): 2561–2569.

[28]   Varadi M, *et al.* Functional advantages of conserved intrinsic disorder in RNA-binding proteins. *PLoS One,* 2015. **10**(10): e0139731.

[29]   Dyson, HJ. Roles of intrinsic disorder in protein-nucleic acid interactions. *Molecular BioSystems,* 2012. **8**(1): 97–104.

[30]   Peng Z, *et al.* A creature with a hundred waggly tails: Intrinsically disordered proteins in the ribosome. *Cellular and Molecular Life Scien–ces,* 2014. **71**(8): 1477–1504.

[31]   Lieutaud P, *et al.* How disordered is my protein and what is its disorder for? A guide through the "dark side" of the protein universe. *Intrinsically Disordered Proteins,* 2016. **4**(1): e1259708.

[32]   Oldfield CJ, *et al.* Introduction to intrinsically disordered proteins and regions. In *Intrinsically Disordered Proteins,* Salvi N (Ed.), Academic Press, 2019. 1–34.

[33] Habchi J, *et al.* Introducing protein intrinsic disorder. *Chemical Reviews*, 2014. **114**(13): 6561–6588.

[34] Si J, *et al.* An overview of the prediction of protein DNA-binding sites. *International Journal of Molecular Sciences*, 2015. **16**(3): 5194–5215.

[35] Yan J, *et al.* A comprehensive comparative review of sequence-based predictors of DNA- and RNA-binding residues. *Briefings in Bioinformatics*, 2016. **17**(1): 88–105.

[36] Katuwawala A and Kurgan L. Comparative assessment of intrinsic disorder predictions with a focus on protein and nucleic acid-binding proteins. *Biomolecules*, 2020. **10**(12): 1636.

[37] Meng F, *et al.* Comprehensive review of methods for prediction of intrinsic disorder and its molecular functions. *Cellular and Molecular Life Sciences*, 2017. **74**(17): 3069–3090.

[38] Si J, *et al.* Computational prediction of RNA-binding proteins and binding sites. *International Journal of Molecular Sciences*, 2015. **16**(11): 26303–26317.

[39] Zhao H, *et al.* Prediction of RNA binding proteins comes of age from low resolution to high resolution. *Molecular BioSystems*, 2013. **9**(10): 2417–2425.

[40] Walia RR, *et al.* Protein-RNA interface residue prediction using machine learning: An assessment of the state of the art. *BMC Bioinformatics*, 2012. **13**: 89.

[41] Gromiha MM and Nagarajan R. Computational approaches for predicting the binding sites and understanding the recognition mechanism of protein-DNA complexes. *Advances in Protein Chemistry and Structural Biology*, 2013. **91**: 65–99.

[42] Wang K, *et al.* Comprehensive survey and comparative assessment of RNA-binding residue predictions with analysis by RNA type. *International Journal of Molecular Sciences*, 2020. **21**(18): 6879.

[43] Li W, *et al.* RefSeq: Expanding the prokaryotic genome annotation pipeline reach with protein family model curation. *Nucleic Acids Research*, 2021. **49**(D1): D1020–D1028.

[44] Katuwawala A, *et al.* Computational prediction of functions of intrinsically disordered regions. *Progress in Molecular Biology and Translational Science*, 2019. **166**: 341–369.

[45] Kurgan L, *et al.* The methods and tools for intrinsic disorder prediction and their application to systems medicine. In *Systems Medicine*, Wolkenhauer O (Ed.), Academic Press: Oxford, 2021. 159–169.

[46] Zhang F, *et al.* DeepDISOBind: Accurate prediction of RNA-, DNA- and protein-binding intrinsically disordered residues with deep multi-task learning. *Briefings in Bioinformatics*, 2022. **23**(1): bbab521.

[47] Georgel PT, *et al.* Sir3-dependent assembly of supramolecular chromatin structures in vitro. *Proceedings of the National Academy of Sciences of the United States of America*, 2001. **98**(15): 8584–8589.

[48] McBryant SJ, *et al*. Chromatin architectural proteins. *Chromosome Research*, 2006. **14**(1): 39–51.

[49] Quaglia F, *et al*. DisProt in 2022: Improved quality and accessibility of protein intrinsic disorder annotation. *Nucleic Acids Research*, 2022. **50**(D1): D480–D487.

[50] Yang J, *et al*. BioLiP: A semi-manually curated database for biologically relevant ligand-protein interactions. *Nucleic Acids Research*, 2013. **41**(Database issue): D1096–D1103.

[51] Sickmeier M, *et al*. DisProt: The database of disordered proteins. *Nucleic Acids Research*, 2007. **35**(Database issue): D786–D793.

[52] Berman HM, *et al*. The protein data bank. *Nucleic Acids Research*, 2000. **28**(1): 235–242.

[53] Jeong E, *et al*. A neural network method for identification of RNA-interacting residues in protein. *Genome informatics*, 2004. **15**(1): 105–116.

[54] Ahmad S, *et al*. Analysis and prediction of DNA-binding proteins and their binding residues based on composition, sequence and structural information. *Bioinformatics*, 2004. **20**(4): 477–486.

[55] Oldfield CJ, *et al*. Disordered RNA binding region prediction with DisoRDPbind. *Methods in Molecular Biology*, 2020. **2106**: 225–239.

[56] Peng Z, *et al*. Prediction of disordered RNA, DNA, and protein binding regions using DisoRDPbind. *Methods in Molecular Biology*, 2017. **1484**: 187–203.

[57] Peng Z and Kurgan L. High-throughput prediction of RNA, DNA and protein binding regions mediated by intrinsic disorder. *Nucleic Acids Research*, 2015. **43**(18): e121.

[58] Zhang J, *et al*. Prediction of protein-binding residues: Dichotomy of sequence-based methods developed using structured complexes versus disordered proteins. *Bioinformatics*, 2020. **36**(18): 4729–4738.

[59] Wang L and Brown SJ. BindN: A web-based tool for efficient prediction of DNA and RNA binding sites in amino acid sequences. *Nucleic Acids Research*, 2006. **34**(Web Server issue): W243–W248.

[60] Su H, *et al*. Improving the prediction of protein–nucleic acids binding residues via multiple sequence profiles and the consensus of complementary methods. *Bioinformatics*, 2019. **35**(6): 930–936.

[61] Qiu J, *et al*. ProNA2020 predicts protein–DNA, protein–RNA, and protein–protein binding proteins and residues from sequence. *Journal of Molecular Biology*, 2020. **432**(7): 2428–2443.

[62] Zhang J, *et al*. NCBRPred: Predicting nucleic acid binding residues in proteins based on multilabel learning. *Briefings in Bioinformatics*, 2021. **22**(5): bbaa397.

[63] Sun Z, *et al*. To improve the predictions of binding residues with DNA, RNA, carbohydrate, and peptide via multi-task deep neural networks. *IEEE/ACM Transactions on Computational Biology and Bioinformatics*, 2021. doi: 10.1109/TCBB.2021.3118916.

[64] Nguyen BP, *et al.* iProDNA-CapsNet: Identifying protein-DNA binding residues using capsule neural networks. *BMC Bioinformatics*, 2019. **20**(23): 1–12.

[65] Ahmad S and Sarai A. PSSM-based prediction of DNA binding sites in proteins. *BMC Bioinformatics*, 2005. **6**(1): 1–6.

[66] Yan C, *et al.* Predicting DNA-binding sites of proteins from amino acid sequence. *BMC Bioinformatics*, 2006. **7**(1): 1–10.

[67] Jeong E and Miyano S. A weighted profile based method for protein-RNA interacting residue prediction. In *Transactions on Computational Systems Biology IV*, Priami C, *et al.* (eds.), Springer, 2006.

[68] Ho S-Y, *et al.* Design of accurate predictors for DNA-binding sites in proteins using hybrid SVM–PSSM method. *Biosystems*, 2007. **90**(1): 234–241.

[69] Hwang S, *et al.* DP-Bind: A web server for sequence-based prediction of DNA-binding residues in DNA-binding proteins. *Bioinformatics*, 2007. **23**(5): 634–636.

[70] Ofran Y, *et al.* Prediction of DNA-binding residues from sequence. *Bioinformatics*, 2007. **23**(13): i347–i353.

[71] Wang Y, *et al.* PRINTR: Prediction of RNA binding sites in proteins using SVM and profiles. *Amino Acids*, 2008. **35**(2): 295–302.

[72] Tong J, *et al.* RISP: A web-based server for prediction of RNA-binding sites in proteins. *Computer Methods and Programs in Biomedicine*, 2008. **90**(2): 148–153.

[73] Kumar M, *et al.* Prediction of RNA binding sites in a protein using SVM and PSSM profile. *Proteins: Structure, Function, and Bioinformatics*, 2008. **71**(1): 189–194.

[74] Cheng C-W, *et al.* Predicting RNA-binding sites of proteins using support vector machines and evolutionary information. *BMC Bioinformatics*, 2008. **9**(12): 1–19.

[75] Wang L, *et al.* Prediction of DNA-binding residues from protein sequence information using random forests. *BMC Genomics*, 2009. **10**(1): 1–9.

[76] Wu J, *et al.* Prediction of DNA-binding residues in proteins from amino acid sequences using a random forest model with a hybrid feature. *Bioinformatics*, 2009. **25**(1): 30–35.

[77] Gao M and Skolnick J. A threading-based method for the prediction of DNA-binding proteins with application to the human genome. *PLoS Computational Biology*, 2009. **5**(11): e1000567.

[78] Wang L, *et al.* BindN+ for accurate prediction of DNA and RNA-binding residues from protein sequence features. *BMC Systems Biology*, 2010. **4**(1): 1–9.

[79] Carson MB, *et al.* NAPS: A residue-level nucleic acid-binding prediction server. *Nucleic Acids Research*, 2010. **38**(Web Server issue): W431–W435.

[80] Murakami Y, *et al.* PiRaNhA: A server for the computational prediction of RNA-binding residues in protein sequences. *Nucleic Acids Research*, 2010. **38**(Suppl 2): W412–W416.

[81]	Huang Y-F, *et al.* Predicting RNA-binding residues from evolutionary information and sequence conservation. *BMC Genomics*, 2010. **11**(Suppl 4): S2.

[82]	Liu Z-P, *et al.* Prediction of protein–RNA binding sites by a random forest method with combined features. *Bioinformatics*, 2010. **26**(13): 1616–1622.

[83]	Wang C-C, *et al.* Identification of RNA-binding sites in proteins by integrating various sequence information. *Amino Acids*, 2011. **40**(1): 239–248.

[84]	Ma X, *et al.* Prediction of RNA RNA-binding sites in proteins by integrating variouusing an enriched random forest model with a novel hybrid feature. *Proteins: Structure, Function, and Bioinformatics*, 2011. **79**(4): 1230–1239.

[85]	Choi S and Han K. Prediction of RNA-binding amino acids from protein and RNA sequences. *BMC Bioinformatics*, 2011. **12**(Suppl 13): S7.

[86]	Terribilini M, *et al.* RNABindR: A server for analyzing and predicting RNA-binding sites in proteins. *Nucleic Acids Research*, 2007. **35**(Suppl 2): W578–W584.

[87]	Ma X, *et al.* Sequence-based prediction of DNA-binding residues in proteins with conservation and correlation information. *IEEE/ACM Transactions on Computational Biology and Bioinformatics*, 2012. **9**(6): 1766–1775.

[88]	Dey S, *et al.* Characterization and prediction of the binding site in DNA-binding proteins: improvement of accuracy by combining residue composition, evolutionary conservation and structural parameters. *Nucleic Acids Research*, 2012. **40**(15): 7150–7161.

[89]	Liu R and Hu J. DNABind: A hybrid algorithm for structure of the binding site in DNA-binding proteins: Improvement of accuracy by combining residue composition *Proteins: Structure, Function, and Bioinformatics*, 2013. **81**(11): 1885–1899.

[90]	Walia RR, *et al.* RNABindRPlus: A predictor that combines machine learning and sequence homology-based methods to improve the reliability of predicted RNA-binding residues in proteins. *PloS One*, 2014. **9**(5): e97725.

[91]	Zhao H, *et al.* Predicting DNA-binding proteins and binding residues by complex structure prediction and application to human proteome. *PloS One*, 2014. **9**(5): e96694.

[92]	Li S, *et al.* Quantifying sequence and structural features of protein–RNA interactions. *Nucleic Acids Research*, 2014. **42**(15): 10086–10098.

[93]	Yang X, *et al.* SNBRFinder: A sequence-based hybrid algorithm for enhanced prediction of nucleic acid-binding residues. *PloS One*, 2015. **10**(7): e0133260.

[94]	Ren H and Shen Y. RNA-binding residues prediction using structural features. *BMC Bioinformatics*, 2015. **16**(1): 1–10.

[95]	Tuvshinjargal N, *et al.* PRIdictor: Protein–RNA interaction predictor. *Biosystems*, 2016. **139**: 17–22.

[96]	Sun M, *et al.* Accurate prediction of RNA-binding protein residues with two discriminative structural descriptors. *BMC Bioinformatics*, 2016. **17**(1): 1–14.

[97]   Zhou J, *et al.* PDNAsite: Identification of DNA-binding site from protein sequence by incorporating spatial and sequence context. *Scientific Reports,* 2016. **6**(1): 1–15.

[98]   Yan J and Kurgan L. DRNApred, fast sequence-based method that accurately predicts and discriminates DNA- and RNA-binding residues. *Nucleic Acids Research,* 2017. **45**(10): e84.

[99]   Hu J, *et al.* Predicting protein-DNA binding residues by weightedly combining sequence-based features and boosting multiple SVMs. *IEEE/ACM Transactions on Computational Biology and Bioinformatics,* 2017. **14**(6): 1389–1398.

[100]   Tang Y, *et al.* A boosting approach for prediction of protein-RNA binding residues. *BMC Bioinformatics,* 2017. **18**(Suppl 13): 465.

[101]   Zhu Y-H, *et al.* DNAPred: Accurate identification of DNA-binding sites from protein sequence by ensembled hyperplane-distance-based support vector machines. *Journal of Chemical Information and Modeling,* 2019. **59**(6): 3057–3071.

[102]   Zhou J, *et al.* EL_LSTM: Prediction of DNA-binding residue from protein sequence by combining long short-term memory and ensemble learning. *IEEE/ACM Transactions on Computational Biology and Bioinformatics,* 2018. **17**(1): 124–135.

[103]   Amirkhani A, *et al.* Prediction of DNA-binding residues in local segments of protein sequences with Fuzzy Cognitive Maps. *IEEE/ACM Transactions on Computational Biology and Bioinformatics,* 2018. **17**(4): 1372–1382.

[104]   Zhang J, *et al.* DNAgenie: Accurate prediction of DNA-type-specific binding residues in protein sequences. *Briefings in Bioinformatics,* 2021. **22**(6): bbab336.

[105]   Hu G and Kurgan L. Sequence Similarity Searching. *Current Protocols in Protein Science,* 2019. **95**(1): e71.

[106]   Altschul SF, *et al.* Gapped BLAST and PSI-BLAST: A new generation of protein database search programs. *Nucleic Acids Research,* 1997. **25**(17): 3389–3402.

[107]   Remmert M, *et al.* HHblits: Lightning-fast iterative protein sequence searching by HMM-HMM alignment. *Nature Methods,* 2012. **9**(2): 173–175.

[108]   Buchan DWA and Jones DT. The PSIPRED protein analysis workbench: 20 years on. *Nucleic Acids Research,* 2019. **47**(W1): W402–W407.

[109]   Faraggi E, *et al.* Accurate single-sequence prediction of solvent accessible surface area using local and global features. *Proteins,* 2014. **82**(11): 3170–3176.

[110]   Faraggi E, *et al.* SPINE X: Improving protein secondary structure prediction by multistep learning coupled with prediction of solvent accessible surface area and backbone torsion angles. *Journal of Computational Chemistry,* 2012. **33**(3): 259–267.

[111] Singh J, *et al.* SPOT-1D-single: Improving the single-sequence-based prediction of protein secondary structure, backbone angles, solvent accessibility and half-sphere exposures using a large training set and ensembled deep learning. *Bioinformatics*, 2021. **37**(20): 3464–3472.

[112] Yang Y, *et al.* SPIDER2: A package to predict secondary structure, accessible surface area, and main-chain torsional angles by deep neural networks. *Methods in Molecular Biology*, 2017. **1484**: 55–63.

[113] Zhang H, *et al.* Critical assessment of high-throughput standalone methods for secondary structure prediction. *Briefings in Bioinformatics*, 2011. **12**(6): 672–688.

[114] Ho HK, *et al.* A survey of machine learning methods for secondary and supersecondary protein structure prediction. *Methods in Molecular Biology*, 2013. **932**: 87–106.

[115] Oldfield CJ, *et al.* Computational prediction of secondary and supersecondary structures from protein sequences. *Methods in Molecular Biology*, 2019. **1958**: 73–100.

[116] Meng F and Kurgan L. Computational prediction of protein secondary structure from sequence. *Current Protocols in Protein Science*, 2016. **86**: 2.3.1–2.3.10.

[117] Chen K and Kurgan L. Computational prediction of secondary and supersecondary structures. *Methods in Molecular Biology*, 2013. **932**: 63–86.

[118] Kurgan L and Disfani FM. Structural protein descriptors in 1-dimension and their sequence-based predictions. *Current Protein & Peptide Science*, 2011. **12**(6): 470–489.

[119] Jiang Q, *et al.* Protein secondary structure prediction: A survey of the state of the art. *Journal of Molecular Graphics & Modelling*, 2017. **76**: 379–402.

[120] Pirovano W and Heringa J. Protein secondary structure prediction. *Methods in Molecular Biology*, 2010. **609**: 327–348.

[121] AlQuraishi, M., AlphaFold at CASP13. *Bioinformatics*, 2019. **35**(22): 4862–4865.

[122] Jumper J, *et al.* Highly accurate protein structure prediction with AlphaFold. *Nature*, 2021. **596**(7873): 583–589.

[123] Schaarschmidt J, *et al.* Assessment of contact predictions in CASP12: Co-evolution and deep learning coming of age. *Proteins*, 2018. **86**(Suppl 1): 51–66.

[124] Guo ZY, *et al.* DNSS2: Improved ab initio protein secondary structure prediction using advanced deep learning architectures. *Proteins-Structure Function and Bioinformatics*, 2021. **89**(2): 207–217.

[125] Lyu Z, *et al.* Protein secondary structure prediction with a reductive deep learning method. *Frontiers in Bioengineering and Biotechnology*, 2021. **9**: 687426.

[126] Hanson J, *et al.* Improving prediction of protein secondary structure, backbone angles, solvent accessibility and contact numbers by using predicted

contact maps and an ensemble of recurrent and residual convolutional neural networks. *Bioinformatics*, 2019. **35**(14): 2403–2410.

[127]   Zhang F, *et al.* DeepFunc: A deep learning framework for accurate prediction of protein functions from protein sequences and interactions. *Proteomics*, 2019. **19**(12): e1900019.

[128]   Kulmanov M, *et al.* DeepGO: Predicting protein functions from sequence and interactions using a deep ontology-aware classifier. *Bioinformatics*, 2018. **34**(4): 660–668.

[129]   Littmann M, *et al.* Embeddings from deep learning transfer GO annotations beyond homology. *Scientific Reports*, 2021. **11**(1): 1160.

[130]   Muller C, *et al.* Artificial intelligence, machine learning, and deep learning in real-life drug design cases. *Methods in Molecular Biology*, 2022. **2390**: 383–407.

[131]   Kim J, *et al.* Comprehensive survey of recent drug discovery using deep learning. *International Journal of Molecular Sciences*, 2021. **22**(18): 9983.

[132]   Li F, *et al.* DeepCleave: A deep learning predictor for caspase and matrix metalloprotease substrates and cleavage sites. *Bioinformatics*, 2020. **36**(4): 1057–1065.

[133]   Zhang F, *et al.* PROBselect: Accurate prediction of protein-binding residues from proteins sequences via dynamic predictor selection. *Bioinformatics*, 2020. **36**(Suppl 2): i735–i744.

[134]   Zhang J and Kurgan L. SCRIBER: Accurate and partner type-specific prediction of protein-binding residues from proteins sequences. *Bioinformatics*, 2019. **35**(14): i343–i353.

[135]   Zhang J and Kurgan L. Review and comparative assessment of sequence-based predictors of protein-binding residues. *Briefings in Bioinformatics*, 2018. **19**(5): 821–837.

Chapter 11

Identification of Cancer Hotspot Residues and Driver Mutations Using Machine Learning

Medha Pandey, P. Anoosha, Dhanusha Yesudhas and
M. Michael Gromiha*

*Department of Biotechnology, Bhupat and Jyoti Mehta School of Biosciences,
Indian Institute of Technology Madras, Chennai 600036, India
corresponding author: gromiha@iitm.ac.in

Cancer is one of the most life-threatening diseases and mutations in selected genes are associated with tumorigenesis. Identification of driver mutations, which are responsible for the disease progression, is crucial for precision oncology. Experimentally available information on cancer-causing mutations is accumulated in several databases and the data is utilized to develop computational algorithms that predict the driver mutations. In this chapter, we survey literature to review the available databases and we summarize their key features. We also explore computational methods for identifying disease-causing mutations in specific genes and cancer types as well as more generic predictive methods. Furthermore, we discuss applications of these computational tools that are focused on large scale studies. The databases and methods for identifying driver mutations discussed in this review are helpful for the development of precision medicine and they advance the blueprint for biological and clinical endeavors.

289

1. Introduction

Cancer is one of the deadliest diseases. It results from the proliferation and uncontrolled cell-division of different types of cells. The essential hallmarks of cancer are linked to an altered cancer cell-intrinsic metabolism, either as a consequence or as a cause [1]. As an example, the resistance of cancer mitochondria against apoptosis-associated permeabilization is closely related with the altered contribution to metabolism [1]. Similarly, the constitutive activation of signaling cascades that stimulate cell growth has a profound impact on anabolic metabolism. One of the important tasks in cancer pathology is the distinction between benign and malignant tumors. Benign tumor remains confined to its original location and does not invade the neighboring tissues whereas malignant tumors invade surrounding tissues and are capable of spreading throughout the body. Radiations and many chemical carcinogens are the key factors that damage DNA and cause mutations that underly cancer.

Genome sequencing has revolutionized our understanding of somatic mutations (genetic alteration acquired by a cell that are not inherited) in cancer, providing a detailed view of the mutational processes and genes that drive cancer. Mutational hotspots indicate selective pressure across a population of tumor samples and understanding their prevalence within and across cancer types is an active area of research. Somatically acquired mutations, which are highly recurring and facilitate cellular growth are known as driver mutations [2]. Novel approaches to detect significantly mutated residues, rather than identifying recurrently mutated genes may uncover new biologically and therapeutically relevant driver mutations [3].

Among the best-studied therapeutic targets in human cancers are proteins encoded by genes with tumor-specific mutational hotspots, such as KRAS, NRAS, BRAF, KIT, TP53 and EGFR [4,5]. The acquisition of somatic mutations is one of the major factors responsible for the dysregulation of proliferation, invasion, and apoptosis. Comprehensive genomic characterization of tumors has produced significant insights into somatic aberrations in individual cancer types as well as dysfunctional molecular pathways that govern tumor initiation, progression, and maintenance. All this data has spurred the development of computational algorithms to identify cancer driver genes and mutations. Mutations that contribute to cancer developments are called driver mutations and the genes that harbor those mutations are called driver genes.

In this chapter, we list comprehensive databases of cancer mutations that we have found via literature search as well as computational tools for identifying driver mutations. Furthermore, we discuss large scale applications of these computational resources.

2. Experimental and computational studies on cancer mutations

One of the major challenges in cancer research is to identify and prioritize gene mutations that are beneficial to cancer cells for driving tumorigenesis. Several investigations have been carried out to understand the influence of mutations in cancer cells [6]. These mutations are experimentally studied using targeted gene sequencing, cytogenetic techniques, systematic mutagenesis, and genetic linkage analysis [7,8]. It is evident from experiments that specific gene families, such as protein kinases and phosphatases, are prominent among cancer genes and these cancer genes cluster on specific signaling pathways. Proteins associated with these gene families include EGFR, RET, FLT3, PIK3CA and KIT, which are mainly involved in thyroid, Acute myeloid leukaemia (AML), lung and breast cancers [9]. For example, in the classical MAPK/ERK pathway, upstream mutations are found in cell-membrane-bound receptor tyrosine kinases, such as EGFR, ERBB2, FGFR1, FGFR2, FGFR3, PDGFRA and PDGFRB, and in the downstream cytoplasmic components NF1, PTPN11, HRAS, KRAS, NRAS and BRAF [10].

In addition, several cancer mutations can be targeted therapeutically. Sorafenib is used as a multikinase inhibitor primarily in kidney and liver cancer to target extracellular signal-regulated kinases (ERK) and other signaling pathways, which are important for tumor cell proliferation and angiogenesis [6]. Another example is Crizotinib, an oral tyrosine kinase inhibitor that targets anaplastic lymphoma kinase (ALK) in ALK-positive lung cancer patients [11]. Anoosha *et al.* [12] collected a set of anti-cancer compounds with known biological activities against EGFR driver mutations and developed Quantitative Structure-Activity Relationship (QSAR) models to relate structural parameters with biological activity. Recently, significant progress has been made for detecting cancer driver genes/mutations due to the advancements in next generation sequencing technology and big data analyses [13]. In the following section, we discuss the development on computational tools for the analysis and prediction of driver mutations along with associated resources.

3. Databases of the cancer-causing mutations

Several databases have been developed to accumulate cancer mutations based on experimental studies and patient samples. Table 1 lists a set of currently available databases.

Tumor Alterations Relevant for Genomics-driven Therapy (TARGET) is a systematic collection and analysis of various human cancers associated with important variants at gene [14] and variant levels [15,16]. It also contains a curated set of clinically and biologically relevant variants, which could be helpful for clinical annotation and computational data analysis.

Large-scale cancer genomics discovery projects, such as The Cancer Genome Atlas (TCGA) [17] and the International Cancer Genome Consortium (ICGC) [16] systematically collect information on human cancer genomes, which are helpful in precision cancer medicine. TCGA is a landmark cancer genomics program jointly managed by NCI and National Human Genome Research Institute, which characterizes over 20,000 tumors and maps the samples from 33 different cancer types. ICGC was developed to analyze the genomic, transcriptomic, and epigenomic changes in 50 different tumor types or subtypes that are of clinical and societal importance across the globe.

Database of Curated Mutations (DoCM) [18] is a curated repository that facilitates the aggregation of gene and variant information for variants with prognostic, diagnostic, predictive or functional roles, which are retrieved from sources such as TCGA [17], ICGC [16] and cBioPortal [19] as well as research articles published in the literature.

Catalogue of Somatic Mutations in Cancer (COSMIC) is a database that holds somatic mutation data and associated information and provides a graphical or tabular view of data with several export options [15]. It includes additional information such as non-coding mutations, gene fusions, copy-number variation, three-dimensional protein structures with mutations, functional impacts, and implications for druggability.

OncoKB [20] is a comprehensive and curated precision oncology knowledge base about individual somatic mutations and structural alterations present in patient tumors. It includes biological, clinical, and therapeutic information curated from multiple information resources. It also provides the level of evidence system, where every level includes genes for which the alterations have been recognized to have response to an FDA-approved drug in a specific cancer type.

Kulandaisamy *et al.* [21] developed MutHTP, a database for disease-causing and neutral mutations in human transmembrane proteins. It also

Table 1. Cancer mutation databases with their properties*

Name	Link	Important Features	Reference
TARGET	https://software.broadinstitute.org/cancer/cga/target	Database of genes linked to clinical action in cancer cases	[14]
TCGA	https://cancergenome.nih.gov/	Publicly accessible epigenome, transcriptome, and proteome data of tumor samples	[17]
ICGC	https://dcc.icgc.org/	Consortium with data from worldwide cancer projects representing somatic mutations and molecular data. Easy visualization, analysis, and download	[16]
DoCM	http://docm.info	Curated repository to track biologically important cancer variants	[18]
CBioPortal	https://www.cbioportal.org/	Integrates somatic mutation, mRNA and microRNA expression, DNA copy-number alterations (CNAs) and methylation, protein, and phosphoprotein RPPA data Graphical summaries	[19,34]
COSMIC	https://cancer.sanger.ac.uk/cosmic	Largest source of manually curated somatic mutations	[15]
OncoKB	https://www.oncokb.org/	Curated biological effect, prevalence and prognostic information and treatment implications for each reported alteration	[20]
MutHTP	http://www.iitm.ac.in/bioinfo/MutHTP/	Disease-causing and neutral mutations from transmembrane proteins	[21]
HuVarBase	https://www.iitm.ac.in/bioinfo/huvarbase	Integration of variant data from all publicly available resources	[25]
dbNSFP	http://database.liulab.science/dbNSFP	Database for functional annotation of all potential variants in human genome	[26–28]
dbCPM	http://bioinfo.ahu.edu.cn:8080/dbCPM/index.jsp	Benchmark dataset for cancer passenger mutations	[30]
IARC TP53 Database	https://p53.iarc.fr/	Various types of data and information on human TP53 gene variations related to cancer	[32]
OncoBase	http://www.oncobase.biols.ac.cn/	Provides annotations for somatic mutations; Interactions between target genes and regulatory elements	[33]

*Last accessed on 11 November 2021

includes conservation scores, neighboring residue information, 3D structure and disease class, and cross linked with other relevant databases such as Gene Cards [22], UniProt [23] and PDB, Protein Data Bank [24].

Ganesan *et al.* [25] developed HUmanVARiant dataBASE (HuVarBase), which has disease-causing and neutral variants at protein and gene levels. Protein-level data include amino acid sequence, secondary structure of the mutant residue, domain, function, subcellular location, and post-translational modification. Gene level data contain gene name, chromosome number and genome position, DNA mutation, mutation type, origin and rs ID number.

dbNSFP is a database for functional annotation for single nucleotide variants (SNVs) discovered in exome sequencing studies [26–28]. This database compiles conservation from PhyloP [29] and functional predictions from multiple algorithms such as SIFT, Polyphen2, LRT and MutationTaster.

Yue *et al.* [30] developed dbCPM, a specific database for cancer passenger mutations, which can be used as a benchmark dataset for improving and evaluating prediction algorithms in cancer research community. For each entry, it provides evidence from functional experiments *in vivo*, *in vitro* and recurrence in healthy controls. Also, the features of missense passenger mutations were evaluated on gene, DNA, and protein levels. In the case of gene level, the association of possible passenger mutations and genes were evaluated using Cancer Gene Census (CGC) whereas, in the case of DNA level, conservation scores and location of mutation sites were identified using PhyloP [29] and Encylcopedia of DNA Elements [31]. Solvent Accessible Surface Area (SASA) and presence of the substituted amino acid within a protein domain were used to evaluate at protein level.

IARC TP53 compiles data and information on human TP53 gene variations related to cancer [32]. OncoBase is an integrated database for annotating approximately 81 million somatic mutations in 68 cancer types from more than 120 cancer projects [33].

Recent advancements in genome and related information have accelerated the development of data repositories as well as computational tools to assist the progress of clinical applications. TCGA and ICGC are being used by many researchers to identify new targets and develop novel biomarker. Databases such as COSMIC and cBioPortal serve as an encyclopedia and provide information on multiomic data. The methods listed in Table 1 offer data analysis and integration tools, interlinking between the different resources as well as user-friendly interface. Furthermore, gene specific database, such as

IARC TP53 database, provides the annotations on tumor phenotype, patient characteristics, and structural and functional impact on mutations. Development of databases with their specific features will have a crucial role in collecting, structuring, and annotating the reported data, hence allowing the interpretation of mutations and their use in molecular pathology. The important features of these databases are presented in Table 1.

4. Identification of hotspot residues

Mutation hotspots are defined as amino acid positions in a protein-coding gene, which are mutated more frequently than would be expected in the absence of selection using background frequency [3]. These are characterized by genes, cancer types, mutation types and sequence contexts. Knowledge of these hotspots helps to identify frequently mutated genes as well as biologically and therapeutically relevant driver mutations. Well-known hotspots are identified in several cancer specific proteins such as KRAS (G12, G13) in lung cancer, NRAS (Q61) in melanoma, BRAF (V600) in colorectal and lung cancers, IDH1 (R132) in gliomas and so on. Anoosha *et al.* [35] reported that driver mutations are highly accumulated in hotspot regions, whereas the number of passenger mutations is relatively low. A set of important mutational hotspots is reported in Table 2.

Table 2. Examples of mutational hotspots in different proteins

Cancer Type	Proteins with Mutational Hotspots						
	CTNNB1	TP53	APC	BRAF	KRAS	IDH1	NRAS
Colorectal cancer	S33, T41, S45, D32, G34, S37	P278	R2714, Y159, R1788	D211, S273, R726, V600	G12, G13, Q61, K117, A146		G12, G13, Q61
Gastric cancer	D32, G34, S45	V272			G12		
Lung cancer	S33, T41, S45	R248, D259	P865, D2796	V600	G12, G13		G12, G13, Q61
Endometrial cancer	D32, S37, D207, S45, T41	P151	P1233		G12, V160, K117, A146		G12, G13, Q61
Liver cancer	G34, S37	N239, A159					
Gliomas						R132	

Several computational methods are developed to identify mutation hotspots using clustering-based algorithms in which hotspots are considered as a cluster of mutations within a small peptide or DNA region. Further, mutations which are close in the three-dimensional structure but distant in protein sequence are also considered as clusters. The features used to identify hotspots include expression, pathways, replication time, epigenetic context, location of mutations, and GC content [36]. Chang *et al.* [37] developed a binomial statistical model for identifying hotspots using nucleotide context mutability, gene-specific mutation rates and patterns of hotspot mutation sites. In addition, several evidence-based criteria are used for eliminating probable false positive hotspots.

Tamborero *et al.* [38] developed OncoDriveClust, a program which calculates a score based on the fraction of mutations composing the cluster of hotspots and the maximal distance between these mutations. HotMaps [39] and Hotspot3D [40] utilize interaction networks for identification of mutation hotspots. Meyer *et al.*, [41] developed Mutation3D for identifying hotspots, which is based on hierarchical clustering, and it encapsulates all α-carbons in a specified linkage distance. This method has identified several hotspots with high precision in TP53, KRAS and PIK3CA proteins. Trevino *et al.*, [42], developed HotSpotAnnotations, a database which contains hotspots data for 33 cancer types that are derived with two statistical methods, APOBEC3A hairpins and dN/dS (N: non-synonymous; S: synonymous) ratio. Pandey *et al.* [43] developed CanProSite, a computational method to identify disease prone sites in lung cancer using deep neural networks, which is based on specific motifs in the neighborhood of the sites. These methods can be used to identify disease prone and neutral sites for specific cancer types.

5. Methods for predicting disease-causing mutations

Several computational algorithms have been developed to identify disease-causing mutations using statistical methods and machine learning techniques. Statistical methods such as DUET [44], PackPred [45] and MuTect [46] predict the mutational effects by calculating scores using graph-based signatures, statistical contact potentials, surface accessibility, electrostatic interactions, and hydrophobicity. Machine learning methods use protein sequences, physicochemical properties, structural information, protein annotation such as information on active site, ligand binding domain and disulfide bridges, and protein-protein interactions as features to build the model [47]. Moreover, conservation information has been used to identify disease-causing mutations [48]. Kulandaisamy *et al.* [49] developed Pred-MutHTP, a sequence-based computational method for predicting disease-causing mutations in transmembrane proteins using evolutionary information, contact potentials, substitution matrices and parameters specific to membrane proteins.

6. Machine learning techniques for predicting cancer-causing mutations

Xie *et al.* [50] developed GDP (Group lasso regularized Deep learning for cancer Prognosis), a method using TCGA data, for survival prediction of cancer patients using molecular, clinical and multi-omic features such as gene expression data, copy number variations, variant call format, and somatic mutation data. The method contains three components including a fully connected deep learning framework, Cox proportional hazard (CPH) module connected to the previous framework and Lasso regularization methods to regularize the coefficients of the input layer of the network. Group Lasso was used to avoid overfitting in the network while training when compared with lasso and no regularization. This method was tested for different cancer types obtained from TCGA such as Liver hepatocellular carcinoma (LIHC), Acute myeloid leukemia (LAML), Stomach adenocarcinoma (STAD), glioblastoma multiforme (GBM), kidney renal clear cell carcinoma (KIRC), Skin cutaneous melanoma (SKCM), and bladder urothelial carcinoma (BLCA). It was observed that group Lasso regularization performed better in cancer types such as glioblastoma multiforme (GBM), kidney renal clear cell carcinoma (KIRC) and bladder urothelial carcinoma (BLCA).

Coudray *et al.* [51] developed DeepPATH, a method to assess the stage, type and subtype of lung tumors using cancer histopathological images. This method applies a convolution neural network, Google's inception v3 using whole slide images to classify them into lung adenocarcinoma (LUAD), lung squamous carcinoma (LUSC) and normal tissues. A set of commonly mutated genes in LUAD, namely STK11, EGFR, SETBP1, TP53, FAT1, KRAS, KEAP1, LRP1B, FAT4 and NF1, is selected for the study and among them STK11, EGFR, FAT1, SETBP1, KRAS and TP53 are predicted from pathology images. This method helps in predicting gene mutational status using whole-slide images. These findings suggest that deep-learning models can assist pathologists in the detection of cancer subtypes or gene mutations.

Several methods such as CHASM, FATHMM, and CanDrA are developed for predicting disease-causing mutation in general as well as distinguishing between driver and passenger mutations in melanoma, breast and lung cancers. FATHMM [52] is a species-independent method, which incorporates pathogenicity weights and is capable of recognizing protein domains sensitive to missense mutations. CHASM [53] is a machine-learning system trained using COSMIC and other cancer-related databases, and utilizes a set of physicochemical properties, conservation score etc. Cancer-Related Analysis of VAriants Toolkit (CRAVAT) is a web-based application for CHASM that provides a simple interface to prioritize genes and variants important for specific cancer tissue types [54]. CanDrA is meta-predictor based on support vector machines, which utilizes the prediction scores obtained from ten prediction algorithms [55].

Anoosha *et al.* [56] developed a machine learning model using sequence and structure-based features of the mutation site to discriminate driver and passenger mutations observed in epidermal growth factor receptor (EGFR), considering its importance in cancer research. Utilizing this method, the highly probable driver mutations are identified by screening all the possible point mutations in EGFR and these results are available at http://www.iitm.ac.in/bioinfo/EGFR_Driver/.

Further, genomic data-based tools, such as Mutational Significance in Cancer (MuSiC) and Maftools [57,58], offer multiple analyses in a single software package. MuSic requires sequence-based inputs along with multiple types of clinical data to establish correlation among mutation sites. It requires large alignment files and significant computational resources and it is available at http://gmt.genome.wustl.edu/. On the other hand, MAFTools provides analysis, visualization and annotation modules and use a single input text file containing somatic variants in the Mutation Annotation Format (MAF). It is used for identifying cancer genes, clinical enrichment analysis and somatic interactions. MAFTools can be accessed at https://github.com/PoisonAlien/Maftools.

7. Large-scale annotation of cancer-causing mutations

Development of efficient sequencing technologies provides promising strategies for analyzing mutations, which include point mutations, indels, copy number alterations and structural variants (fusion genes) in cancer genomics [59,60]. Single cell RNA-seq has been widely used to understand SNV and allele-specific expression (ASE) as well as to predict cancer types and subtypes. On the other hand, somatic mutations are highly heterogeneous, lead to genomic variation, and provide baseline data for several types of analyses such as driver gene detection, pathway analysis, mutational signatures, and estimation of tumor heterogeneity [61]. Consequently, most of the associated computational methods target predicting the effect of singe point mutations, insertions and deletions [62,63]. We summarize the widely used prediction methods in Table 3.

Genomic deep learning (GDL) is a classification method to study the relationship between genomic variations and traits based on deep neural networks as well as identifying cancer risk. The architecture of the model contains feature selection, feature quantization, data filters and deep neural networks, which have multiple hidden layers between input and output layers. The identification of cancer based on genomic information using this method offers a new direction for disease diagnosis [64].

NeoMutate transforms the raw reads from sequencing technology platforms into a list of prioritized somatic variants with high degree of scalability and portability. It combines a set of machine learning models (MuTect2, Strelka2, SomaticSniper, VarScan2, VarDict, Lancet, Freebayes) and train them using ensemble calling and a set of biological relevant features. NeoMutate outperformed with a high rule-based filtering and a range of variant allele frequencies and mutation types [65].

Schulte-Sasse *et al.* [66] developed a machine learning method (EMOGI) based on graph convolutional networks for predicting cancer genes by combining multi-omics pan-cancer data (such as mutations, copy number changes, DNA methylation and gene expression together with protein-protein interaction networks) and it is used in precision oncology and for predicting biomarkers for complex diseases.

Rogers *et al.* [67] developed CScape-somatic, a method to discriminate driver mutations and benign variants using genomic, evolutionary features and allelic consequences. Zhao *et al.* [69] developed CanDriS, a statistical prediction method to identify driver sites using somatic mutations from TCGA and ICGC. Zeng *et al.* [69] developed Deep learning for disease Classification using exome sequencings (DeepCues) to explore the genetic variants including germline variants, insertion and deletion to predict the cancer types.

Leiserson *et al.* [70] proposed HotNet2, an integrative approach for identifying cancer driver genes with high mutation signatures, which

Table 3. Computational methods for mutation analysis based on genomic data

Name	Link	Important Features	References
Genome deep learning	https://github.com/Sunysh/Genome-Deep-Learning	Deep learning method to study the relationship between genomic variations and traits	[64]
NeoMutate	Access upon reasonable requests	An ensemble machine learning framework for the prediction of somatic mutations in cancer using biological and sequence features	[65]
EMOGI	https://github.com/schulter/EMOGI	A method to integrate multi-omics data with graph convolutional networks for identification of cancer genes	[66]
CScape-somatic	http://CScape-somatic.biocompute.org.uk/	Integrative classifier for discriminating between recurrent and rare variants in the human cancer genome	[67]
CanDriS	http://biopharm.zju.edu.cn/candrisdb/	To profile the potential cancer-driving sites for tumor cells	[68]
DeepCues*	https://github.com/zexian/DeepCues	Deep learning method, which uses whole genome sequencing features, germline variants and somatic mutations to predict cancer types.	[69]
HotNet2	http://compbio.cs.brown.edu/projects/hotnet2/	To identify significantly mutated groups of interacting genes from sequencing studies	[70]
MTGCN	https://github.com/weiba/MTGCN	To identify cancer driver genes using convolutional neutral network and protein-protein interaction networks	[71]
SWnet	https://github.com/zuozhaorui/SWnet.git	Deep learning method to integrate gene expression, genetic mutation, and chemical structure of compounds	[72]
PMCE	https://github.com/BIMIB-DISCo/PMCE	Identification of mutation expression profiles and survival analysis	[73]
dbWGFP*	http://bioinfo.au.tsinghua.edu.cn/dbwgfp	A database and web server of human who-genome single nucleotide variants and their functional predictions	[74]

Last accessed on 11 November, 2021
*Not accessible

utilizes the dataset from genomic studies of 12 cancer types from TCGA, combines different types of genomic alterations (point mutations, indel and copy number alterations) from different cancers and maps the mutated genes onto protein-protein interaction network. Recently, Peng *et al.* [71] developed a Multi-Task learning method based on the Graph Convolutional Network (MTGCN) to identify cancer driver genes by integrating biological and network features.

Zuo *et al.* [72] developed SWnet (self-attention gene weight layer network), a deep-learning predictive method which integrates genomics and cheminformatics for predicting the drug efficacy (IC50). It combines genomic signatures and molecular graphs and uses Graph Neural Networks to identify the interactions between genetics and structural similarity of the compounds. Angaroni *et al.* [73] developed PMCE, which uses mutational profiles from cross-sectional sequencing data and provides the correlation between evolutionary paths and overall survival in several cancer types including SKCM, KIRC, GBM and LIHC.

dbWGFP is a comprehensive database of functional predictions for SNVs by integrating 18 scoring methods used in different machine learning-based approaches for predicting nonsynonymous SNVs [74].

In essence, multilayer experiments using genomic and proteomic methods are developed for understanding the mutations, which govern tumor progression and their clinical implications. These methods are focused on pathogenic changes in the genome sequence as well as epigenetic perturbations or dysregulated gene activity, which may lead to cancer. The potential variants and hidden genes are identified using the combination of different approaches such as bioinformatics analysis, machine learning algorithms and Artificial Intelligence methods.

8. Conclusions

Identification of cancer-causing mutations and hotspot residues are very important for precision oncology. In this chapter, we identify and summarize databases of the cancer-causing mutations. The experimental and biomedical data available for cancer mutations are utilized to relate them with protein sequence and structural characteristics. We systematically discuss computational methods for identifying hotspot residues as well as the influence of mutations in specific genes and cancer types. Furthermore, we list and explore computational methods for predicting the driver mutations. In particular, we examine machine learning algorithms that are used to develop these methods. This comprehensive review aims to introduce readers to this area of research and help experimentalists to identify tools that assist with prioritizing mutations, and in more general sense, to advance the field of precision medicine.

9. Acknowledgements

The authors thank IIT Madras for computational facilities and Professor Lukasz Kurgan for providing the opportunity to contribute to this book. The work is partially supported by the Department of Science and Technology, Government of India to MMG (DST/INT/SWD/P-05/2016). Medha Pandey acknowledges the Ministry of Human Resource and Development for HTRA fellowship.

References

[1] Kroemer G and Pouyssegur J. Tumor cell metabolism: cancer's Achilles' heel. *Cancer Cell*, 2008. **13**(6): 472–482.

[2] Nesta AV, *et al.* Hotspots of human mutation. *Trends in Genetics*, 2021. **37**(8): 717–729.

[3] Chang MT, *et al.* Identifying recurrent mutations in cancer reveals widespread lineage diversity and mutational specificity. *Nature Biotechnology*, 2016. **34**: 155–163.

[4] Zhang J, *et al.* Molecular spectrum of KRAS, NRAS, BRAF and PIK3CA mutations in Chinese colorectal cancer patients: analysis of 1,110 cases. *Scientific Reports*, 2015. **5**: 18678.

[5] Jordan EJ, *et al.* Prospective comprehensive molecular characterization of lung adenocarcinomas for efficient patient matching to approved and emerging therapies. *Cancer Discovery*, 2017. **7**(6): 596–609.

[6] Pinter M, *et al.* Sorafenib in unresectable hepatocellular carcinoma from mild to advanced stage liver cirrhosis. *The Oncologist*, 2009. **14**(1): 70–76.

[7] Stratton MR, *et al.* The cancer genome. *Nature*, 2009. **458**(7239): 719–724.

[8] Touw IP and van de Geijn GJ. Granulocyte colony-stimulating factor and its receptor in normal myeloid cell development, leukemia and related blood cell disorders. *Frontiers in Bioscience*, 2007. **12**(1): 800–815.

[9] Bardelli A, *et al.* Mutational analysis of the tyrosine kinome in colorectal cancers. *Science*, 2003. **300**(5621): 949.

[10] Johnson GL and Lapadat R. Mitogen-activated protein kinase pathways mediated by ERK, JNK, and p38 protein kinases. *Science*, 2002. **298**(5600): 1911–1912.

[11] Shaw AT, *et al.* Crizotinib versus chemotherapy in advanced ALK-positive lung cancer. *New England Journal of Medicine*, 2013. **368**: 2385–2394.

[12] Anoosha P, *et al.* Investigating mutation-specific biological activities of small molecules using quantitative structure-activity relationship for epidermal growth factor receptor in cancer. *Mutation Research*, 2017. **806**: 19–26.

[13] Dlamini Z, *et al.* Artificial intelligence (AI) and big data in cancer and precision oncology. *Computational and Structural Biotechnology Journal*, 2020. **18**: 2300–2311.

[14] Van EM, *et al.* Whole-exome sequencing and clinical interpretation of formalin-fixed, paraffin-embedded tumor samples to guide precision cancer medicine. *Nature Medicine*, 2014. **20**: 682–688.

[15] Forbes SA, *et al.* COSMIC: Exploring the world's knowledge of somatic mutations in human cancer. *Nucleic Acids Research*, 2015. **43**(D1): D805–D811.

[16] Zhang J, *et al.* International cancer genome consortium data portal — a one-stop shop for cancer genomics data. *Database*, 2011. **2011**: bar026.

[17] Tomczak K, *et al.* The cancer genome atlas (TCGA): An immeasurable source of knowledge. *Contemporary Oncology*, 2015. **19**(1A): A68.

[18] Ainscough BJ, *et al.* DoCM: A database of curated mutations in cancer. *Nature Methods*, 2016. **13**(10): 806–807.

[19] Cerami E, *et al.* The cBio cancer genomics portal: An open platform for exploring multidimensional cancer genomics data. *Cancer Discovery*, 2012. **2**(5): 401–404.

[20] Chakravarty D, *et al.* OncoKB: A precision oncology knowledge base. *JCO Precision Oncology*, 2017. **1**: 1–16.

[21] Kulandaisamy A, *et al.* MutHTP: Mutations in human transmembrane proteins. *Bioinformatics*, 2018. **34**(13): 2325–2326.

[22] Rebhan M, *et al.* GeneCards: A novel functional genomics compendium with automated data mining and query reformulation support. *Bioinformatics*, 1998. **14**(8): 656–664.

[23] The UniProt Consortium. UniProt: The universal protein knowledgebase in 2021. *Nucleic Acids Research*, 2021. D480–D489.

[24] Burley SK, *et al.* RCSB Protein Data Bank: Powerful new tools for exploring 3D structures of biological macromolecules for basic and applied research and education in fundamental biology, biomedicine, biotechnology, bioengineering and energy sciences. *Nucleic Acids Research*, 2021. **49**(D1): D437–D451.

[25] Ganesan K, *et al.* HuVarBase: A human variant database with comprehensive information at gene and protein levels. *PLoS One*, 2019. **14**(1): e0210475.

[26] Liu X, *et al.* dbNSFP: A lightweight database of human nonsynonymous SNPs and their functional predictions. *Human Mutation*, 2011. **32**(8): 894–899.

[27] Liu X, *et al.* dbNSFP v2. 0: A database of human non–synonymous SNVs and their functional predictions and annotations. *Human Mutation*, 2013. **34**(9): E2393–E2402.

[28] Liu X, *et al.* dbNSFP v3. 0: A one–stop database of functional predictions and annotations for human nonsynonymous and splice-site SNVs. *Human Mutation*, 2016. **37**(3): 235–241.

[29] Siepel A, *et al.* Evolutionarily conserved elements in vertebrate, insect, worm, and yeast genomes. *Genome Research*, 2005. **15**(8): 1034–1050.

[30] Yue Z, *et al.* dbCPM: A manually curated database for exploring the cancer passenger mutations. *Briefings in Bioinformatics*, 2020. **21**(1): 309–317.

[31] ENCODE Project Consortium. An integrated encyclopedia of DNA elements in the human genome. *Nature*, 2012. **489**(7414): 57.

[32] Olivier M, *et al.* The IARC TP53 database: New online mutation analysis and recommendations to users. *Human Mutation*, 2002. **19**(6): 607–614.

[33] Li X, *et al.* OncoBase: A platform for decoding regulatory somatic mutations in human cancers. *Nucleic Acids Research*, 2019. **47**(D1): D1044–D1055.

[34] Gao J, *et al.* Integrative analysis of complex cancer genomics and clinical profiles using the cBioPortal. *Science Signaling*, 2013. **6**(269): pl1.

[35] Anoosha P, *et al.* Exploring preferred amino acid mutations in cancer genes: Applications to identify potential drug targets. *Biochimica et Biophysica Acta*, 2016. **1862**(2): 155–165.

[36] Martinez-Ledesma E, *et al.* Computational methods for detecting cancer hotspots. *Computational and Structural Biotechnology Journal*, 2020. **18**: 3567–3576.

[37] Chang *et al.* Accelerating discovery of functional mutant alleles in cancer. *Cancer discovery*, 2018. **8**(2): 174–183.

[38] Tamborero D, *et al.* OncodriveCLUST: Exploiting the positional clustering of somatic mutations to identify cancer genes. *Bioinformatics*, 2013. **29**: 2238–2244.

[39] Tokheim C, *et al.* Exome-scale discovery of hotspot mutation regions in human cancer using 3D protein structure. *Cancer Research*, 2016. **76**(13): 3719–3731.

[40] Chen S, *et al.* HotSpot3D web server: An integrated resource for mutation analysis in protein 3D structures. *Bioinformatics*, 2020. **36**(12): 3944–3946.

[41] Meyer MJ, *et al.* Mutation3D: Cancer gene prediction through atomic clustering of coding variants in the structural proteome. *Human Mutation*, 2016. **37**(5): 447–456.

[42] Trevino V, *et al.* HotSpotAnnotations-a database for hotspot mutations and annotations in cancer. *Database*, **2020**: baaa025.

[43] Pandey M and Gromiha MM. Predicting potential residues associated with lung cancer using deep neural networks. *Mutation Research*, 2020. **822**: 111737.

[44] Pandurangan AP and Blundell TL. Prediction of impacts of mutations on protein structure and interactions: SDM, a statistical approach, and mCSM, using machine learning. *Protein Science*, 2020. **29**(1): 247–257.

[45] Tan KP, *et al.* Packpred: Predicting the functional effect of missense mutations. *Frontiers in Molecular Biosciences*, 2021. **8**: 646288.

[46] Cibulskis K, *et al.* Sensitive detection of somatic point mutations in impure and heterogeneous cancer samples. *Nature Biotechnology*, 2013. **31**(3): 213–219.

[47] Choi Y, *et al.* Predicting the functional effect of amino acid substitutions and indels. *PLoS One*, 2012. **7**(10): e46688.

[48] Ramani R, *et al.* PhastWeb: A web interface for evolutionary conservation scoring of multiple sequence alignments using phastCons and phyloP. *Bioinformatics*, 2019. **35**(13): 2320–2322.

[49] Kulandaisamy A, *et al.* Pred-MutHTP: Prediction of disease-causing and neutral mutations in human transmembrane proteins. *Human Mutation*, 2020. **41**(3): 581–590.

[50] Xie G, *et al.* Group lasso regularized deep learning for cancer prognosis from multi-omics and clinical features. *Genes*, 2019. **10**(3): 240.

[51] Coudray N, *et al.* Classification and mutation prediction from non–small cell lung cancer histopathology images using deep learning. *Nature Medicine*, 2018. **24**(10): 1559–1567.

[52] Shihab HA, *et al.* Predicting the functional, molecular, and phenotypic consequences of amino acid substitutions using hidden Markov models. *Human Mutation*, 2013. **34**: 57–65.

[53] Carter H, *et al.* Cancer-specific high-throughput annotation of somatic mutations: computational prediction of driver missense mutations. *Cancer Research*, 2009. **69**: 6660–6667.

[54] Douville C, *et al.* CRAVAT: Cancer-related analysis of variants toolkit. *Bioinformatics*, 2013. **29**(5): 647–648.

[55] Mao Y, *et al.* CanDrA: Cancer-specific driver missense mutation annotation with optimized features. *PLoS One*, 2013. **8**(10): e77945.

[56] Anoosha P, *et al.* Discrimination of driver and passenger mutations in epidermal growth factor receptor in cancer. *Mutation Research*, 2015. **780**: 24–34.

[57] Dees ND, *et al.* MuSiC: Identifying mutational significance in cancer genomes. *Genome Research*, 2012. **22**(8): 1589–1598.

[58] Mayakonda A, *et al.* Maftools: Efficient and comprehensive analysis of somatic variants in cancer. *Genome Research*, 2018. **28**(11), 1747–1756.

[59] Fan J, *et al.* Single-cell transcriptomics in cancer: Computational challenges and opportunities. *Experimental & Molecular Medicine*, 2020. **52**(9): 1452–1465.

[60] Vu TN, *et al.* Cell-level somatic mutation detection from single-cell RNA sequencing. *Bioinformatics*, 2019. **35**(22): 4679–4687.

[61] Alexandrov LB and Stratton MR. Mutational signatures: the patterns of somatic mutations hidden in cancer genomes. *Current Opinion in Genetics & Development*, 2014. **24**: 52–60.

[62] Soh KP, *et al.* Predicting cancer type from tumour DNA signatures. *Genome Medicine*, 2017. **8**(1): 104.

[63] Madsen BE and Browning SR. A groupwise association test for rare mutations using a weighted sum statistic. *PLOS Genetics*, 2009. **5**(2): e1000384.

[64] Sun Y, *et al.* Identification of 12 cancer types through genome deep learning. *Scientific Reports*, 2019. **9**: 17256.

[65] Anzar I, *et al.* NeoMutate: An ensemble machine learning framework for the prediction of somatic mutations in cancer. *BMC Medical Genomics*, 2019. **12**: 63.

[66] Schulte-Sasse R, *et al.* Integration of multiomics data with graph convolutional networks to identify new cancer genes and their associated molecular mechanisms. *Nature Machine Intelligence*, 2021. **3**(6): 513–526.

[67] Rogers MF, *et al.* CScape-somatic: Distinguishing driver and passenger point mutations in the cancer genome. *Bioinformatics*, 2020. **36**(12): 3637–3644.

[68] Zhao W, *et al.* CanDriS: Posterior profiling of cancer-driving sites based on two-component evolutionary model. *Briefings in Bioinformatics*, 2021. **22**(5): bbab131.

[69] Zeng Z, *et al.* Deep learning for cancer type classification and driver gene identification. *BMC Bioinformatics*, 2021. **22**(Suppl 4): 491.

[70] Leiserson MD, *et al.* Pan-cancer network analysis identifies combinations of rare somatic mutations across pathways and protein complexes. *Nature Genetics*, 2015. **47**(2): 106–114.

[71] Peng W, *et al.* Improving cancer driver gene identification using multi-task learning on graph convolutional network. *Briefings in Bioinformatics*, 2022. **23**: bbab432.

[72] Zuo Z, *et al.* SWnet: A deep learning model for drug response prediction from cancer genomic signatures and compound chemical structures. *BMC Bioinformatics*, 2021. **22**(1): 434.

[73] Angaroni F, *et al.* PMCE: Efficient inference of expressive models of cancer evolution with high prognostic power. *Bioinformatics*, 2021. btab717.

[74] Wu J, *et al.* dbWGFP: A database and web server of human whole-genome single nucleotide variants and their functional predictions. *Database*, 2016. **2016**: baw024.

Part IV

Practical Resources

https://doi.org/10.1142/9789811258589_0012

Chapter 12

Designing Effective Predictors of Protein Post-Translational Modifications Using iLearnPlus

Zhen Chen*, Fuyi Li[†,‡], Xiaoyu Wang[†,‡], Yanan Wang[†,‡], Lukasz Kurgan[¶,††] and Jiangning Song[†,‡,§,**]

*Collaborative Innovation Center of Henan Grain Crops, Henan Agricultural University, Zhengzhou 450046, China;
[†]Monash Biomedicine Discovery Institute and Department of Biochemistry and Molecular Biology, Monash University, Melbourne, VIC 3800, Australia;
[‡]Monash Data Futures Institute, Faculty of Information Technology, Monash University, Melbourne, VIC 3800, Australia;
[§]Department of Veterinary Biosciences, Melbourne Veterinary School, The University of Melbourne, Parkville, Victoria 3010, Australia;
[¶]Department of Computer Science, Virginia Commonwealth University, Richmond, VA, USA.
**corresponding author: Jiangning.Song@monash.edu;
[††]corresponding author: lkurgan@vcu.edu.

Posttranslational modifications (PTMs) have vital roles in a myriad of biological processes, such as metabolism, DNA damage response, transcriptional regulation, protein-protein interactions, cell death, immune response, signaling pathways and aging. Identification of PTM sites is a crucial first step for biochemical, pathological and pharmaceutical studies

309

associated with the functional characterization of proteins. However, experimental approaches for identifying PTM sites are relatively expensive, labor-intensive and time-consuming, partly due to the dynamics and reversibility of PTMs. In this context, computational methods that accurately predict PTMs serve as a useful alternative, especially when targeting large-scale whole-proteome annotations. We briefly summarize and review existing predictors of PTM sites in protein sequences. Moreover, we introduce the iLearnPlus platform that facilitates the development of new predictive methods and apply it to generate a new PTM predictor. We elaborate a detailed procedure for the development of predictive models, particularly focusing on the deep learning (DL) techniques. We assess predictive performance of the developed DL model and demonstrate how to compare it against other machine learning algorithms. While we use iLearnPlus in the context of the PTM prediction, we emphasize that this platform can be used to design predictive systems for a broad spectrum of other related problems that cover prediction of structural and functional characteristics of proteins and nucleic acids from their sequences.

1. Introduction

Posttranslational modifications (PTMs) refer to the reversible or irreversible chemical changes that some proteins undergo after translation [1]. PTMs play vital roles in a broad array of cellular processes, such as metabolism, signal transduction, stability, structural state, and localization of proteins [2–8]. For example, phosphorylation is implicated in orchestrating signal transduction, cytoskeleton rearrangement, and cell cycle progression [9,10]. Moreover, ubiquitination mediates protein degradation by the Ub-proteasome system in eukaryotic cells [11] while malonylation plays vital roles for metabolic reprogramming in determining the function of immune cells [12]. To date, advances in experimental techniques have significantly assisted biologists in identifying various types of PTMs. Currently, over 680 types of PTMs have been characterized experimentally (see http://www.uniprot.org/docs/ptmlist.txt).

Given the prevalence and importance of PTMs, aberrant modifications are shown to be associated with various human diseases [4,5,13–16]. Systematic identification of different types of PTM substrates and PTM sites in proteomic data is becoming an urgent issue. To date, numerous efforts have been dedicated to the investigation of cellular mechanisms that underly PTMs, which is based on accurate identification of corresponding PTM substrates and sites. Advances in the PTM research benefit from computational studies that accurately predict PTM sites, significantly reducing the time and effort involved in the experimental identification. Compared to the labor-intensive and time-consuming experimental characterization of PTMs, computational prediction of PTMs in proteins provides a valuable and complementary approach to shortlist likely candidates for subsequent experimental validation. Thus, a variety of computational methods for PTM identification have been developed using various protein sequence features and state-of-the-art machine learning (ML) techniques [17–21]. These methods predict new PTM sites by learning features of the sequence context of experimentally verified PTM sites primarily using ML algorithms. We briefly overview existing computational predictors of PTM sites.

We describe an innovative and comprehensive platform for the development of new predictive methods, iLearnPlus [22], and we apply it to generate a new PTM predictor for lysine malonylation. We detail the procedure for development of predictive models based on iLearnPlus, focusing on the DL techniques. This includes benchmark dataset preparation, feature extraction, model construction, and performance evaluation. In particular, we compare the results produced by the DL model against other ML algorithms.

While here we apply iLearnPlus for the lysine malonylation prediction, this software can be used to design, implement, and comparatively validate predictive systems for many other related problems. These application areas broadly cover prediction of structural and functional characteristics of proteins and nucleic acids, such as secondary structure, intrinsic disorder, protein-ligand and nucleic acid-ligand binding, and many others.

2. Brief review of computational PTM site prediction

Recent years have witnessed the development and proliferation of computational approaches for the prediction of PTM and cleavage sites [17,18,20,21,23–39]. These methods differ in a variety of aspects, including the dataset collection and preprocessing, feature descriptors and feature selection techniques employed, classification algorithms used, and performance evaluation strategies utilized.

Generally, the current models for the PTM sites prediction could be divided into three main categories based on the adopted techniques. The first category is based on peptide similarity. Methods in this group usually calculate a similarity score between the peptide that is being predicted and peptides with experimentally annotated PTM sites [38,40]. The similarities are computed using a number of measures, such as the BLOcks SUbstitution Matrix (BLOSUM62) matrix [41] and position-specific scoring matrix (PSSM) [42]. Representative methods in this category include the Group-based Prediction System (GPS) series approaches for predicting phosphorylation [43], methylation [44], sumoylation [45], as well as the acetylation set enrichment-based (ASEB) approach [46] for the acetylation sites prediction [47].

The secondary category relies on conventional ML algorithms using sequence-derived features. Here, ML algorithms are used to derive predictive models from experimentally annotated training data. Authors of these methods manually develop an approach to transform the input sequences into a fixed-size numeric feature set that is subsequently input into the ML model. The fixed-size feature vector is required by these types of models. Fortunately, development of these feature sets from biological sequences (i.e., protein and nucleic acid sequences) is supported by a variety of convenient tools, such as ProtrWeb [48], iFeature [49], BioSeq-Analysis [50], iLearn [51] and iLearnPlus [22]. Conventional ML algorithms (i.e., ML algorithms that exclude deep learners) are employed to use the extracted features to build an accurate predictive model. Conventional ML algorithms that are commonly used to predict PTMs include support vector machine [18,52,53], random forest [54–56], shallow artificial neural network [57,58], k-nearest neighbors [59,60], logistic regression [32], and their ensembles [31].

The third category covers end-to-end approaches that rely on deep learning techniques. For the end-to-end approaches, the protein sequence is not encoded into a feature vector but rather used directly as an input to a deep neural network. These deep learners extract latent features by themselves. Many of the recent PTM predictors belong to this category. Examples include DeepNitro [34], CapsNet_PTM [30], DeepPTM [28], DeepSuccinylSite [35], MusiteDeep [29], DeepPPSite [36], MultiLyGAN [37], and nhKcr [39].

As described above, dozens of computational methods have been developed for the prediction of various types of PTM sites. However, dedicated predictors are missing for numerous PTM types, and the rapid accumulation of the experimental data motivate the need to develop many more predictors in a near future. We use the iLearnPlus platform to demonstrate how easy and convenient it is to design, develop and validate a new PTM predictor, focusing on the recently popular deep learning/end-to-end methods.

3. Design of novel predictive methods using iLearnPlus

3.1 *iLearnPlus*

iLearnPlus is an advanced software package that provides a comprehensive platform to analyze various structural and functional characteristics of the DNA, RNA and protein sequences and to efficiently conceptualize, design, implement and comparatively evaluate ML-based solutions for prediction of these characteristics [22]. This platform includes four modules:

1) iLearnPlus-Basic for analysis and prediction using feature-based representation of protein/RNA/DNA sequences and a selected ML classifier;
2) iLearnPlus-Estimator that facilitates comprehensive feature extraction from protein and nucleic acid sequences;
3) iLearnPlus-AutoML that provides automated benchmarking and optimization of predictive accuracy by considering different ML algorithms and features; and
4) iLearnPlus-LoadModel that enables uploading, deploying and testing models on user's own data.

Altogether, iLearnPlus supports a broad spectrum of activities including feature extraction and analysis, rational design of ML models, training and empirical assessment of ML classifiers, comparative statistical analysis of classifiers, and visualisation of data and predictive results. As a highlight, iLearnPlus covers 21 ML algorithms including seven types of popular and modern deep learners.

iLearnPlus can be utilized by users with limited bioinformatics expertise, such as biologists and biochemists, who can take advantage of the easy and convenient to use webserver version at http://ilearnplus.erc.monash.edu/. All activities, including generation of models and testing are performed on the server side. More experienced bioinformaticians should use the command line and/or GUI (Graphical User Interface) versions that can be downloaded from https://github.com/Superzchen/iLearnPlus/.

Here, we illustrate how to use this platform to conceptualize, design and test a deep learning-based predictor of protein lysine malonylation sites.

3.2 Data collection and preprocessing

Lysine malonylation (Kmal) is a recently discovered PTM type [61] that is associated with several important cellular processes [62–65]. Only a few methods can predict the Kmal sites [31,66–69]. Experimental data on the lysine-malonylated proteins are retrieved from mice and humans in two proteomic assays [70,71]. Based on work in [69], the following data pre-processing steps are used to derive datasets needed for the development of predictive models:

1) Kmal-containing proteins are retrieved from the UniProt database [72] and protein sequences with sequence identities greater than 30% are removed using the CD-HIT tool [73];
2) The annotated Kmal sites are considered as positive samples and the remaining lysine residues on the same proteins are considered as negative samples;
3) 31-residue long peptides (–15 to + 15) with the lysine site in the center are extracted for each sample. If the positive peptides are identical to the negative peptides then the negative peptides are removed;
4) All the samples are randomly divided into two parts. About 80% of all samples are subjected to five-fold cross-validation, and the remaining are used as an independent test dataset (i.e., dataset excluded from the classifier training procedure). The finalized version of the training dataset contains 4,242 positive peptides and 71,809 negative peptides, while the independent test dataset has 1,046 positive peptides and 16,827 negative peptides.

3.3 Model construction and performance evaluation

The iLearnPlus platform is used to construct the model and assess the model performance. Figure 1 shows the main GUI interface of iLearnPlus. The predictive model is built using the following sequence of nine steps:

1) Transform the positive and negative peptides into FASTA format. The FASTA header consists of three parts: part 1, part 2 and part 3, which

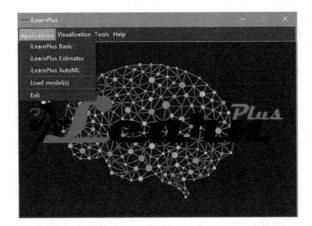

Figure 1.   The main interface of iLearnPlus.

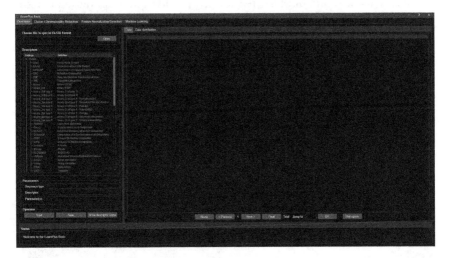

Figure 2.   An example of extracting feature descriptors using the iLearnPlus-Basic module.

are separated by the "|" symbol (Fig. 2). Part 1 is the sequence name. Part 2 is the sample category information, which can be filled with any integer. For instance, users may use "1" to indicate the positive samples and "0" to represent the negative samples for a binary classification task, or use "0, 1, 2, ..." to represent different classes in a multiclass classification task. Part 3 indicates the role of the sample, where for instance "training" would indicate that the corresponding sequence would be used as part of the training set in the *k*-fold cross-validation test, and "testing" that the sequence would be used as part of the independent test dataset;

2) Click "iLearnPlus Basic" (Fig. 1) to launch the iLearnPlus-Basic module that is shown in Figure 2; This module facilitates generation of features from the sequences.

3) Click the "Open" button in the "Descriptor" panel, and select the file with the protein sequences. The biological sequence type (i.e., DNA, RNA or protein) is automatically detected based on the input sequences. Click the "Save" button to save the feature descriptors to a file named "binary.csv";

4) Click the "binary" descriptor. We use the default parameters here;

5) Click the "Start" button to calculate the descriptor. This initiates a procedure to compute numeric features from the input peptide sequences, which is needed to subsequently use the ML algorithm. The feature encoding and graphical presentation are displayed in the "Data" and "Data distribution" areas, respectively;

6) Switch to the "Machine Learning" panel and load data through the "Select" button (Fig. 3). Select "Descriptor data" in the data selection dialog box and click the "OK" button; This step moves the process to the production of the ML model from the already prepared feature-based dataset.

7) Select "Net_1_CNN" and set the "Input channels" as 20 (Fig. 4). This denotes that we select a modern deep convolutional neural network (CNN) model. The default values are used for the remaining parameters;

Figure 3.   Load data using the data selection explorer.

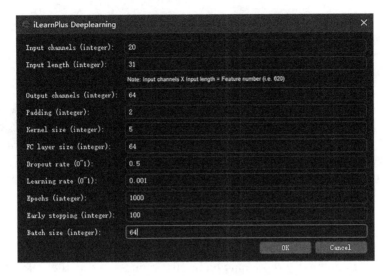

Figure 4. An example of parameter setting for the deep convolutional neural network (CNN) algorithm.

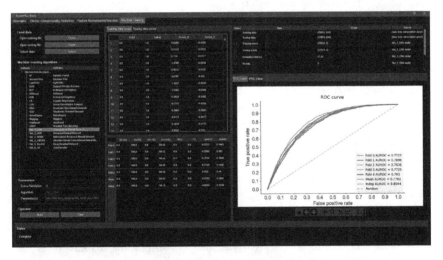

Figure 5. An example of prediction model construction using the CNN algorithm.

8) Set the *K* number as 5; This sets up the test procedure as the 5-fold cross validation.

9) Click the "Start" button to start the modeling. This starts the process of generating the CNN model using the training dataset in the 5-fold

cross-validation setting. The resulting prediction score, performance evaluation metrics for the cross-validation and independent test and the Receiver Operating Characteristic (ROC) curves are displayed in Figure 5;

We observe that iLearnPlus produces a relatively accurate predictor of the Kmal sites, with the average AUROC (area under the ROC curve) over the 5 cross-validation folds of about 0.78 (Fig. 5). This model achieves a similar AUROC of 0.80 on the independent test dataset, which suggests that the trained model is robust and did not overfit the cross-validation experiment.

3.4 *Comparison with other ML algorithms*

So far, we designed and evaluated a predictive model using the CNN algorithm. Now, we use the iLearnPlus-AutoML module to compare the predictive performance between different ML algorithms. This requires executing the following sequence of five steps:

1) Open the "data.csv" file;
2) Select ML algorithms. Here, we consider five ML algorithms including "RF" (random forest), "DecisionTree" (decision tree), "LightGBM" (gradient boosted forest), "LR" (logistic regression), and "CNN" (deep convolutional neural network).
3) Set the K number as 5; This defines the test procedure to be the 5-fold cross validation.
4) Click the "Start" button to train the five models. For each of the selected ML algorithms, the program will build the predictive model automatically, one by one, and correspondingly test them via the cross-validation protocol.
5) The results are displayed using four panels:
   • The evaluation metrics for the five classifiers are displayed in the table widget (Fig. 6);
   • The correlation matrix of the five classifiers is displayed in the form of a heatmap (Fig. 7); This heatmap quantifies the degree of similarity between the results/predictions produced by the five models.
   • Boxplots for the evaluation metrics (Fig. 8); These plots quantify the spread of the values of the metrics over the five cross-validation folds.

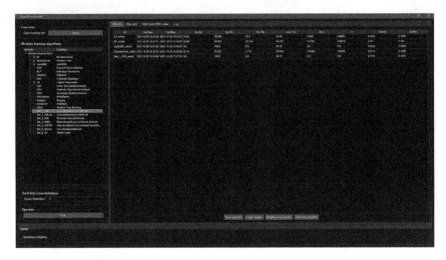

Figure 6. Performance evaluation metrics for the five models. The metrics include sensitivity (Sn), specificity (Sp), precision (Pre), accuracy (Acc), Matthew's correlation coefficient (MCC), F1, AUROC, and AUPRC (area under the precision-recall curve).

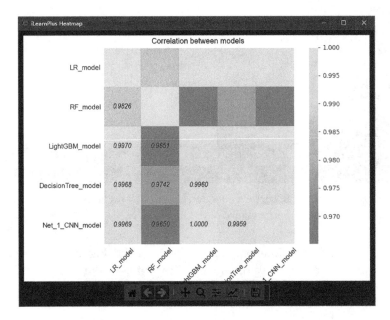

Figure 7. The correlation matrix generated by the iLearnPlus-AutoML module.

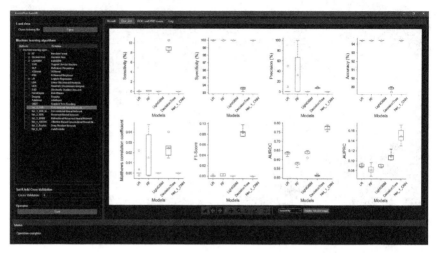

Figure 8. The boxplots generated by the iLearnPlus-AutoML module.

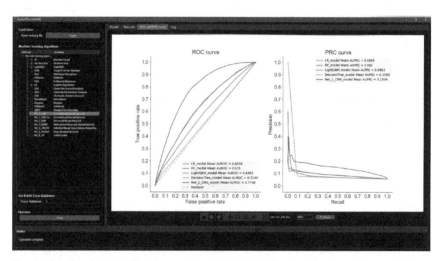

Figure 9. ROC and PR curves generated by the iLearnPlus-AutoML module to evaluate the predictive performance of the five models.

- ROC and precision-recall curve (PRC) curves (Fig. 9). These curves quantify the quality of the putative scores produced by the five predictors.

The five-step process automates numerous activities that include loading of the data, selection of ML algorithms, setting up test protocol, running training, and test experiments across the five selected algorithms, and generation of a wide range of helpful metrics and plots that summarize and compare the corresponding results of the five ML algorithms. From Figure 7, we learn that the results generated by the various ML algorithms are highly correlated, with Pearson correlation coefficients ranging between 0.98 and 1. This is not surprising since these models solve the same problem using the same training dataset. While the five results are correlated, Figure 6 and Figure 9 reveal that the corresponding predictive performance is substantially different. The most accurate solution that relies on the CNN model achieves AUROC of 0.77, which agrees with the results in Figure 5. The other ML algorithms are not as accurate as the CNN model, with the second-best algorithm (i.e., the gradient boosted forest) securing AUROC of 0.64 and the simplest decision tree obtaining AUROC of 0.51. This type of analysis allows the users to compare different solutions easily and conveniently, in this case, by relying on different ML algorithms to select the one that generates favorable levels of predictive quality.

4. Summary

Prediction of the PTM sites from the protein sequences is an active research area that requires the development of novel methods that would provide results for the many PTM types that lack predictors and that would take advantage of the newly released experimental data to improve over the current solutions. Generation of the predictive systems is a relatively complex process that involves collection of training and test data, various data conversions that include feature encoding, extraction and selection, modeling that covers setup and generation of predictive models typically using multiple ML algorithms, and comparative analysis to select the best solution. The execution of this entire process can be automated and facilitated with modern software platforms, such as iLearnPlus [22]. We use the example of the prediction of the lysine malonylation sites to demonstrate how iLearnPlus is used to develop accurate models and to perform comparative analysis. We find that predictors that rely on deep neural networks outperform more classical ML algorithms for this predictive task.

Importantly, we highlight the fact that iLearnPlus can be utilized to conceptualize, design, test, and deploy predictive solutions for many other related problems that extend beyond the PTM predictions. These problems cover prediction of functional and structural annotations from the proteins and nucleic acid sequences. Examples include prediction of the protein secondary structure [74–78] and other structural features of proteins [79], RNA secondary structure [80,81], protein-nucleic acids interactions [82–84], protein-protein interactions [85–88], intrinsic disorder and its functions [89–97], cleavage sites [98], and many other annotations.

References

[1]   Uversky VN. Posttranslational modification. In *Brenner's Encyclopedia of Genetics*. Maloy S & Hughes K (eds.), Academic Press: San Diego, 2013. 425–430.

[2]   Mann M and Jensen ON. Proteomic analysis of post-translational modifications. *Nature Biotechnology*, 2003. **21**(3): 255–261.

[3]   Hendriks IA, *et al.* Site-specific mapping of the human SUMO proteome reveals co-modification with phosphorylation. *Nature Structural & Molecular Biology*, 2017. **24**(3): 325–336.

[4]   Xu H, *et al.* PTMD: A database of human disease-associated post-translational modifications. *Genomics Proteomics Bioinformatics*, 2018. **16**(4): 244–251.

[5]   Li F, *et al.* PRISMOID: A comprehensive 3D structure database for post-translational modifications and mutations with functional impact. *Briefings in Bioinformatics*, 2020. **21**(3): 1069–1079.

[6]   Duan G and Walther D. The roles of post-translational modifications in the context of protein interaction networks. *PLoS Computational Biology*, 2015. **11**(2): e1004049–e1004049.

[7]   Hu G, *et al.* Functional analysis of human hub proteins and their interactors involved in the intrinsic disorder-enriched interactions. *International Journal of Molecular Science*, 2017. **18**(12): 2761.

[8]   Zhou JH, *et al.* Intrinsically disordered proteins link alternative splicing and post-translational modifications to complex cell signaling and regulation. *Journal of Molecular Biology*, 2018. **430**(16): 2342–2359.

[9]   Manning G, *et al.* The protein kinase complement of the human genome. *Science*, 2002. **298**(5600): 1912–1934.

[10] Kuntz EM, *et al.* Targeting mitochondrial oxidative phosphorylation eradicates therapy-resistant chronic myeloid leukemia stem cells. *Nature Medicine*, 2017. **23**(10): 1234–1240.

[11] Hagai T and Levy Y. Ubiquitin not only serves as a tag but also assists degradation by inducing protein unfolding. *Proceedings of the National Academy of Sciences of the United States of America*, 2010. **107**(5): 2001–2006.

[12] Galvan-Pena S, *et al.* Malonylation of GAPDH is an inflammatory signal in macrophages. *Nature Communications*, 2019. **10**(1): 338.

[13] D'Amore C and Salvi M. Editorial of Special Issue "Protein post-translational modifications in signal transduction and diseases". *International Journal of Molecular Science*, 2021. **22**(5): 2232.

[14] Li F, *et al.* GlycoMine: A machine learning-based approach for predicting N-, C- and O-linked glycosylation in the human proteome. *Bioinformatics*, 2015. **31**(9): 1411–1419.

[15] Li F, *et al.* GlycoMine(struct): A new bioinformatics tool for highly accurate mapping of the human N-linked and O-linked glycoproteomes by incorporating structural features. *Scientific Reports*, 2016. **6**: 34595.

[16] Li F, *et al.* Positive-unlabelled learning of glycosylation sites in the human proteome. *BMC Bioinformatics*, 2019. **20**(1): 112.

[17] Gianazza E, *et al.* In silico prediction and characterization of protein post-translational modifications. *Journal of Proteomics*, 2016. **134**: 65–75.

[18] Wang B, *et al.* Prediction of post-translational modification sites using multiple kernel support vector machine. *PeerJ*, 2017. **5**: e3261.

[19] Zhou F, *et al.* A general user interface for prediction servers of proteins' post-translational modification sites. *Nature Protocols*, 2006. **1**(3): 1318–1321.

[20] Audagnotto M and Dal Peraro M. Protein post-translational modifications: In silico prediction tools and molecular modeling. *Computational and Structural Biotechnology Journal*, 2017. **15**: 307–319.

[21] He W, *et al.* Research progress in protein posttranslational modification site prediction. *Briefings in Functional Genomics*, 2018. **18**(4): 220–229.

[22] Chen Z, *et al.* iLearnPlus: A comprehensive and automated machine-learning platform for nucleic acid and protein sequence analysis, prediction and visualization. *Nucleic Acids Research*, 2021. **49**(10): e60.

[23] Li F, *et al.* Quokka: A comprehensive tool for rapid and accurate prediction of kinase family-specific phosphorylation sites in the human proteome. *Bioinformatics*, 2018. **34**(24): 4223–4231.

[24] Song J, *et al.* PROSPERous: High-throughput prediction of substrate cleavage sites for 90 proteases with improved accuracy. *Bioinformatics*, 2018. **34**(4): 684–687.

[25] Li F, *et al.* DeepCleave: A deep learning predictor for caspase and matrix metalloprotease substrates and cleavage sites. *Bioinformatics*, 2020. **36**(4): 1057–1065.

[26] Li F, *et al.* Procleave: Predicting protease-specific substrate cleavage sites by combining sequence and structural information. *Genomics Proteomics Bioinformatics*, 2020. **18**(1): 52–64.

[27] Ahmed S, *et al.* DeepPPSite: A deep learning-based model for analysis and prediction of phosphorylation sites using efficient sequence information. *Analytical Biochemistry*, 2020. **612**: 113955.

[28] Baisya DR and Lonardi S. Prediction of histone post-translational modifications using deep learning. *Bioinformatics*, 2020. **36**(24): 5610–5617.

[29]   Wang D, *et al.* MusiteDeep: A deep-learning based webserver for protein post-translational modification site prediction and visualization. *Nucleic Acids Research*, 2020. **48**(W1): W140–W146.

[30]   Wang D, *et al.* Capsule network for protein post-translational modification site prediction. *Bioinformatics*, 2019. **35**(14): 2386–2394.

[31]   Zhang Y, *et al.* Computational analysis and prediction of lysine malonylation sites by exploiting informative features in an integrative machine-learning framework. *Briefings in Bioinformatics*, 2019. **20**(6): 2185–2199.

[32]   Li F, *et al.* Quokka: A comprehensive tool for rapid and accurate prediction of kinase family-specific phosphorylation sites in the human proteome. *Bioinformatics*, 2018. **34**(24): 4223–4231.

[33]   Lo Monte M, *et al.* ADPredict: ADP-ribosylation site prediction based on physicochemical and structural descriptors. *Bioinformatics*, 2018. **34**(15): 2566–2574.

[34]   Xie Y, *et al.* DeepNitro: Prediction of protein nitration and nitrosylation sites by deep learning. *Genomics Proteomics Bioinformatics*, 2018. **16**(4): 294–306.

[35]   Thapa N, *et al.* DeepSuccinylSite: A deep learning based approach for protein succinylation site prediction. *BMC Bioinformatics*, 2020. **21**(Suppl 3): 63.

[36]   Ahmed S, *et al.* DeepPPSite: A deep learning-based model for analysis and prediction of phosphorylation sites using efficient sequence information. *Analytical Biochemistry*, 2021. **612**: 113955.

[37]   Yang Y, *et al.* Prediction and analysis of multiple protein lysine modified sites based on conditional wasserstein generative adversarial networks. *BMC Bioinformatics*, 2021. **22**(1): 171.

[38]   Li F, *et al.* Twenty years of bioinformatics research for protease-specific substrate and cleavage site prediction: A comprehensive revisit and benchmarking of existing methods. *Briefings in Bioinformatics*, 2019. **20**(6): 2150–2166.

[39]   Chen YZ, *et al.* nhKcr: A new bioinformatics tool for predicting crotonylation sites on human nonhistone proteins based on deep learning. *Briefings in Bioinformatics*, 2021. **22**(6): bbab146.

[40]   Chen Z, *et al.* Large-scale comparative assessment of computational predictors for lysine post-translational modification sites. *Briefings in Bioinformatics*, 2019. **20**(6): 2267–2290.

[41]   Henikoff S and Henikoff JG. Amino acid substitution matrices from protein blocks. *Proceedings of the National Academy of Sciences of the United States of America*, 1992. **89**(22): 10915–10919.

[42]   Altschul SF, *et al.* Gapped BLAST and PSI-BLAST: A new generation of protein database search programs. *Nucleic Acids Research*, 1997. **25**(17): 3389–3402.

[43]   Xue Y, *et al.* GPS: A comprehensive www server for phosphorylation sites prediction. *Nucleic Acids Research*, 2005. **33**(Web Server issue): W184–W187.

[44] Deng W, *et al.* Computational prediction of methylation types of covalently modified lysine and arginine residues in proteins. *Briefings in Bioinformatics,* 2017. **18**(4): 647–658.

[45] Zhao Q, *et al.* GPS-SUMO: A tool for the prediction of sumoylation sites and SUMO-interaction motifs. *Nucleic Acids Research,* 2014. **42**(Web Server issue): W325–W330.

[46] Li T, *et al.* Characterization and prediction of lysine (K)-acetyl-transferase specific acetylation sites. *Molecular & Cellular Proteomics,* 2012. **11**(1): M111 011080.

[47] Liu Z, *et al.* GPS-PUP: Computational prediction of pupylation sites in prokaryotic proteins. *Molecular BioSystems,* 2011. **7**(10): 2737–2740.

[48] Xiao N, *et al.* protr/ProtrWeb: R package and web server for generating various numerical representation schemes of protein sequences. *Bioinformatics,* 2015. **31**(11): 1857–1859.

[49] Chen Z, *et al.* iFeature: A Python package and web server for features extraction and selection from protein and peptide sequences. *Bioinformatics,* 2018. **34**(14): 2499–2502.

[50] Liu B, *et al.* BioSeq-Analysis2.0: An updated platform for analyzing DNA, RNA and protein sequences at sequence level and residue level based on machine learning approaches. *Nucleic Acids Research,* 2019. **47**(20): e127.

[51] Chen Z, *et al.* iLearn: An integrated platform and meta-learner for feature engineering, machine-learning analysis and modeling of DNA, RNA and protein sequence data. *Briefings in Bioinformatics,* 2020. **21**(3): 1047–1057.

[52] Cao W, *et al.* Prediction of N-myristoylation modification of proteins by SVM. *Bioinformation,* 2011. **6**(5): 204–206.

[53] Chen X, *et al.* Proteomic analysis and prediction of human phosphorylation sites in subcellular level reveal subcellular specificity. *Bioinformatics,* 2015. **31**(2): 194–200.

[54] Chang CC, *et al.* SUMOgo: Prediction of sumoylation sites on lysines by motif screening models and the effects of various post-translational modifications. *Scientific Reports,* 2018. **8**(1): 15512.

[55] Hamby SE and Hirst JD. Prediction of glycosylation sites using random forests. *BMC Bioinformatics,* 2008. **9**: 500.

[56] Wang YG, *et al.* Accurate prediction of species-specific 2-hydroxyisobutyrylation sites based on machine learning frameworks. *Analytical Biochemistry,* 2020. **602**: 113793.

[57] Basu S and Plewczynski D. AMS 3.0: Prediction of post-translational modifications. *BMC Bioinformatics,* 2010. **11**: 210.

[58] Dewhurst HM and Torres MP. Systematic analysis of non-structural protein features for the prediction of PTM function potential by artificial neural networks. *PLoS One,* 2017. **12**(2): e0172572.

[59]   Feng KY, *et al.* Using WPNNA classifier in ubiquitination site prediction based on hybrid features. *Protein & Peptide Letters*, 2013. **20**(3): 318–323.

[60]   Niu S, *et al.* Prediction of tyrosine sulfation with mRMR feature selection and analysis. *Journal of Proteome Research*, 2010. **9**(12): 6490–6497.

[61]   Peng C, *et al.* The first identification of lysine malonylation substrates and its regulatory enzyme. *Molecular & Cellular Proteomics*, 2011. **10**(12): M111 012658.

[62]   Nie LB, *et al.* Global proteomic analysis of lysine malonylation in toxoplasma gondii. *Frontiers in Microbiology*, 2020. **11**: 776.

[63]   Ma Y, *et al.* Malonylome analysis reveals the involvement of lysine malonylation in metabolism and photosynthesis in cyanobacteria. *Journal of Proteome Research*, 2017. **16**(5): 2030–2043.

[64]   Qian L, *et al.* Global profiling of protein lysine malonylation in Escherichia coli reveals its role in energy metabolism. *Journal of Proteome Research*, 2016. **15**(6): 2060–2071.

[65]   Hirschey MD and Zhao Y. Metabolic regulation by lysine malonylation, succinylation, and glutarylation. *Molecular & Cellular Proteomics*, 2015. **14**(9): 2308–2315.

[66]   Chung CR, *et al.* Incorporating hybrid models into lysine malonylation sites prediction on mammalian and plant proteins. *Scientific Reports*, 2020. **10**(1): 10541.

[67]   Ahmad W, *et al.* Mal-Light: Enhancing lysine malonylation sites prediction problem using evolutionary-based features. *IEEE Access*, 2020. **8**: 77888–77902.

[68]   Xiang Q, *et al.* Prediction of lysine malonylation sites based on pseudo amino acid. *Combinatorial Chemistry & High Throughput Screening*, 2017. **20**(7): 622–628.

[69]   Chen Z, *et al.* Integration of a deep learning classifier with a random forest approach for predicting malonylation sites. *Genomics Proteomics Bioinformatics*, 2018. **16**(6): 451–459.

[70]   Nishida Y, *et al.* SIRT5 regulates both cytosolic and mitochondrial protein malonylation with glycolysis as a major target. *Molecular Cell*, 2015. **59**(2): 321–332.

[71]   Colak G, *et al.* Proteomic and biochemical studies of lysine malonylation suggest its malonic aciduria-associated regulatory role in mitochondrial function and fatty acid oxidation. *Molecular & Cellular Proteomics*, 2015. **14**(11): 3056–3071.

[72]   UniProt C. UniProt: The universal protein knowledgebase in 2021. *Nucleic Acids Research*, 2021. **49**(D1): D480–D489.

[73]   Fu L, *et al.* CD-HIT: Accelerated for clustering the next-generation sequencing data. *Bioinformatics*, 2012. **28**(23): 3150–3152.

[74]   Kashani-Amin E, *et al.* A systematic review on popularity, application and characteristics of protein secondary structure prediction tools. *Current Drug Discovery Technologies*, 2018. **16**(2): 159–172.

[75]   Zhang H, *et al.* Critical assessment of high-throughput standalone methods for secondary structure prediction. *Briefings in Bioinformatics*, 2011. **12**(6): 672–688.

[76]   Oldfield CJ, *et al.* Computational prediction of secondary and supersecondary structures from protein sequences. *Methods in Molecular Biology*, 2019. **1958**: 73–100.

[77]   Meng F and Kurgan L. Computational prediction of protein secondary structure from sequence. *Current Protocols in Protein Science*, 2016. **86**: 2 3 1–2 3 10.

[78]   Jiang Q, *et al.* Protein secondary structure prediction: A survey of the state of the art. *Journal of Molecular Graphics and Modelling*, 2017. **76**: 379–402.

[79]   Kurgan L and Disfani FM. Structural protein descriptors in 1-dimension and their sequence-based predictions. *Current Protein & Peptide Science*, 2011. **12**(6): 470–489.

[80]   Zhao Q, *et al.* Review of machine learning methods for RNA secondary structure prediction. *PLoS Computational Biology*, 2021. **17**(8): e1009291.

[81]   Tahi F, *et al.* In silico prediction of RNA secondary structure. *Methods in Molecular Biology*, 2017. **1543**: 145–168.

[82]   Yan J, *et al.* A comprehensive comparative review of sequence-based predictors of DNA- and RNA-binding residues. *Briefings in Bioinformatics*, 2016. **17**(1): 88–105.

[83]   Zhang J, *et al.* Comprehensive review and empirical analysis of hallmarks of DNA-, RNA- and protein-binding residues in protein chains. *Briefings in Bioinformatics*, 2019. **20**(4): 1250–1268.

[84]   Miao Z and Westhof E. A Large-Scale Assessment of Nucleic Acids Binding Site Prediction Programs. *PLoS Computational Biology*, 2015. **11**(12): e1004639.

[85]   Zhang J, *et al.* Prediction of protein-binding residues: dichotomy of sequence-based methods developed using structured complexes versus disordered proteins. *Bioinformatics*, 2020. **36**(18): 4729–4738.

[86]   Katuwawala A, *et al.* Computational prediction of MoRFs, short disorder-to-order transitioning protein binding regions. *Computational and Structural Biotechnology Journal*, 2019. **17**: 454–462.

[87]   Chen H, *et al.* Systematic evaluation of machine learning methods for identifying human-pathogen protein-protein interactions. *Briefings in Bioinformatics*, 2020. **22**(3): bbaa068.

[88]   Fernandez-Recio J. Prediction of protein binding sites and hot spots. *Wiley Interdisciplinary Reviews-Computational Molecular Science*, 2011. **1**(5): 680–698.

[89] Meng F, *et al.* Comprehensive review of methods for prediction of intrinsic disorder and its molecular functions. *Cellular and Molecular Life Sciences*, 2017. **74**(17): 3069–3090.

[90] He B, *et al.* Predicting intrinsic disorder in proteins: An overview. *Cell Research*, 2009. **19**(8): 929–949.

[91] Zhao B and Kurgan L. Surveying over 100 predictors of intrinsic disorder in proteins. *Expert Review of Proteomics*, 2021. **18**(12): 1019–1029.

[92] Katuwawala A, *et al.* Computational prediction of functions of intrinsically disordered regions. *Progress in Molecular Biology and Translational Science*, 2019. **166**: 341–369.

[93] Kurgan L, *et al.* The methods and tools for intrinsic disorder prediction and their application to systems medicine. In *Systems Medicine*, Wolkenhauer O (Ed.), Academic Press: Oxford, 2021. 159–169.

[94] Dosztanyi Z, *et al.* Bioinformatical approaches to characterize intrinsically disordered/unstructured proteins. *Briefings in Bioinformatics*, 2010. **11**(2): 225–243.

[95] Dosztányi Z and Tompa P. Bioinformatics approaches to the structure and function of intrinsically disordered proteins. In *From Protein Structure to Function with Bioinformatics*, Rigden DJ (Ed.), Springer Netherlands: Dordrecht, 2017. 167–203.

[96] Liu Y, *et al.* A comprehensive review and comparison of existing computational methods for intrinsically disordered protein and region prediction. *Briefings in Bioinformatics*, 2019. **20**(1): 330–346.

[97] Atkins JD, *et al.* Disorder prediction methods, their applicability to different protein targets and their usefulness for guiding experimental studies. *International Journal of Molecular Science*, 2015. **16**(8): 19040–19054.

[98] Bao Y, *et al.* Toward more accurate prediction of caspase cleavage sites: A comprehensive review of current methods, tools and features. *Briefings in Bioinformatics*, 2019. **20**(5): 1669–1684.

Chapter 13

Databases of Protein Structure and Function Predictions at the Amino Acid Level

Bi Zhao and Lukasz Kurgan*

Department of Computer Science, Virginia Commonwealth University, Richmond, Virginia, United States
* corresponding author: lkurgan@vcu.edu

The rapid growth of the number of protein sequences greatly exceeds the pace of efforts to annotate these proteins functionally and structurally. The closing of the ensuing large and growing gap in the amino acid (AA)-level annotations of protein structure and function can be facilitated using accurate and fast computational predictors. Hundreds of sequence-based predictors of the AA-level annotations have been developed, making it challenging for the end users to identify suitable/good predictors and collect their results. One convenient solution is to obtain pre-computed predictions from large-scale databases, which include MobiDB, D^2P^2 and DescribePROT. These databases provide access to a diverse set of structural and functional characteristics, such as domains, secondary structures, solvent accessibility, intrinsic disorder, posttranslational modifications (PTMs), protein/DNA/RNA-binding AAs, disordered linkers and signal peptides. We motivate and introduce these databases, discuss and compare their contents, and comment on their applications and limitations. We find that these databases provide complementary scope and services, with D^2P^2 delivering comprehensive annotations of domains and PTMs, MobiDB focusing on the intrinsic disorder and being highly-connected to other resources, and DescribePROT covering the most diverse set of structural and functional features. We briefly examine practical applications for some of the structural predictions covered by these databases. We also concisely discuss modern predictive webservers that can be used when users need to collect the AA-level annotations for proteins that are not included in these databases.

1. Introduction

We face an enormous challenge to characterize hundreds of millions of protein sequences functionally and structurally [1,2]. The current 2021_04 version UniProt includes 225.01 million of proteins and has more than tripled in size compared to the version 2016_04 from just five years ago that featured 63.69 million proteins [2,3]. These annotations are done at three levels: atomic, amino acid (AA) and whole protein. The arguably most popular atomic-level database, Protein Data Bank (PDB) [4], covers 185,000 protein structures. The most popular protein-level database, UniProt, has 565,000 manually curated proteins (Swiss-Prot) and close to 225 million proteins with alignment-generated/predicted annotations (TrEMBL) [2]. The AA-level annotations bridge the gap between the atomic and protein-level annotations. They are computed from the PDB files and extracted from a sparsely populated subset of the UniProt records. However, only a small fraction of AAs was annotated so far. Computational methods that predict the AA-level annotations from protein sequences (i.e., sequence-based predictors), many of which are described in this book, are widely used to assist with closing the huge and rapidly growing gap in the AA-level annotations.

The sequence-based predictors output AA-level annotations using predictive models trained and validated/tested using the ground truth generated by experimental methods, are typically collected from PDB or related/derived databases, such as BioLip [5] or DisProt [6]. They often rely on models produced by machine learning (ML) algorithms. ML algorithms utilize experimentally annotated training datasets to parametrize models to "optimally" differentiate between AAs that have a given function/structure and the remaining non-functional/non-structural AAs. The training sets are two orders of magnitude larger than the corresponding set of training proteins since they concern AAs; average protein sequence has around 300 AAs [7]. Consequently, the amount of the experimentally annotated training data is sufficient to train and test accurate predictive models using sophisticated ML algorithms, such as deep neural networks. We stress that these models are optimized to provide accurate predictions for proteins that share low levels of similarity/homology with the proteins in the training dataset, typically less than 30% similarity. In essence, the sequence-based *ab initio* methods can be used to make AA-level predictions for any of the 225 million of the sequenced proteins.

Hundreds of the sequence-based predictors of the AA-level annotations have been developed. They can be divided into two major groups: (1) methods that target prediction of functional AAs; and (2) methods that predict structural characteristics of AAs. The first group covers a broad spectrum of functions including prediction of AAs that interact with RNA, DNA, lipids and proteins, catalytic residues, cleavage and post-translational modification sites (PTMs), and intrinsic disorder. Selected, popular examples include DP-Bind [8,9] and DBS-PSSM [10] that predict DNA-binding AAs; RNABindR [11–13] and Pprint [14] that identify putative RNA-binding residues; BindN+ [15], DRNApred [16] and NucBind [17] that predict DNA and RNA binding AAs; SPPIDER [18], PSIVER [19] and SCRIBER [20] that find putative protein-binding residues; DisoLipPred [21] that predicts lipid-binding AAs; PROSPERous [22] and DeepCleave [23] that generate putative cleavage sites; INTREPID [24,25], PREvaIL [26] and CRpred [27] that produce putative catalytic residues; NetPhosK [28], SUMOsp [29,30] and UbPred [31] that find putative PTMs; SignalP [32–35] and ChloroP [36] that identify putative signal peptides; IUPred [37–40], DISOPRED [41,42] and flDPnn [43] that predict intrinsic disorder; and DisoRDPbind [44–46] and DeepDISObind [47] that generate putative disordered residues that interact with DNA, RNA and proteins. The second category targets prediction of various structural features of the AAs including their secondary structure, torsion angles, solvent accessibility, flexibility and residue-residue contacts. Example popular predictors include PSIPRED [48,49], PHD [50,51] and JPRED [52–54] that predict secondary structure; PHDacc [55] and ACCpro [56] that predict solvent accessibility; PROFbval [57,58] and FlexRP [59] that generate putative flexibility; and PSICOV [60], GREMLIN [61], ContactMap [62] and SVMcon [63] that produce putative residue-residue contacts. There are many more methods that target prediction of each of these structural and functional characteristics. For instance, there are over 100 predictors of intrinsic disorder [64–67], over 60 tools for the prediction of secondary structure [68–71], close to 40 predictors of AAs that interact with DNA and/or RNA [72–74], and over 30 that predict protein-binding AAs [75,76]. The sheer number and diversity of these methods make it rather challenging for the end users to select suitable/good predictors and collect their predictions.

The predictive quality of the sequence-based predictors of the AA-level function and structure annotations is evaluated on benchmark datasets. While authors of individual predictors compare their methods to a selected collection of other tools, arguably, the more reliable information source belongs to community-driven assessments. In the latter case, a large collection of methods competes in a blind prediction task on a common dataset (unknown to the authors of methods) under guidance of an independent group assessors (excluding authors). Examples include the Critical Assessment of Structure Prediction (CASP) [77–79] that evaluates the disorder and contact maps predictions [80,81], the Critical Assessment of PRotein Interactions (CAPRI) [82–84], Critical Assessment of Intrinsic protein Disorder (CAID) [85], and the discontinued Critical Assessment of Fully Automated Structure Prediction (CAFASP) [86]. The AA-level predictions that do not have community assessments can be reliably compared by utilizing large-scale comparative surveys. Recent examples for the prediction of secondary structure [69] (which was discontinued in CASP after 2002) include AAs that interact with RNA and DNA [72,74,87,88], protein-binding AAs [75,89], and disordered protein-binding AAs [76]. The community assessments and the comparative survey give useful guidance for the selection of well-performing predictors.

The collection of predictions could be difficult and time-consuming, particularly for less computer savvy users. Users interested in collecting several types of putative annotations have to navigate multiple websites and/or software, correspondingly adjust the format of the input protein sequences, and parse and standardize the diverse formats of outputs that different predictors use. One convenient alternative is to use platforms that provide multiple and diverse predictions. Several platforms that integrate predictions of multiple AA-level descriptors are currently available including PredictProtein [90], PSIPRED workbench [91], MULTICOM [92], Distill [93], and DEPICTER [94]. However, these platforms require a significant amount of runtime to collect results, particularly in scenarios when users require to predict a large number of proteins, and typically focus on a specific annotation type (structural versus functional) and structural state (disordered versus structured). Moreover, they are relatively inefficient since the same protein sequence that is being input by different users is typically predicted repeatedly.

The ultimate solution to these two prediction problems (selection and collection) are found in databases that offer convenient access to pre-computed AA-level predictions for a broad collection of predictors. This chapter describes, compares and analyzes these databases in the effort to disseminate and popularize their use.

2. Databases of the AA-level predictions

Three databases of the sequence-based AA-level predictions were released to date: MobiDB [95–98], D^2P^2 [99], and DescribePROT [100]. They provide instantaneous access to results generated by several disorder predictors for large datasets of proteins ranging from 1.35 million proteins in DescribePROT, through 10.43 million proteins in D^2P^2, to 219.74 million proteins in MobiDB. The first two databases focus on annotations associated with intrinsic disorder while DescribePROT offers a more holistic collection of putative annotations. The disorder is defined by the lack of stable structure under physiological conditions [101,102]. It was bioinformatically shown to be common across all kingdoms of life [103–107] and distributed across cellular compartments [108,109]. The focus on intrinsic disorder can be explained by its functional importance [110–117], association with human diseases [118] and defining contribution to poorly functionally/structurally characterized dark proteomes [119–121]. The prediction that underly these three databases consist of a numeric propensity (higher value signifies higher likelihood for a given annotation) and a binary value (annotated versus lacking a given annotation). The binary prediction is typically generated from the propensities, where AAs associated with the propensities higher than a threshold are classified as annotated with a given structural/functional characteristic. We summarize key characteristics of MobiDB, D^2P^2 and DescribePROT in Table 1.

2.1 *MobiDB*

MobiDB was developed by the Silvio Tosatto's group at the University of Padua. It was first released around 2012 [95] and continues to advance and expand along the years, with version 2 published around 2015 [98], version 3 in 2017 [97], and version 4 in 2020 [96].

Table 1. Summary of databases of the sequence-based AA-level predictions of protein structure and function.

Database	References	Year Released	Size [Millions of Proteins]	Predicted Properties: Structural (S) and Functional (F)	Predictors Included	Databases Linked
MobiDB version 4.1	[95–98]	2012	219.74	Intrinsic disorder (S) Disordered protein-binding residues (F) Secondary structure (S) Low complexity regions (S) Domains (F)	AlphaFold2 [122] ANCHOR [123] DisEMBL [124] DynaMine [125] ESpritz [126] FeSS [127] Gene3D [128] GlobPlot [129] IUPred2A [38] JRONN [130] MobiDB-lite [131] Pfilt [132] PONDR VSL2B [133,134] SEG [135]	CoDNaS [136] DIBS [137] DisProt [6] ELM [138] FuzDB [139] IDEAL [140] MFIB [141] PDBe [142] PhasePro [143] UniProt [2]
D^{2}P^2 version 1.0	[99]	2013	10.43	Intrinsic disorder (S) Disordered protein-binding residues (F) Domains (F)	PONDR VL-XT [144] PONDR VSL2B [133,134] PrDOS [145] PV2 [146] ESpritz [126] IUPred [40]	IDEAL [140] DisProt [6] PhosphoSitePlus [148]
DescribePROT version 1.4	[100]	2021	1.37	Solvent accessibility (S) Secondary structure (S) Disordered and structured protein-binding (F) Disordered and structured RNA-binding (F) Disordered and structured DNA-binding (F) Intrinsic disorder (S) Disordered linkers (F) Signal peptides (F)	SUPERFAMILY [147] ASAquick [149] DFLpred [150] DRNApred [16] DisoRDPbind [44–46] MoRFchibi [151] PONDR VSL2B [133,134] PSIPRED [48,152] SCRIBER [20,75] SignalP [34,153]	UniProt [2]

Availability: https://mobidb.bio.unipd.it/ [95–98]

Advantages: This is by far the largest database that aims to cover the UniProt-size collection of proteins, which currently totals to 219.7 million. Another key highlight is its linkage to 10 external databases (Table 1) and inclusion of experimental data that was collected from these databases. MobiDB features results generated by 14 predictors, including eight methods that predict intrinsic disorder. The primary annotation of putative disorder is produced using a meta/consensus method, MobiDB-lite [131]. The meta-predictors input multiple disorder predictions to produce a new disorder prediction that improves over the input predictions. This approach is motivated by empirical works that conclude that well-designed meta-methods in fact produce predictions with favorable accuracy [154,155].

Disadvantages: MobiDB almost exclusively focuses on annotations of intrinsic disorder. Moreover, it provides only the binary values for the disorder predictions, lacking the corresponding putative propensities.

2.2 D^2P^2

D^2P^2 was released around 2012 by Julian Gough's team at the University of Bristol. His research group has recently moved to the MRC Laboratory of Molecular Biology at Cambridge and D^2P^2 is no longer supported. The release of this resource was supported by a large international group of researchers including Drs Takeshi Ishida (Tokyo Institute of Technology), Bin Xue and Vladimir Uversky (University of South Florida), Zsuzsanna Dosztanyi (Eotvos Lorand University), Zoran Obradovic (Temple University), Lukasz Kurgan (Virginia Commonwealth University), and A. Keith Dunker (Indiana University).

Availability: https://d2p2.pro/ [99]

Advantages: D^2P^2 offers access to the results produced by a diverse collection of six disorder predictors (Table 1). It also combines these predictions using a 75% consensus approach, i.e., a residue is predicted as disorder if at least 75% of methods predicts it as disordered in binary. The use of this meta/consensus approach is motivated by the past empirical studies [154,155]. Moreover, D^2P^2 provides arguably the most comprehensive annotations of protein domains and PTMs.

Disadvantages: Similar to MobiDB, D^2P^2 almost fully focuses on the intrinsic disorder annotations. Furthermore, this resource was last updated in 2013 and is no longer maintained.

2.3 DescribePROT

DescribePROT was produced by Lukasz Kurgan's lab at the Virginia Commonwealth University and made available to the public in 2020. Similar to D^2P^2, DescribePROT was a collaborative effort that involved a big team of researchers including Drs A. Keith Dunker (Indiana University), Andrzej Kloczkowski (Ohio State University), Jorg Gsponer (University of British Columbia), Johannes Soding (Max Planck Institute for Biophysical Chemistry), Zoran Obradovic (Temple University), Martin Steinegger (Seoul National University), and Yaoqi Zhou (Shenzhen Bay Laboratory).

Availability: http://biomine.cs.vcu.edu/servers/DESCRIBEPROT/ [100]

Advantages: The strongest point of DescribePROT is the diversity of its predictions that cover several structural and functional characteristics including solvent accessibility, secondary structure, protein-, RNA- and DNA-binding AAs, intrinsic disorder, disordered linkers, and signal peptides. Consequently, DescribePROT stores over 7.8 billion AA-level predictions. Moreover, it provides access to position specific scoring matrices (PSSMs) generated from protein sequences using MMSeqs2 [156–158] and the relative entropy-based conservation scores that are produced from PSSMs [159,160]. Furthermore, this is the only database that combines complementary predictions of DNA, RNA and protein interactions that are trained using structured versus disordered data [75], which results in a more complete coverage of these interactions.

Disadvantages: The main downside of DescribePROT is a relatively low number of proteins that it covers (1.37 million), which spans over 83 complete proteomes/species. It also has insufficient linkage to external resources. However, both of these issues should be resolved in the subsequent releases.

2.4 Example results

Figure 1(A) shows experimental annotations of structure and function for the SIR3 protein, a transcriptional repressor from *Saccharomyces cerevisiae*

(a) Experimental annotation from the DisProt database

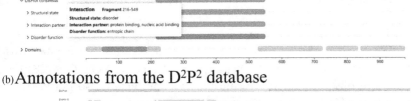

(b) Annotations from the D²P² database

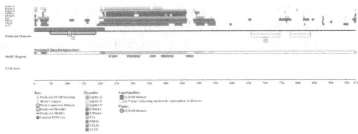

(c) Annotations from the MobiDB database

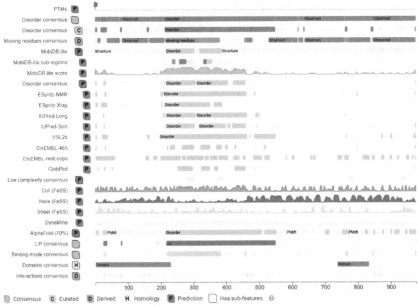

Figure 1. Experimental and predicted disorder annotations for the SIR3 protein (UniProt ID: P06701, DisProt ID: DP00533). Panel A shows the experimental annotations collected from DisProt (https://www.disprot.org/). Panel B shows the results generated by the D2P2 database (https://d2p2.pro/). Panel C presents the results produced by the MobiDB database (https://mobidb.bio.unipd.it/). The legends included in panels B and C explain the encoding of the presented data.

(UniProt ID: P06701) which we extract from the DisProt database (DisProt ID: DP00533) [6]. SIR3 modulates chromatin structure and correspondingly includes a long intrinsically disordered region (positions 216 to 549) that interacts with proteins and DNA [161].

We compare these annotations against the results that we collect from the D^2P^2 (Fig. 1B), MobiDB (Fig. 1C) and DescribePROT (Fig. 2B) databases. We observe that the location of the predicted disordered AAs in these three databases agrees to a large degree with the experimental data. This suggests that the corresponding disorder predictors produce accurate results, which concurs with recent empirical assessments that similarly conclude that disorder predictions are generally done accurately [85,162,163]. We emphasize that these resources provide well-designed and color-coded visualizations of the predictions and annotations, each using its own format. D^2P^2 groups all disorder predictions together and presents an "agreement" line that compares them against experimental annotations, if available (Fig. 1B). This is accompanied with the location of identified domains and PTMs. MobiDB similarly clusters several disorder predictions together with the corresponding consensus result (Fig. 1C). It also provides annotations of domains and protein interactions at the bottom of the panel. DescribePROT divides the panel into two parts where the top aggregates information at the protein level and the bottom provides complete AA-level results (Fig. 2B). The residue-level annotations supplied by DescribePROT include both binary predictions (horizontal bars) and numeric propensities (thin solid lines). We note that MobiDB and DescribePROT provide interactive interfaces where users can select specific functional/structural characteristics, zoom in and out on selected parts of the sequence, and are shown convenient and informative callouts that display additional detail and which appear on the mouse hover.

3. Conclusions, impact and limitations

Three large-scale databases that we introduce and discuss in this chapter, MobiDB, D^2P^2 and DescribePROT, facilitate easy and free access to large collections of the AA-level annotations of protein structure and function. We demonstrate that they provide complementary scope and services. D^2P^2 arguably delivers the most comprehensive set of annotations of protein domains and PTMs. However, this database was last updated in 2013 and is no longer actively supported. MobiDB focuses primarily on the intrinsic disorder and is by far the largest and most externally connected resource,

(a) Experimental annotation from the DisProt database

(b) Annotations from the DescribePROT database

Figure 2. Experimental and predicted disorder annotations for the SIR3 protein (UniProt ID: P06701, DisProt ID: DP00533). Panel A shows the experimental annotations collected from DisProt (https://www.disprot.org/). Panel B gives the outputs from the DescribePROT database (http://biomine.cs.vcu.edu/ servers/ DESCRIBEPROT/). The legend included in panel B explains the encoding of the presented data.

while DescribePROT covers the most diverse collection of the structural and functional features. Thus, we recommend the latter two resources as the most valuable, current, and complete solutions to conveniently collect the AA-level annotations.

The data available in these databases is utilized for numerous practical applications. We briefly summarize the impact of one of the structural aspects covered by these resources, the intrinsic disorder. Just in 2021, the disorder predictions of the popular IUPred [37–40], which are available via D^2P^2 and MobiDB databases, were used to analyze the SARS-CoV-2 proteins [164–167], link mutations in the intrinsically disordered sequence regions to cancer [168,169], investigate liquid-liquid phase separation [170–172], localize disorder across compartments of the human cell [108], and develop a wide range of predictive tools [173–178], among many other applications. Similarly, a long list of diverse uses can be attributed to the results produced by DisoRDPbind [44–46], which covers putative disordered protein/DNA/RNA binding AAs and which are available via DescribePROT. These predictions were utilized to investigate several viral genomes including SARS-CoV-2 [179], porcine astrovirus type 3 [180], and hepatitis E [181], and to decipher functions of genes from animal pathogens [182]. They were also applied to investigate several specific proteins, such as CS-like zinc finger (FLZ) [183], nonstructural nsP2 protein from Salmonid alphavirus [184], spindle-defective protein 2 (SPD-2) [185], heat shock factor 1 (Hsf1) [186], and Mixed Lineage Leukemia 4 (MLL4) [187], some of which are connected to cancers and neurodegenerative and viral diseases. More broadly, we find that the intrinsic disorder predictions are utilized across many research and development areas, such as drug design [188–192], molecular and systems medicine [193,194], and structural genomics [124,195]. These examples and studies clearly demonstrate the significant impact of the use of the putative AA-level annotations, which are directly facilitated by the databases described here.

Lastly, we emphasize that the use of these databases is limited to the proteins that they include. Users who like to collect the AA-level data outside the protein sets covered in these resources, e.g., for a novel protein sequence, have the option of applying for one of the freely available predictive platforms. These platforms include PredictProtein (https://predictprotein.org/) [90], PSIPRED workbench (http://bioinf.cs.ucl.ac.uk/psipred/) [91], MULTICOM (http://sysbio.rnet.missouri.edu/multicom_cluster/) [92], Distill (http://distillf.ucd.ie/distill/) [93], and

DEPICTER (http://biomine.cs.vcu.edu/servers/DEPICTER/) [94]. We briefly discuss details of the DisorderEd PredictIon CenTER (DEPICTER) webserver, which is closest to the scope of the three databases. This webserver conveniently generates the AA-level predictions on the server side, covering a broad selection of disorder and disorder function predictions. It produces consensus/meta prediction of disordered AAs using results output by the fast UPred-short [40], IUPred-long [40] and SPOT-Disorder-Single [196] methods. It also predicts disordered linkers using DFLpred [150], disordered AAs that bind proteins, DNA and/or RNA by combining results of fMoRFpred [197], DisoRDPbind [46] and ANCHOR2 [38] methods, and putative disordered multifunctional (moonlighting) AAs generated by DMRpred [198]. The predictions are visualized and delivered as a parsable text file in the browser window and sent to the user's email, if the email was provided as one of the inputs.

4. Funding

This work was funded in part by the National Science Foundation (grants 2146027 and 2125218) and the Robert J. Mattauch Endowment funds to L.K.

References

[1] Li W, *et al.* RefSeq: Expanding the prokaryotic genome annotation pipeline reach with protein family model curation. *Nucleic Acids Research*, 2021. **49**(D1): D1020–D1028.

[2] UniProt C. UniProt: The universal protein knowledgebase in 2021. *Nucleic Acids Research*, 2021. **49**(D1): D480–D489.

[3] UniProt C. UniProt: A hub for protein information. *Nucleic Acids Research*, 2015. **43**(Database issue): D204–D212.

[4] wwPDB consortium. Protein data bank: The single global archive for 3D macromolecular structure data. *Nucleic Acids Research*, 2019. **47**(D1): D520–D528.

[5] Yang J, *et al.* BioLiP: A semi-manually curated database for biologically relevant ligand-protein interactions. *Nucleic Acids Research*, 2013. **41**(Database issue): D1096–D1103.

[6] Hatos A, *et al.* DisProt: Intrinsic protein disorder annotation in 2020. *Nucleic Acids Research*, 2020. **48**(D1): D269–D276.

[7] Brocchieri L and Karlin S. Protein length in eukaryotic and prokaryotic proteomes. *Nucleic Acids Research*, 2005. **33**(10): 3390–3400.

[8]   Kuznetsov IB, *et al.* Using evolutionary and structural information to predict DNA-binding sites on DNA-binding proteins. *Proteins-Structure Function and Bioinformatics*, 2006. **64**(1): 19–27.

[9]   Hwang S, *et al.* DP-Bind: A Web server for sequence-based prediction of DNA-binding residues in DNA-binding proteins. *Bioinformatics*, 2007. **23**(5): 634–636.

[10] Ahmad S and Sarai A. PSSM-based prediction of DNA binding sites in proteins. *BMC Bioinformatics*, 2005. **6**: 33.

[11] Walia RR, *et al.* RNABindRPlus: A predictor that combines machine learning and sequence homology-based methods to improve the reliability of predicted RNA-binding residues in proteins. *PLoS One*, 2014. **9**(5): e97725.

[12] Terribilini M, *et al.* Prediction of RNA binding sites in proteins from amino acid sequence. *RNA-a Publication of the RNA Society*, 2006. **12**(8): 1450–1462.

[13] Terribilini M, *et al.* RNABindR: A server for analyzing and predicting RNA-binding sites in proteins. *Nucleic Acids Research*, 2007. **35**: W578–W584.

[14] Kumar M, *et al.* Prediction of RNA binding sites in a protein using SVM and PSSM profile. *Proteins-Structure Function and Bioinformatics*, 2008. **71**(1): 189–194.

[15] Wang L, *et al.* BindN+ for accurate prediction of DNA and RNA-binding residues from protein sequence features. *BMC Systems Biology*, 2010. **4**(Suppl 1): S3.

[16] Yan J and Kurgan L. DRNApred, fast sequence-based method that accurately predicts and discriminates DNA- and RNA-binding residues. *Nucleic Acids Research*, 2017. **45**(10): e84.

[17] Su H, *et al.* Improving the prediction of protein-nucleic acids binding residues via multiple sequence profiles and the consensus of complementary methods. *Bioinformatics*, 2019. **35**(6): 930–936.

[18] Porollo A and Meller J. Prediction-based fingerprints of protein-protein interactions. *Proteins*, 2007. **66**(3): 630–645.

[19] Murakami Y and Mizuguchi K. Applying the Naive Bayes classifier with kernel density estimation to the prediction of protein-protein interaction sites. *Bioinformatics*, 2010. **26**(15): 1841–1848.

[20] Zhang J and Kurgan L. SCRIBER: Acurate and partner type-specific prediction of protein-binding residues from proteins sequences. *Bioinformatics*, 2019. **35**(14): i343–i353.

[21] Katuwawala A, *et al.* DisoLipPred: Accurate prediction of disordered lipid binding residues in protein sequences with deep recurrent networks and transfer learning. *Bioinformatics*, 2021. **38**(1): 115–124.

[22] Song J, *et al.* PROSPERous: High-throughput prediction of substrate cleavage sites for 90 proteases with improved accuracy. *Bioinformatics*, 2018. **34**(4): 684–687.

[23]   Li F, *et al.* DeepCleave: A deep learning predictor for caspase and matrix metalloprotease substrates and cleavage sites. *Bioinformatics,* 2020. **36**(4): 1057–1065.

[24]   Sankararaman S and Sjolander K. INTREPID — INformation-theoretic TREe traversal for Protein functional site IDentification. *Bioinformatics,* 2008. **24**(21): 2445–2452.

[25]   Sankararaman S, *et al.* INTREPID: A web server for prediction of functionally important residues by evolutionary analysis. *Nucleic Acids Research,* 2009. **37**(Web Server issue): W390–W395.

[26]   Song JN, *et al.* PREvaIL, an integrative approach for inferring catalytic residues using sequence, structural, and network features in a machine-learning framework. *Journal of Theoretical Biology,* 2018. **443**: 125–137.

[27]   Zhang T, *et al.* Accurate sequence-based prediction of catalytic residues. *Bioinformatics,* 2008. **24**(20): 2329–2338.

[28]   Blom N, *et al.* Prediction of post-translational glycosylation and phosphorylation of proteins from the amino acid sequence. *Proteomics,* 2004. **4**(6): 1633–1649.

[29]   Ren J, *et al.* Systematic study of protein sumoylation: Development of a site-specific predictor of SUMOsp 2.0. *Proteomics,* 2009. **9**(12): 3409–3412.

[30]   Xue Y, *et al.* SUMOsp: A web server for sumoylation site prediction. *Nucleic Acids Research,* 2006. **34**(Web Server issue): W254–W257.

[31]   Radivojac P, *et al.* Identification, analysis, and prediction of protein ubiquitination sites. *Proteins,* 2010. **78**(2): 365–3680.

[32]   Nielsen H, *et al.* Identification of prokaryotic and eukaryotic signal peptides and prediction of their cleavage sites. *Protein Engineering,* 1997. **10**(1): 1–6.

[33]   Bendtsen JD, *et al.* Improved prediction of signal peptides: SignalP 3.0. *Journal of Molecular Biology,* 2004. **340**(4): 783–795.

[34]   Almagro Armenteros JJ, *et al.* SignalP 5.0 improves signal peptide predictions using deep neural networks. *Nature Biotechnology,* 2019. **37**(4): 420–423.

[35]   Petersen TN, *et al.* SignalP 4.0: Discriminating signal peptides from transmembrane regions. *Nature Methods,* 2011. **8**(10): 785–786.

[36]   Emanuelsson O, *et al.* ChloroP, a neural network-based method for predicting chloroplast transit peptides and their cleavage sites. *Protein Science,* 1999. **8**(5): 978–984.

[37]   Erdos G, *et al.* IUPred3: Prediction of protein disorder enhanced with unambiguous experimental annotation and visualization of evolutionary conservation. *Nucleic Acids Research,* 2021. **49**(W1): W297–W303.

[38]   Meszaros B, *et al.* IUPred2A: Context-dependent prediction of protein disorder as a function of redox state and protein binding. *Nucleic Acids Research,* 2018. **46**(W1): W329–W337.

[39]   Dosztanyi Z. Prediction of protein disorder based on IUPred. *Protein Science,* 2018. **27**(1): 331–340.

[40] Dosztanyi Z, *et al.* IUPred: Web server for the prediction of intrinsically unstructured regions of proteins based on estimated energy content. *Bioinformatics*, 2005. **21**(16): 3433–3434.

[41] Jones DT and Cozzetto D. DISOPRED3: Precise disordered region predictions with annotated protein-binding activity. *Bioinformatics*, 2015. **31**(6): 857–863.

[42] Ward JJ, *et al.* The DISOPRED server for the prediction of protein disorder. *Bioinformatics*, 2004. **20**(13): 2138–2139.

[43] Hu G, *et al.* flDPnn: Accurate intrinsic disorder prediction with putative propensities of disorder functions. *Nature Communications*, 2021. **12**(1): 4438.

[44] Oldfield CJ, *et al.* Disordered RNA-binding region prediction with DisoRDPbind. *Methods in Molecular Biology*, 2020. **2106**: 225–239.

[45] Peng Z, *et al.* Prediction of disordered RNA, DNA, and protein binding regions using DisoRDPbind. *Methods in Molecular Biology*, 2017. **1484**: 187–203.

[46] Peng Z and Kurgan L. High-throughput prediction of RNA, DNA and protein binding regions mediated by intrinsic disorder. *Nucleic Acids Research*, 2015. **43**(18): e121.

[47] Zhang F, *et al.* DeepDISOBind: Accurate prediction of RNA-, DNA- and protein-binding intrinsically disordered residues with deep multi-task learning. *Briefings in Bioinformatics*, 2022. **23**(1): bbab521.

[48] McGuffin LJ, *et al.* The PSIPRED protein structure prediction server. *Bioinformatics*, 2000. **16**(4): 404–405.

[49] Jones DT. Protein secondary structure prediction based on position-specific scoring matrices. *Journal of Molecular Biology*, 1999. **292**(2): 195–202.

[50] Rost B and Sander C. Prediction of protein secondary structure at better than 70% accuracy. *Journal of Molecular Biology*, 1993. **232**(2): 584–599.

[51] Rost B, *et al.* Phd — an automatic mail server for protein secondary structure prediction. *Computer Applications in the Biosciences*, 1994. **10**(1): 53–60.

[52] Cuff JA, *et al.* JPred: A consensus secondary structure prediction server. *Bioinformatics*, 1998. **14**(10): 892–893.

[53] Cole C, *et al.* The Jpred 3 secondary structure prediction server. *Nucleic Acids Research*, 2008. **36**(Web Server issue): W197–W201.

[54] Drozdetskiy A, *et al.* JPred4: A protein secondary structure prediction server. *Nucleic Acids Research*, 2015. **43**(W1): W389–W394.

[55] Rost B and Sander C. Conservation and prediction of solvent accessibility in protein families. *Proteins*, 1994. **20**(3): 216–226.

[56] Pollastri G, *et al.* Prediction of coordination number and relative solvent accessibility in proteins. *Proteins*, 2002. **47**(2): 142–153.

[57] Schlessinger A, *et al.* PROFbval: Predict flexible and rigid residues in proteins. *Bioinformatics*, 2006. **22**(7): 891–893.

[58]   Schlessinger A and Rost B. Protein flexibility and rigidity predicted from sequence. *Proteins*, 2005. **61**(1): 115–126.

[59]   Chen K, *et al.* Prediction of flexible/rigid regions from protein sequences using k-spaced amino acid pairs. *BMC Structural Biology*, 2007. **7**: 25.

[60]   Jones DT, *et al.* PSICOV: Precise structural contact prediction using sparse inverse covariance estimation on large multiple sequence alignments. *Bioinformatics*, 2012. **28**(2): 184–190.

[61]   Kamisetty H, *et al.* Assessing the utility of coevolution-based residue-residue contact predictions in a sequence- and structure-rich era. *Proceedings of the National Academy of Sciences of the United States of America*, 2013. **110**(39): 15674–15679.

[62]   Wang S, *et al.* Accurate De Novo prediction of protein contact map by ultra-deep learning model. *PLoS Computational Biology*, 2017. **13**(1): e1005324.

[63]   Cheng JL and Baldi P. Improved residue contact prediction using support vector machines and a large feature set. *BMC Bioinformatics*, 2007. **8**: 113.

[64]   Zhao B and Kurgan L. Surveying over 100 predictors of intrinsic disorder in proteins. *Expert Review of Proteomics*, 2021. **18**(12): 1019–1029.

[65]   Liu Y, *et al.* A comprehensive review and comparison of existing computational methods for intrinsically disordered protein and region prediction. *Briefings in Bioinformatics*, 2019. **20**(1): 330–346.

[66]   Meng F, *et al.* Comprehensive review of methods for prediction of intrinsic disorder and its molecular functions. *Cellular and Molecular Life Sciences*, 2017. **74**(17): 3069–3090.

[67]   Meng F, *et al.* Computational prediction of intrinsic disorder in proteins. *Current Protocols in Protein Science*, 2017. **88**: 2 16 1–2 16 14.

[68]   Kashani-Amin E, *et al.* A systematic review on popularity, application and characteristics of protein secondary structure prediction tools. *Current Drug Discovery Technologies*, 2018. **16**(2): 159–172.

[69]   Zhang H, *et al.* Critical assessment of high-throughput standalone methods for secondary structure prediction. *Briefings in Bioinformatics*, 2011. **12**(6): 672–88.

[70]   Oldfield CJ, *et al.* Computational prediction of secondary and supersecondary structures from protein sequences. *Methods in Molecular Biology*, 2019. **1958**: 73–100.

[71]   Meng F and Kurgan L. Computational prediction of protein secondary structure from sequence. *Current Protocols in Protein Science*, 2016. **86**: 2 3 1–2 3 10.

[72]   Yan J, *et al.* A comprehensive comparative review of sequence-based predictors of DNA- and RNA-binding residues. *Briefings in Bioinformatics*, 2016. **17**(1): 88–105.

[73]   Zhang J, *et al.* Comprehensive review and empirical analysis of hallmarks of DNA-, RNA- and protein-binding residues in protein chains. *Briefings in Bioinformatics*, 2019. **20**(4): 1250–1268.

[74]   Miao Z and Westhof E. A large-scale assessment of nucleic acids binding site prediction programs. *PLoS Computational Biology*, 2015. **11**(12): e1004639.

[75]   Zhang J, *et al.* Prediction of protein-binding residues: dichotomy of sequence-based methods developed using structured complexes versus disordered proteins. *Bioinformatics*, 2020. **36**(18): 4729–4738.

[76]   Katuwawala A, *et al.* Computational prediction of MoRFs, short disorder-to-order transitioning protein binding regions. *Computational and Structural Biotechnology Journal*, 2019. **17**: 454–462.

[77]   Alexander LT, *et al.* Target highlights in CASP14: Analysis of models by structure providers. *Proteins*, 2021. **89**(12): 1647–1672.

[78]   Kinch LN, *et al.* Target classification in the 14th round of the critical assessment of protein structure prediction (CASP14). *Proteins*, 2021. **89**(12): 1618–1632.

[79]   Moult J. A decade of CASP: Progress, bottlenecks and prognosis in protein structure prediction. *Current Opinion in Structural Biology*, 2005. **15**(3): 285–289.

[80]   Monastyrskyy B, *et al.* Assessment of protein disorder region predictions in CASP10. *Proteins*, 2014. **82**(Suppl 2): 127–137.

[81]   Ruiz-Serra V, *et al.* Assessing the accuracy of contact and distance predictions in CASP14. *Proteins*, 2021. **89**(12): 1888–1900.

[82]   Lensink MF, *et al.* Prediction of protein assemblies, the next frontier: The CASP14-CAPRI experiment. *Proteins*, 2021. **89**(12): 1800–1823.

[83]   Lensink MF, *et al.* Modeling protein-protein, protein-peptide, and protein-oligosaccharide complexes: CAPRI 7th edition. *Proteins*, 2020. **88**(8): 916–938.

[84]   Dapkunas J, *et al.* Template-based modeling of diverse protein interactions in CAPRI rounds 38–45. *Proteins*, 2020. **88**(8): 939–947.

[85]   Necci M, *et al.* Critical assessment of protein intrinsic disorder prediction. *Nature Methods*, 2021. **18**(5): 472–481.

[86]   Eyrich VA, *et al.* CAFASP3 in the spotlight of EVA. *Proteins*, 2003. **53**(Suppl 6): 548–560.

[87]   Walia RR, *et al.* Protein-RNA interface residue prediction using machine learning: An assessment of the state of the art. *BMC Bioinformatics*, 2012. **13**: 89.

[88]   Jung Y, *et al.* Partner-specific prediction of RNA-binding residues in proteins: A critical assessment. *Proteins*, 2019. **87**(3): 198–211.

[89]   Zhang J and Kurgan L. Review and comparative assessment of sequence-based predictors of protein-binding residues. *Briefings in Bioinformatics*, 2018. **19**(5): 821–837.

[90] Bernhofer M, *et al.* PredictProtein — predicting protein structure and function for 29 years. *Nucleic Acids Research*, 2021. **49**(W1): W535–W540.

[91] Buchan DWA and Jones DT. The PSIPRED protein analysis workbench: 20 years on. *Nucleic Acids Research*, 2019. **47**(W1): W402–W407.

[92] Hou J, *et al.* The MULTICOM protein structure prediction server empowered by deep learning and contact distance prediction. In *Protein Structure Prediction*, Kihara D (ed.), Humana, New York, NY, 2020. **2165**: 13–26.

[93] Bau D, *et al.* Distill: A suite of web servers for the prediction of one-, two- and three-dimensional structural features of proteins. *BMC Bioinformatics*, 2006. **7**: 402.

[94] Barik A, *et al.* DEPICTER: Intrinsic disorder and disorder function prediction server. *Journal of Molecular Biology*, 2020. **432**(11): 3379–3387.

[95] Di Domenico T, *et al.* MobiDB: A comprehensive database of intrinsic protein disorder annotations. *Bioinformatics*, 2012. **28**(15): 2080–2081.

[96] Piovesan D, *et al.* MobiDB: Intrinsically disordered proteins in 2021. *Nucleic Acids Research*, 2021. **49**(D1): D361–D367.

[97] Piovesan D, *et al.* MobiDB 3.0: More annotations for intrinsic disorder, conformational diversity and interactions in proteins. *Nucleic Acids Research*, 2018. **46**(D1): D471–D476.

[98] Potenza E, *et al.* MobiDB 2.0: An improved database of intrinsically disordered and mobile proteins. *Nucleic Acids Research*, 2015. **43**(Database issue): D315–D320.

[99] Oates ME, *et al.* D(2)P(2): Database of disordered protein predictions. *Nucleic Acids Research*, 2013. **41**(Database issue): D508–D516.

[100] Zhao B, *et al.* DescribePROT: Database of amino acid-level protein structure and function predictions. *Nucleic Acids Research*, 2021. **49**(D1): D298–D308.

[101] Lieutaud P, *et al.* How disordered is my protein and what is its disorder for? A guide through the "dark side" of the protein universe. *Intrinsically Disordered Proteins*, 2016. **4**(1): e1259708.

[102] Oldfield CJ, *et al.* Introduction to intrinsically disordered proteins and regions. In *Intrinsically Disordered Proteins*, Salvi N (ed.), Academic Press, 2019. 1–34.

[103] Xue B, *et al.* Orderly order in protein intrinsic disorder distribution: Disorder in 3500 proteomes from viruses and the three domains of life. *Journal of Biomolecular Structure and Dynamics*, 2012. **30**(2): 137–149.

[104] Peng Z, *et al.* Exceptionally abundant exceptions: Comprehensive characterization of intrinsic disorder in all domains of life. *Cellular and Molecular Life Sciences*, 2015. **72**(1): 137–151.

[105] Wang C, *et al.* Disordered nucleiome: Abundance of intrinsic disorder in the DNA- and RNA-binding proteins in 1121 species from Eukaryota, Bacteria and Archaea. *Proteomics*, 2016. **16**(10): 1486–1498.

[106] Peng Z, *et al.* Genome-scale prediction of proteins with long intrinsically disordered regions. *Proteins*, 2014. **82**(1): 145–158.

[107] Ward JJ, *et al.* Prediction and functional analysis of native disorder in proteins from the three kingdoms of life. *Journal of Molecular Biology*, 2004. **337**(3): 635–645.

[108] Zhao B, *et al.* IDPology of the living cell: Intrinsic disorder in the subcellular compartments of the human cell. *Cellular and Molecular Life Sciences*, 2021. **78**(5): 2371–2385.

[109] Meng F, *et al.* Compartmentalization and functionality of nuclear disorder: Intrinsic disorder and protein-protein interactions in intra-nuclear compartments. *International Journal of Molecular Sciences*, 2015. **17**(1): 24.

[110] Uversky VN, *et al.* Showing your ID: Intrinsic disorder as an ID for recognition, regulation and cell signaling. *Journal of Molecular Recognition*, 2005. **18**(5): 343–384.

[111] Liu J, *et al.* Intrinsic disorder in transcription factors. *Biochemistry*, 2006. **45**(22): 6873–6888.

[112] Peng Z, *et al.* A creature with a hundred waggly tails: Intrinsically disordered proteins in the ribosome. *Cellular and Molecular Life Sciences*, 2014. **71**(8): 1477–1504.

[113] Babu MM. The contribution of intrinsically disordered regions to protein function, cellular complexity, and human disease. *Biochemical Society Transactions*, 2016. **44**(5): 1185–1200.

[114] Peng ZL, *et al.* More than just tails: Intrinsic disorder in histone proteins. *Molecular Biosystems*, 2012. **8**(7): 1886–1901.

[115] Hu G, *et al.* Functional analysis of human hub proteins and their interactors involved in the intrinsic disorder-enriched interactions. *International Journal of Molecular Sciences*, 2017. **18**(12): 2761.

[116] Na I, *et al.* Autophagy-related intrinsically disordered proteins in intra-nuclear compartments. *Molecular Omics*, 2016. **12**(9): 2798–2817.

[117] Peng Z, *et al.* Resilience of death: Intrinsic disorder in proteins involved in the programmed cell death. *Cell Death & Differentiation*, 2013. **20**(9): 1257–1267.

[118] Uversky VN, *et al.* Unfoldomics of human diseases: Linking protein intrinsic disorder with diseases. *BMC Genomics*, 2009. **10**(Suppl 1): S7.

[119] Bhowmick A, *et al.* Finding our way in the dark proteome. *Journal of the American Chemical Society*, 2016. **138**(31): 9730–9742.

[120] Hu G, *et al.* Taxonomic landscape of the dark proteomes: Whole-proteome scale interplay between structural darkness, intrinsic disorder, and crystallization propensity. *Proteomics*, 2018. **18**(21–22): e1800243.

[121] Kulkarni P and Uversky VN. Intrinsically disordered proteins: The dark horse of the dark proteome. *Proteomics*, 2018. **18**(21–22): e1800061.

[122] Jumper J, *et al.* Highly accurate protein structure prediction with AlphaFold. *Nature*, 2021. **596**(7873): 583–589.

[123] Dosztanyi Z, *et al.* ANCHOR: Web server for predicting protein binding regions in disordered proteins. *Bioinformatics*, 2009. **25**(20): 2745–2746.

[124] Linding R, *et al.* Protein disorder prediction: Implications for structural proteomics. *Structure*, 2003. **11**(11): 1453–1459.

[125] Cilia E, *et al.* From protein sequence to dynamics and disorder with DynaMine. *Nature Communications*, 2013. **4**: 2741.

[126] Walsh I, *et al.* ESpritz: Accurate and fast prediction of protein disorder. *Bioinformatics*, 2012. **28**(4): 503–509.

[127] Piovesan D, *et al.* FELLS: Fast estimator of latent local structure. *Bioinformatics*, 2017. **33**(12): 1889–1891.

[128] Lewis TE, *et al.* Gene3D: Extensive prediction of globular domains in proteins. *Nucleic Acids Research*, 2018. **46**(D1): D435–D439.

[129] Linding R, *et al.* GlobPlot: Exploring protein sequences for globularity and disorder. *Nucleic Acids Research*, 2003. **31**(13): 3701–3708.

[130] Yang ZR, *et al.* RONN: The bio-basis function neural network technique applied to the detection of natively disordered regions in proteins. *Bioinformatics*, 2005. **21**(16): 3369–3376.

[131] Necci M, *et al.* MobiDB-lite: Fast and highly specific consensus prediction of intrinsic disorder in proteins. *Bioinformatics*, 2017. **33**(9): 1402–1404.

[132] Jones DT and Swindells MB. Getting the most from PSI-BLAST. *Trends in Biochemical Sciences*, 2002. **27**(3): 161–164.

[133] Peng K, *et al.* Length-dependent prediction of protein intrinsic disorder. *BMC Bioinformatics*, 2006. **7**: 208.

[134] Obradovic Z, *et al.* Exploiting heterogeneous sequence properties improves prediction of protein disorder. *Proteins*, 2005. **61**(Suppl 7): 176–182.

[135] Wootton JC. Non-globular domains in protein sequences: automated segmentation using complexity measures. *Computational Chemistry*, 1994. **18**(3): 269–285.

[136] Monzon AM, *et al.* CoDNaS 2.0: A comprehensive database of protein conformational diversity in the native state. *Database (Oxford)*, 2016. **2016**: baw038.

[137] Schad E, *et al.* DIBS: A repository of disordered binding sites mediating interactions with ordered proteins. *Bioinformatics*, 2018. **34**(3): 535–537.

[138] Dinkel H, *et al.* ELM 2016 — data update and new functionality of the eukaryotic linear motif resource. *Nucleic Acids Research*, 2016. **44**(D1): D294–D300.

[139] Miskei M, *et al.* FuzDB: Database of fuzzy complexes, a tool to develop stochastic structure-function relationships for protein complexes and higher-order assemblies. *Nucleic Acids Research*, 2017. **45**(D1): D228–D235.

[140] Fukuchi S, *et al.* IDEAL in 2014 illustrates interaction networks composed of intrinsically disordered proteins and their binding partners. *Nucleic Acids Research*, 2014. **42**(Database issue): D320–D325.

[141] Ficho E, *et al.* MFIB: A repository of protein complexes with mutual folding induced by binding. *Bioinformatics*, 2017. **33**(22): 3682–3684.

[142] consortium PD-K. PDBe-KB: Collaboratively defining the biological context of structural data. *Nucleic Acids Research*, 2022. **50**(D1): D534–D542.

[143] Meszaros B, *et al.* PhaSePro: The database of proteins driving liquid-liquid phase separation. *Nucleic Acids Research*, 2020. **48**(D1): D360–D367.

[144] Romero P, *et al.* Sequence complexity of disordered protein. *Proteins*, 2001. **42**(1): 38–48.

[145] Ishida T and Kinoshita K. Prediction of disordered regions in proteins based on the meta approach. *Bioinformatics*, 2008. **24**(11): 1344–1348.

[146] Ghalwash MF, *et al.* Uncertainty analysis in protein disorder prediction. *Molecular Omics*, 2012. **8**(1): 381–391.

[147] Gough J, *et al.* Assignment of homology to genome sequences using a library of hidden Markov models that represent all proteins of known structure. *Journal of Molecular Biology*, 2001. **313**(4): 903–919.

[148] Hornbeck PV, *et al.* PhosphoSitePlus: A comprehensive resource for investigating the structure and function of experimentally determined post-translational modifications in man and mouse. *Nucleic Acids Research*, 2012. **40**(Database issue): D261–D270.

[149] Faraggi E, *et al.* Fast and accurate accessible surface area prediction without a sequence profile. *Prediction of Protein Secondary Structure*, 2017. **1484**: 127–136.

[150] Meng F and Kurgan L. DFLpred: High-throughput prediction of disordered flexible linker regions in protein sequences. *Bioinformatics*, 2016. **32**(12): i341–i350.

[151] Malhis N, *et al.* MoRFchibi SYSTEM: Software tools for the identification of MoRFs in protein sequences. *Nucleic Acids Research*, 2016. **44**(W1): W488–W493.

[152] Buchan DWA and Jones DT. The PSIPRED protein analysis workbench: 20 years on. *Nucleic Acids Research*, 2019. **47**(W1): W402–W407.

[153] Teufel F, *et al.* SignalP 6.0 predicts all five types of signal peptides using protein language models. *Nature Biotechnology*, 2022. https://doi.org/10.1038/s41587-021-01156-3

[154] Fan X and Kurgan L. Accurate prediction of disorder in protein chains with a comprehensive and empirically designed consensus. *Journal of Biomolecular Structure and Dynamics*, 2014. **32**(3): 448–464.

[155] Peng Z and Kurgan L. On the complementarity of the consensus-based disorder prediction. *Pacific Symposium on Biocomputing*, 2012. 176–187.

[156] Mirdita M, *et al.* MMseqs2 desktop and local web server app for fast, interactive sequence searches. *Bioinformatics*, 2019. **35**(16): 2856–2858.

[157] Steinegger M and Soding J. MMseqs2 enables sensitive protein sequence searching for the analysis of massive data sets. *Nature Biotechnology*, 2017. **35**(11): 1026–1028.

[158] Hu G and Kurgan L. Sequence similarity searching. *Current Protocols in Protein Science*, 2019. **95**(1): e71.

[159] Wang K and Samudrala R. Incorporating background frequency improves entropy-based residue conservation measures. *BMC Bioinformatics*, 2006. **7**: 385.

[160] Fischer J, *et al.* Prediction of protein functional residues from sequence by probability density estimation. *Bioinformatics*, 2008. **24**(5): 613–620.

[161] McBryant SJ, *et al.* Domain organization and quaternary structure of the Saccharomyces cerevisiae silent information regulator 3 protein, Sir3p. *Biochemistry*, 2006. **45**(51): 15941–15948.

[162] Katuwawala A and Kurgan L. Comparative assessment of intrinsic disorder predictions with a focus on protein and nucleic acid-binding proteins. *Biomolecules*, 2020. **10**(12): 1636.

[163] Katuwawala A, *et al.* Accuracy of protein-level disorder predictions. *Briefings in Bioinformatics*, 2020. **21**(5): 1509–1522.

[164] Schiavina M, *et al.* The highly flexible disordered regions of the SARS-CoV-2 nucleocapsid N protein within the 1–248 residue construct: Sequence-specific resonance assignments through NMR. *Biomolecular NMR Assignments*, 2021. **15**(1): 219–227.

[165] Zhao M, *et al.* GCG inhibits SARS-CoV-2 replication by disrupting the liquid phase condensation of its nucleocapsid protein. *Nature Communications*, 2021. **12**(1): 2114.

[166] Cubuk J, *et al.* The SARS-CoV-2 nucleocapsid protein is dynamic, disordered, and phase separates with RNA. *Nature Communications*, 2021. **12**(1): 1936.

[167] Akbayrak IY, *et al.* Structures of MERS-CoV macro domain in aqueous solution with dynamics: impacts of parallel tempering simulation techniques and CHARMM36m and AMBER99SB force field parameters. *Proteins*, 2021. **89**(10): 1289–1299.

[168] Meszaros B, *et al.* Mutations of intrinsically disordered protein regions can drive cancer but lack therapeutic strategies. *Biomolecules*, 2021. **11**(3): 381.

[169] Wong ETC, *et al.* Protein-protein interactions mediated by intrinsically disordered protein regions are enriched in missense mutations. *Biomolecules*, 2020. **10**(8): 1097.

[170] Agarwal A, *et al.* An intrinsically disordered pathological prion variant Y145Stop converts into self-seeding amyloids via liquid-liquid phase separation. *Proceedings of the National Academy of Sciences of the United States of America*, 2021. **118**(45): e2100968118.

[171] Li J, *et al.* Protein phase separation and its role in chromatin organization and diseases. *Biomedicine & Pharmacotherapy*, 2021. **138**: 111520.

[172] Agarwal A and Mukhopadhyay S. Prion protein biology through the lens of liquid-liquid phase separation. *Journal of Molecular Biology*, 2022. **434**(1): 167368.

[173] Katuwawala A, *et al.* QUARTERplus: Accurate disorder predictions integrated with interpretable residue-level quality assessment scores. *Computational and Structural Biotechnology Journal*, 2021. **19**: 2597–2606.

[174] Katuwawala A, *et al.* DISOselect: Disorder predictor selection at the protein level. *Protein Science*, 2020. **29**(1): 184–200.

[175] Ghadermarzi S, *et al.* XRRpred: Accurate predictor of crystal structure quality from protein sequence. *Bioinformatics*, 2021. **37**(23): 4366–4374.

[176] Zhang J, *et al.* DNAgenie: Accurate prediction of DNA-type-specific binding residues in protein sequences. *Briefings in Bioinformatics*, 2021. **22**(6): bbab336.

[177] van Mierlo G, *et al.* Predicting protein condensate formation using machine learning. *Cell Reports*, 2021. **34**(5): 108705.

[178] Lei Y, *et al.* A deep-learning framework for multi-level peptide-protein interaction prediction. *Nature Communications*, 2021. **12**(1): 5465.

[179] Giri R, *et al.* Understanding COVID-19 via comparative analysis of dark proteomes of SARS-CoV-2, human SARS and bat SARS-like coronaviruses. *Cellular and Molecular Life Sciences*, 2021. **78**(4): 1655–1688.

[180] Matias Ferreyra F, *et al.* Comparative analysis of novel strains of Porcine Astrovirus type 3 in the USA. *Viruses*, 2021. **13**(9): 1859.

[181] Zoya S, *et al.* Shedding light on the dark proteome of Hepatitis E Virus. *Network Biology*, 2021. **11**(4): 295–314.

[182] Oliva Chavez AS, *et al.* Mutational analysis of gene function in the Anaplasmataceae: Challenges and perspectives. *Ticks and Tick-Borne Diseases*, 2019. **10**(2): 482–494.

[183] Jamsheer KM, *et al.* The FCS-like zinc finger scaffold of the kinase SnRK1 is formed by the coordinated actions of the FLZ domain and intrinsically disordered regions. *Journal of Biological Chemistry*, 2018. **293**(34): 13134–13150.

[184] Jami R, *et al.* The C-terminal domain of Salmonid Alphavirus nonstructural protein 2 (nsP2) Is essential and sufficient to block RIG-I pathway induction and interferon-mediated antiviral response. *Journal of Virology*, 2021. **95**(23): e0115521.

[185] Murph M, *et al.* The Centrosomal Swiss Army Knife: A combined in silico and in vivo approach to the structure-function annotation of SPD-2 provides mechanistic insight into its functional diversity. bioRxiv, 2021.

[186] Pujols J, *et al.* The disordered C-terminus of yeast Hsf1 contains a cryptic low-complexity amyloidogenic region. *International Journal of Molecular Sciences*, 2018. **19**(5): 1384.

[187] Szabo B, *et al.* Disordered regions of mixed lineage leukemia 4 (MLL4) protein are capable of RNA binding. *International Journal of Molecular Sciences*, 2018. **19**(11): 3478.

[188] Hu G, *et al.* Untapped potential of disordered proteins in current druggable human proteome. *Current Drug Targets*, 2016. **17**(10): 1198–205.

[189] Hosoya Y and Ohkanda J. Intrinsically disordered proteins as regulators of transient biological processes and as untapped drug targets. *Molecules*, 2021. **26**(8): 2118.

[190] Biesaga M, *et al.* Intrinsically disordered proteins and biomolecular condensates as drug targets. *Current Opinion in Chemical Biology*, 2021. **62**: 90–100.

[191] Ambadipudi S and Zweckstetter M. Targeting intrinsically disordered proteins in rational drug discovery. *Expert Opinion on Drug Discovery*, 2016. **11**(1): 65–77.

[192] Ghadermarzi S, *et al.* Sequence-derived markers of drug targets and potentially druggable human proteins. *Frontiers in Genetics*, 2019. **10**: 1075.

[193] Deng X, *et al.* An overview of practical applications of protein disorder prediction and drive for faster, more accurate predictions. *International Journal of Molecular Sciences*, 2015. **16**(7): 15384–15404.

[194] Kurgan L, *et al.* The methods and tools for intrinsic disorder prediction and their application to systems medicine. In *Systems Medicine*, Wolkenhauer O (Ed.), Academic Press: Oxford, 2021. 159–169.

[195] Oldfield CJ, *et al.* Utilization of protein intrinsic disorder knowledge in structural proteomics. *Biochimica et Biophysica Acta*, 2013. **1834**(2): 487–98.

[196] Hanson J, *et al.* Accurate single-sequence prediction of protein intrinsic disorder by an ensemble of deep recurrent and convolutional architectures. *Journal of Chemical Information and Modeling*, 2018. **58**(11): 2369–2376.

[197] Yan J, *et al.* Molecular recognition features (MoRFs) in three domains of life. *Molecular Omics*, 2016. **12**(3): 697–710.

[198] Meng F and Kurgan L. High-throughput prediction of disordered moonlighting regions in protein sequences. *Proteins*, 2018. **86**(10): 1097–1110.

Index